CONTINENTAL MARGINS

Geological and Geophysical Research Needs and Problems

Ad Hoc Panel to Investigate the
   Geological and Geophysical Research Needs
   and Problems of Continental Margins
Ocean Sciences Board
Assembly of Mathematical and Physical Sciences
National Research Council

NATIONAL ACADEMY OF SCIENCES
Washington, D.C.   1979

NOTICE   The project that is the subject of this report was approved by the Governing Board of the National Research Council, whose members are drawn from the Councils of the National Academy of Sciences, the National Academy of Engineering, and the Institute of Medicine. The members of the Panel responsible for this report were chosen for their special competences and with regard for appropriate balance.

This report has been reviewed by a group other than the authors according to procedures approved by a Report Review Committee consisting of members of the National Academy of Sciences, the National Academy of Engineering, and the Institute of Medicine.

The Ocean Sciences Board is pleased to acknowledge the support of the National Science Foundation, the U.S. Geological Survey, and the Office of Naval Research for the conduct of this study.

Library of Congress Cataloging in Publication Data

National Research Council. Ad Hoc Panel to Investigate the Geological and Geophysical Research
    Needs and Problems of Continental Margins.
    Continental margins.

The report of the Ad Hoc Panel to Investigate the Geological and Geophysical Research Needs
and Problems of Continental Margins.
    Includes bibliographies.
    1. Submarine geology—Research—United States. 2. Continental margins—Research—United
States. 3. Geophysical research—United States. I. National Research Council. Ocean Sciences
Board. II. Title.
QE39.N35   1979   551.4′608′072   79-22
ISBN 0-309-02793-4

Available from

Office of Publications
National Academy of Sciences
2101 Constitution Avenue, N.W.
Washington, D.C. 20418

Printed in the United States of America

Ocean Sciences Board

iii

Ad Hoc Panel to Investigate the Geological and Geophysical Research Needs and Problems of Continental Margins

ALBERT W. BALLY, Shell Oil Company, *Chairman*
WILLIAM A. BERGGREN, Woods Hole Oceanographic Institution
THOMAS A. CLINGAN, JR., University of Miami
JOSEPH R. CURRAY, Scripps Institution of Oceanography
EDGAR S. DRIVER, Gulf Science and Technology Company
JOHN I. EWING, Woods Hole Oceanographic Institution
DONN S. GORSLINE, University of Southern California
THOMAS H. JORDAN, Scripps Institution of Oceanography
CHARLOTTE E. KEEN, Bedford Institute, Dartmouth, Nova Scotia, Canada
LAVERNE D. KULM, Oregon State University
JOHN D. MILLIMAN, Woods Hole Oceanographic Institution
WALTER C. PITMAN, JR., Lamont-Doherty Geological Observatory of Columbia University
JOEL S. WATKINS, JR., Gulf Research and Development Company*
CHARLES L. DRAKE, Dartmouth College, *Ex officio*
JACK E. OLIVER, Cornell University, *Ex officio*

NRC Staff

MARY LOU LINDQUIST, *Principal Staff Officer,* Ocean Sciences Board
RICHARD C. VETTER, *Executive Secretary,* Ocean Sciences Board

*As of October 1, 1977; previously with the University of Texas.

v

W. W. Hay, University of Miami
G. R. Heath, Oregon State University
R. R. Hume, National Research Council
G. L. Johnson, Office of Naval Research
M. K. Johrde, National Science Foundation
I. R. Kaplan, University of California at Los Angeles
T. T. Kato, Shell Development Company
M. J. Keen, Dalhousie University, Halifax, Nova Scotia, Canada
J. Kofoed, National Oceanic and Atmospheric Administration
J. C. Kraft, University of Delaware
N. W. Lauritzen, Gulf Energy and Minerals Company–International
R. Le Blanc, Shell Oil Company
B. T. R. Lewis, University of Washington
M. Lloyd, Consulting Geologist, Houston
G. P. Lohmann, Woods Hole Oceanographic Institution
P. D. Lowman, Jr., National Aeronautics and Space Administration
F. I. Manheim, U.S. Geological Survey, Woods Hole
P. G. Mathieu, Gulf Science and Technology Company
R. D. McIver, Exxon Production Research Company
E. L. Miles, University of Washington
S. Milner, Shell Development Company
R. M. Mitchum, Jr., Exxon Production Research Company
R. J. Mousseau, Gulf Science and Technology Company
R. W. Murphy, Exxon Production Research Company
J. W. Norris, Gulf Science and Technology Company
N. A. Ostenso, National Oceanic and Atmospheric Administration
L. L. Raymer, Schlumberger-Doll Research Center
M. A. Rogers, Exxon Production Research Company
J. B. Sangree, Exxon Production Research Company
C. H. Savit, Western Geophysical Company
D. R. Seely, Exxon Production Research Company
R. E. Sheridan, University of Delaware
J. G. Smith, Gulf Science and Technology Company
J. Thiede, University of Oslo
S. Thompson III, New Mexico Bureau of Mines and Mineral Resources
P. R. Vail, Exxon Production Research Company
T. H. van Andel, Stanford University
R. P. Von Herzen, Woods Hole Oceanographic Institution
R. Von Huene, U.S. Geological Survey
R. E. Wall, National Science Foundation
A. B. Watts, Lamont-Doherty Geological Observatory, Columbia University
J. Whelan, Woods Hole Oceanographic Institution
W. S. Wooster, University of Washington
P. C. Wuenschel, Gulf Science and Technology Company

Substantial portions of this report are based on the work of three IPOD panels during late 1976:

Active Margins Panel

S. Uyeda, University of Tokyo, *Chairman*
R. Blanchet, Centre de Recherche en Géologie, France
C. A. Burk, University of Texas

vi

J. S. CREAGER, University of Washington
K. KOBAYASHI, University of Tokyo
I. P. KOSMINSKAYA, U.S.S.R. Academy of Sciences, Moscow
L. W. KROENKE, Mineral Resources Division, Suva, Fiji
L. D. KULM, Oregon State University
K. KVENVOLDEN, U.S. Geological Survey, Menlo Park, California
W. J. LUDWIG, Lamont-Doherty Geological Observatory of Columbia University
G. PACKHAM, University of Sydney, Australia
A. V. PEYVE, U.S.S.R. Academy of Sciences, Moscow
D. W. SCHOLL, U.S. Geological Survey, Menlo Park, California
R. VON HUENE, U.S. Geological Survey, Menlo Park, California

Passive Margins Panel

J. R. CURRAY, Scripps Institution of Oceanography, *Chairman*
A. W. BALLY, Shell Oil Company
H. BEIERSDORF, Bundesantalt für Geowissenschaften, Federal Republic of Germany
D. BERNOULLI, University of Basel, Switzerland
H. CLOSS, Bundesanstalt für Geowissenschaften, Federal Republic of Germany
J. I. EWING, Lamont-Doherty Geological Observatory of Columbia University
J. GROW, U.S. Geological Survey
K. HINZ, Bundesanstalt für Geowissenschaften, Federal Republic of Germany
J. M. HUNT, Woods Hole Oceanographic Institution
H. KAGAMI, University of Tokyo
L. MONTADERT, Institut Français du Pétrole
D. G. MOORE, Scripps Institution of Oceanography
D. G. ROBERTS, Institute of Oceanographic Sciences, England
E. SEIBOLD, Universität Kiel, Federal Republic of Germany
R. E. SHERIDAN, University of Delaware
J. THIEDE, University of Oslo

Palaeoenvironment Panel

Y. LANCELOT, C.N.E.X.O., France, *Chairman*
W. BERGGREN, Woods Hole Oceanographic Institution
P. L. BEZRUKOV, U.S.S.R. Academy of Sciences, Moscow
P. CEPEK, Bundesanstalt für Geowissenschaften, Federal Republic of Germany
J. DEBYSER, C.N.E.X.O., France
B. M. FUNNELL, University of East Anglia, England
W. W. HAY, University of Miami
J. KENNETT, University of Rhode Island
V. KRASHENNINIKOV, U.S.S.R. Academy of Sciences, Moscow
T. C. MOORE, University of Rhode Island
W. RIEDEL, Scripps Institution of Oceanography
H. SCHRADER, Universität Kiel, Federal Republic of Germany
N. SHACKLETON, University of Cambridge, England
I. P. SILVA, Instituto di Palentologie, Italy
Y. TAKAYANAGI, Tohoku University, Japan
H. THIERSTEIN, Scripps Institution of Oceanography
T. H. VAN ANDEL, Stanford University
T. WORSLEY, University of Washington

Contents

ix

Contents

7   CONTINENTAL MARGINS OF THE ARCTIC OCEAN   102
    A Introduction, 102
    B Structure and Scientific Problems, 102
       Lomonosov Ridge, 102; Eurasian Basin
       and Margins, 103; Amerasian Basin and Margins, 103
    C Current and Future Projects, 104
    D Operational Aspects, 104
    E Recommendations and Priorities, 106
       References and Bibliography, 106

8   ACTIVE MARGINS   108
    A Introduction, 108
    B The Evolving Active-Margin Model, 110
    C The Forearc Region, 111
    D The Arc Region, Its Volcanoes and Geochemistry of Igne-
       ous Rocks, 112
    E The Backarc Region, 115
    F Peri-Sutural Basins, 116
    G Transform Margins, 117
    H Incipient Subduction, 118
    I The Mediterranean and the Caribbean: Two Special
       Cases, 118
    J The Role of Drilling on Active Margins, 120
       Introduction, 120; Concerning Present Drilling Capability, 120;
       Concerning Future Drilling Capabilities, 121
    K Transects on Active Margins, 121
       Introduction, 121; Alaska, 122; Bering Sea Area, 123; U.S. West
       Coast Transects, 123; Middle America Trench Transects, 123;
       Caribbean Transects, 123; South American Transects, 123;
       Northwest Pacific Plate Dynamics Traverses, 124; East and
       Southeast Asia Transects, 127; Seismologic Studies on Active
       Margins, 127
    L Recommendation, 127
       References and Bibliography, 128

9   MOBILE FOLD BELTS AND ANCIENT CONTINENTAL
    MARGINS   130
    Introduction, 130
    The "Basement" of Fold Belts, 130
    Reconstructions of Folded Belts, 133
    Geological Field Observations on Ancient Margins, 135
       Suture Zones—Boundaries of Lithospheric Plates, 135; Border-
       lands, 137; The Western Cordillera of the United States, 137;
       Unconformities, 139; Deep-Sea Turbidites, 139; Study the Oldest
       Rocks in Outcropping Miogeosynclinal Sequences, 139
    References and Bibliography, 139

10 SEISMICITY AND THE DEEP STRUCTURE OF
    CONTINENTAL MARGINS   141
    A Introduction, 141
    B Seismicity of the Margins, 141
    C Tectonic Stresses along the Margins, 142
    D Continent-Ocean Transition, 145
    E The Fate of Subducted Lithosphere, 148
    F Recommendations, 148
       References and Bibliography, 148

Contents

Contents

xiv

List of Figures

List of Figures

xvii

Preface

Starting in the 1930's and continuing until the 1950's, scientists in academic institutions made investigations along the U.S. continental margins using the technologies available during that period. The International Geophysical Year (IGY) in the late 1950's led them to investigate the nature and history of the deep-ocean basins. These investigations led, during the 1960's, to the plate-tectonics model, which has proved remarkably effective in explaining the properties of the seafloor and the processes affecting it. Academic geological and geophysical oceanographers then turned their attention more toward specifics—the processes taking place in seismically active zones and the nature and history of continental margins.

In 1962, a marine-studies program within the U.S. Geological Survey (USGS) was begun, "to identify and evaluate potential mineral resources on and beneath the sea floor, and to aid in the solution of problems of coastal areas—which were mushrooming because of rapid population growth, urbanization, and industrial expansion" (Agnew, 1975). The program involved limited collaboration with nongovernmental oceanographic institutions, such as the Woods Hole Oceanographic Institution and the Scripps Institution of Oceanography. By late 1975, the USGS had substantially increased its activities on U.S. continental margins. Furthermore, accelerated leasing schedules on the Outer Continental Shelf of the United States resulted in a large increase in industrial activity on the continental margins. In the 1970's, when geoscientists in oceanographic institutions began to explore U.S. margins more intensely for answers to some scientific puzzles, they found it difficult to counter the argument that their efforts might better be diverted to other areas because industry and the government agencies would adequately handle the problems of the U.S. margins.

To confront this problem, geoscientists from oceanographic institutions, federal agencies, and industry met in Galveston, Texas, in March 1976. USGS scientists explained their mandate and research plans. Their counterparts from the academic community discussed their research interests in continental margins. All participants recognized that it was timely and necessary to have a national program that placed emphasis and priorities on continental-margins research. The Ocean Sciences Board of the National Research Council responded to this need by establishing an *ad hoc* Panel to Investigate the Geological and Geophysical Research Needs and Problems of Continental Margins, the authors of this report. The Panel was asked to do its work within a year, starting October 1, 1976. Three agencies—the National Science Foundation, the U.S. Geological Survey, and the Office of Naval Research—contributed support for the Panel's activities. The terms of reference for the Panel were framed as follows:

The *ad hoc* Panel to Investigate Geological/Geophysical Research Needs and Problems of Continental Margins will be concerned with solid earth research problems of continental margins (i.e., shelf, slope, and rise). It will include consideration, as required, of the effects of adjacent onshore and deep-sea areas which are in geological continuity with the continental margins, as well as ancient continental margins now outcropping in mountain ranges.

The Panel will not concern itself with the details of exploration and prospecting for nonrenewable resources, nor with the environmental aspects associated with exploitation of such resources. Instead, the Panel will concentrate on geological-geochemical and geophysical research oriented towards gaining new conceptual insights. The main emphasis will be on problem definition and goals for future solid earth research on continental margins.

Concurrently, the Panel will consider alternative plans for solving the most important problems and search for ways to coordinate current and future research efforts in keeping with the respective roles of academic, governmental, and industrial institutions. Such coordination should enhance more effective use of research facilities, currently limited manpower, and financial resources without reducing diversity of creative efforts—one of the strengths of our current research efforts in the field.*

The Panel met in November 1976 at Denver, Colorado, and in May 1977 at Boulder, Colorado. Several members of the Panel also met informally while serving with other groups dealing with more specific aspects of continental-margins research. It soon became obvious that several related efforts paralleled ours. The most important of these were the following:

• The U.S. Geodynamics Committee (National Research Council) focused on national and international solid-earth research in the 1980's. A working document, *Crustal Dynamics, a Framework for Resource Systems* (August 1976, circulated by the U.S. Geodynamics Committee) provides a summary, from which we quote: "...it is time to invoke a program that first supports and encourages fundamental research in fields contributing to the understanding of the principles governing the distribution of the resources of the solid earth, and

---

*Note from the chairman with regard to the last paragraph of the Terms of Reference: The Panel did consider the need for coordination of continental-margins research done by government, industry, and academe, but in the end, we refrained from making significant specific proposals in this area. This was for the following reasons: (1) In the time available, we were not able to obtain reasonably detailed current and past expenditure figures for federally supported earth-science research on continental margins as we perceived them (i.e., including both land and marine portions). (2) We also lacked forecasts that dealt with the number of earth and marine scientists needed to fulfill the needs for resource exploration and environmental management. (3) We became aware that the federal government has begun to reexamine its own organization. This activity includes a search for better ways to coordinate resources and environmental and related research activities.

second provides a bridge between the geoscience community and the public."
The Geophysics Research Board of the National Research Council has requested
the U.S. Geodynamics Committee to propose plans for long-range research in
the solid-earth sciences.

The Inter-Union Commission on Geodynamics (ICG) included many sugges-
tions paralleling those of the U.S. Geodynamics Committee in the report,
*Geodynamics Post 1980—A Successor to the International Geodynamics Pro-
gramme* (ICG, December 1976), which concludes with: "...we recommend...the continued study of the nature of the Earth's surface, including
oceanic and continental structures, continental margins, and mountain ranges,
the state and composition of the crust and the vertical movements and evolution
of sedimentary basins." Task groups of the International Union for Geodesy and
Geophysics (IUGG) and the International Union for Geological Sciences (IUGS)
are currently working to finalize a plan that would implement the preceding
recommendation.

• JOI, Inc. (a consortium of major academic oceanographic institutions)
adopted a specific program for the remainder of this decade and for the 1980's.
This is summarized in *A Program for Ocean Crustal Dynamics* (Talwani, 1977),
which emphasizes (1) studies of the evolution of the lithosphere and asthenosphere, (2) passive margins studies, and (3) active margins studies. The program
contemplates total science and facilities expenditures on the order of $9 million
per year for 1977–1980 and $21.2 million per year for 1980–1990, with ship time
(not included in the preceding figures) adding up to 48 ship-months per year for
1977–1980 and 66 ship-months per year for 1980–1990.

• A JOIDES (Joint Oceanographic Institutions for Deep Earth Sampling) Subcommittee on The Future of Scientific Ocean Drilling prepared a report based
on a series of white papers concerned with ocean-crust drilling, passive and
active continental margins, and paleooceanography. Following a meeting at
Woods Hole in spring 1977, the subcommittee wrote a summary recommendation that proposed a 10-year drilling program to follow the current International
Phase of Ocean Drilling (IPOD) program (scheduled to end late in 1979). The
estimated costs for the 10-year program total $450 million, which include the
costs to convert the *Glomar Explorer* to develop a 12,000-foot riser system; $83
million for geophysical work; and $38 million for analysis, interpretation, and
synthesis of the results. Continental-margin studies constitute a large part of this
program.

• A shelf sediment dynamics workshop, sponsored by the International Decade of Ocean Exploration Program of the National Science Foundation, the
National Oceanic and Atmospheric Administration, the USGS, and the Energy
Research and Development Administration, was held in November 1976. The
results of this workshop are summarized in *Shelf Sediment Dynamics: A National Overview* (Gorsline and Swift, 1977).

• The International Decade of Ocean Exploration (IDOE) Program will end in
1979. The National Science Foundation's IDOE Office held a series of meetings
during the spring and summer of 1977 to identify possible projects for large-
scale, long-term, cooperative ocean studies in the 1980's. These workshops covered physical oceanography, chemical oceanography, biological oceanography,
and geological/geophysical oceanography (including sediment dynamics). The
last topic is of special interest to this Panel. Finally, the NRC's Ocean Sciences
Board organized a major workshop involving all aspects. The workshop was held
in September 1977. The preliminary conclusions of this report were presented at
that workshop. The findings of the relevant workshops were published in Heath
(1977) and Wooster (1977).

At one time or another, individual members of our Panel were involved in most of these activities. In this report, we have freely used material written and discussed by our colleagues for these other meetings.

This report provides an overview and recommendations for continental-margins geoscience research in the 1980's. We concentrate on academic research, but we also review coordination and communication between federal, academic, and industrial research groups. Our focus is on the priorities and the sequence of geoscience research on continental margins.

We have attempted in the report to reconcile the need for a concise summary with the desire for a more comprehensive documentation of our current knowledge. The essence of our thoughts is found in the beginning of the report under Introduction, Summary, and Principal Recommendations. The remaining text and the appendixes are for those who wish to dig deeper. This introductory material is intended for science planners concerned with strategy. It presents an overview of the main scientific problems, our most important recommendations for continental-margins research in the 1980's, and general conclusions.

The scope and costs of the proposed research for the next decade are unusually large. Because of this, we believe that a more detailed perspective is in order—if only to help those readers who want more than a mere summary of our Panel's judgment. This is the aim of Parts I and II.

Fellow scientists and others involved in the details of research planning may want to read Parts I and II. These sections contain recommendations that are presented within their scientific contexts, without regard to priorities.

In Part I, we look at the state of the science of continental margins and pose some questions. From the surface of the seafloor into the depths of the earth, we proceed from physiography to sediments, their geochemistry and diagenesis, and conclude with their spatial and temporal distribution (stratigraphy and paleooceanography). Following this is a discussion of the geophysical and geological evolution of continental margins. Finally, we focus on the deeper structure of continental margins as revealed by seismic studies.

Part II tackles methods and tools. It first deals with some geophysical methods and then discusses remote sensing, drilling, and research vessels.

Following some general remarks in Part III, we reiterate our high-priority recommendations summarized in the introductory material, expand on them, and relate them to some of the second-priority items contained in Parts I and II. We did this to give the reader a look at the array from which we chose our high-priority items.

To differentiate more easily between our priorities, three styles of presentation are used:

1. *High Priority*   In this class, recommendations and conclusions are in **boldface print.**
2. *Second Priority*   Recommendations that are either of secondary importance or subsets of high-priority items are presented in *italics.*
3. *Lower Priority*   High- and second-priority recommendations are repeated in Parts I and II, where they are *italicized.* In addition to these, Parts I and II contain lower-priority items, also printed in *italics.* These are good but less urgent programs, often of smaller scope.

Appendix A contains a review of the organization of and funding modes for continental-margins research in the United States and Canada. The discussions of organization and funding were written in mid-1977 to provide an overview of the agencies involved, some idea as to how they are structured (in terms of continental-margins research), and to give some rough idea of what amounts of funds are allocated to such research. Appendix A evolved as an attempt to respond to the third paragraph of the Panel's terms of reference (see p. xxii and the chairman's footnote, particularly items 1 and 3). A chapter on the Law of the Sea and scientific research is also included to give the reader an indication of some further problems that scientists may face in doing research on continental margins.

In the course of our work, we were offered a number of geophysical papers that document specific areas of interest. Some of these are included in Appendix B. They may prove useful for the reader who is more interested in the technology of continental-margins research.

Appendix C describes a modern geophysical research vessel.

Note that, unless indicated otherwise, all dollar figures used throughout this report are in 1977 dollars. *Thus, no allowance has been made for cost increases due to inflation.*

Acknowledgments: Many individuals participated in the work of our Panel. We are most grateful for all the fine contributions they made, for without their cooperation, we would not have been able to provide such a broad perspective for research needs on continental margins. Scientists from government agencies as well as industry and academe contributed much material for this report. Also, they and other of their federal colleagues spent many hours reviewing the technical and descriptive sections to ensure the accuracy of our reportage. We owe special thanks to William E. Benson, who provided us with the section on the history of research in the field that this report considers. We gleaned much from the work of the panels of the International Phase of Ocean Drilling Project that wrote white papers on active margins, passive margins, and paleooceanography. We have listed the names of the members of these panels in light of their contribution to this report. The names of all participants and contributors precede this Preface. Their listing indicates that they contributed information, time, and attention to this report but does not intend to imply their agreement with the Panel's recommendations or conclusions.

A very sincere and special vote of thanks goes to Mary Lou Lindquist for the charming blend of enthusiasm, stamina, and class she brought to the work of our Panel. Herb Scott prepared a large number of the illustrations, and we thank him and Shell Oil for making them available to this report.

ALBERT W. BALLY

REFERENCES AND BIBLIOGRAPHY

Ad Hoc Subcommittee of the JOIDES Executive Committee (1977). *The Future of Scientific Ocean Drilling*, JOIDES Office, University of Washington, 92 pp.

Agnew, A. F. (1975). *The U.S. Geological Survey*, Congressional Research Service, Library of Congress, Environmental Policy Division, U.S. Government Printing Office, Washington, D.C.

Gorsline, D. S., and D. J. P. Swift (1977). *Shelf Sediment Dynamics: A National Overview*, available from the Office of the International Decade of Ocean Exploration, the National Science Foundation, Washington, D.C., 134 pp.

Heath, G. R. (1977). Geological/Geophysical Oceanography, Section IV in *Ocean Research in the 1980's: Recommendations from a Series of Workshops in Large Scale Oceanographic Research*, Center for Ocean Management Studies, University of Rhode Island, Kingston, Rhode Island, 34 pp.

Talwani, M., ed. (1977). *A Program for Ocean Crustal Dynamics*, from a workshop held at Lamont-Doherty Geological Observatory, Columbia University, N.Y., informal draft.

Wooster, W. S. (1977). Post-IDOE Planning, Section V in *Ocean Research in the 1980's: Recommendations from a Series of Workshops in Large Scale Oceanographic Research*, Center for Ocean Management Studies, University of Rhode Island, Kingston, Rhode Island, 22 pp.

Introduction, Summary, and Principal Recommendations

INTRODUCTION

Today, the earth sciences are especially concerned with continental margins. In the past, people regarded the seashore as the margin of a continent. Since the nineteenth century, marine surveyors have discovered extensive underwater terraces and slopes. These became the continental margins of the oceanographers. However, geologists working on land have long been aware of sea-level changes through the ages. They consider the present level of the sea to be an ephemeral coincidence.

Modern earth scientists are beginning to take a more comprehensive view of continental margins—one that encompasses a wide transition zone that separates oceanic from continental realms. The zone includes the continental shelf, slope, and rise, but it also embraces the landward extension of this geologic province. Thus, continental margins have now become the joint concern of scientists working on land and their colleagues who work in the oceans.

The continental margin is the place where land scientists meet ocean scientists, where sedimentologists meet physical oceanographers, where stratigraphers meet geophysicists, where economic geologists meet environmentalists, where industry meets government, and where government meets academe.

The plate-tectonics hypothesis, developed during the last decade, serves well to define major problem areas. Figure 1 shows how the outer shell of the earth (the lithosphere) is segmented in plates that have convergent and divergent boundaries. We differentiate three types of continental margins, based on their relation to plates and plate boundaries and to the presence or absence of seismic and volcanic activity (Figure 2).

1

Passive continental margins are those without significant concentrations of seismic and volcanic activity. They are located within a plate on the transition between continental and oceanic crust. These margins form at divergent plate boundaries. With time, they move away from those boundaries and become sites of massive subsidence and thick accumulations of sediment.

Cratonic margins lie entirely on continental crust. Strictly speaking, they do not qualify as transition between continental and oceanic crust. They do, however, occupy large areas covered by seas. Cratonic margins also may contain thick sediment accumulations.

Active margins are associated with intensive earthquake activity and spectacular volcanism. These margins form at convergent plate boundaries, that is, where rigid lithospheric plates are sinking deep in the more viscous asthenosphere of the earth or where plates move laterally with respect to each other. Differential uplifts and downwarps on active margins lead to the formation of mountain ranges and small, but deep, sedimentary basins. Most characteristic of conver-

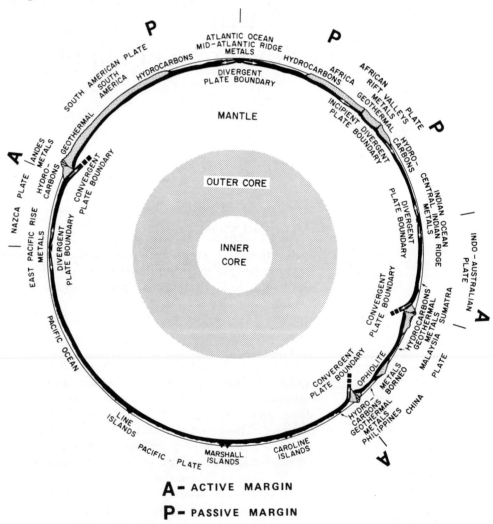

A - ACTIVE MARGIN

P - PASSIVE MARGIN

FIGURE 1 Schematic great-circle section through the equator of the earth, showing lithospheric plates, plate boundaries, and occurrences of energy and mineral resources. For clarity, the thickness of the lithosphere is expanded by a factor of approximately 3. From P. A. Rona, 1977. Plate tectonics, energy and mineral resources: Basic research leading to payoff. *EOS, Trans. Am. Geophys. Union* 58(8): 629–639. Copyrighted by the American Geophysical Union.

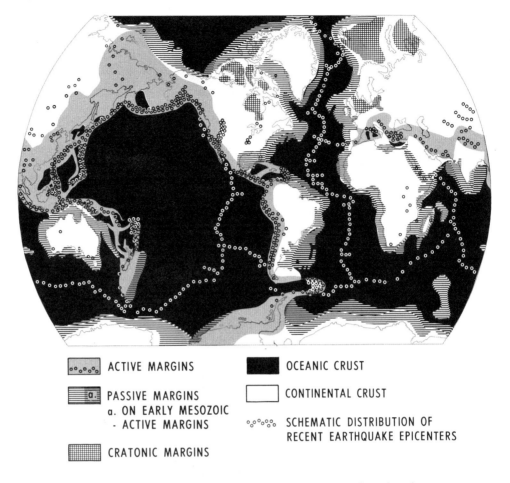

⊙°⊙°⊙° ACTIVE MARGINS	■ OCEANIC CRUST
═a.═ PASSIVE MARGINS a. ON EARLY MESOZOIC - ACTIVE MARGINS	☐ CONTINENTAL CRUST
▦ CRATONIC MARGINS	°⊙°⊙°⊙ SCHEMATIC DISTRIBUTION OF RECENT EARTHQUAKE EPICENTERS

FIGURE 2   Classification of continental margins. Active margins are also referred to as converging margins, and passive margins as diverging margins. The detailed earthquake distribution is shown on Figure 1.4.

gent active margins is the formation of island arcs, with marginal basins developing on their concave side.

Sedimentary basins belonging to all continental-margin types are sites of oil and gas accumulations. Significant amounts of oil and gas trapped in these basins remain to be discovered. Other mineral resources of continental margins include sand and gravel, heavy minerals (such as cassiterite, diamonds, gold, and barite), phosphates in shallow water depths, and manganese nodules at greater depths. Metallic minerals are associated with active margins as indicated in Figure 1.

A stepped-up effort in solid-earth research on continental margins is desirable, because without scientific underpinnings we cannot properly evaluate and manage resources of these margins. Because all resource exploitation affects the surrounding environment, we should also try to understand that environment.

Erosion and sedimentation mold the surface of continental margins. On coasts, these processes have been studied for many years. In the past, sedimentologists have made surveys of the sediment distribution on continental margins. The next task is to go beyond these studies and try to understand the processes that dictate the entrainment, transport, and deposition of sediments in coastal areas and on the shelf, slope, and rise. We know there are complex currents sweeping over the continental margins, but we know little of their spatial distributions.

3

Physical oceanographers have made great advances in modeling these currents, but many theoretical aspects are not fully understood. Most of the concepts have not been tested. We need to measure the capacity of these currents to erode and to carry sedimentary particles in order to understand the details of sediment settling and deposition.

In order to apply what we learn of sediment settling and deposition, we must reconstruct conditions of the geological past. We now better understand how sea level has changed by amplitudes of more than a hundred meters during glacial versus nonglacial epochs of the Pleistocene. With these changes of sea level, continental shelves have been alternately exposed and inundated. The last rise of sea level was so recent that sediment distributions have not yet come into equilibrium with environmental conditions. To apply the principles of distribution and deposition of sediment on continental shelves, we ought to understand fully how the processes we observe today differ from those that were prevalent during other geological times.

Recently, geological and physical oceanographers have paid more attention to the benthic boundary layer, that ill-defined layer involving some 20 cm of seawater over the seafloor and about 20 cm of sediments below the seafloor. Physical, chemical, and biological processes occurring in this layer are only dimly perceived; yet these processes ultimately determine the "soil" properties so important to seafloor engineers. Viewed in a different light, the benthic boundary layer is where sediments, through early diagenesis, start on their way to becoming solid sedimentary rock. As these rocks are buried, they undergo many more diagenetic changes and losses in porosity, which determine the quality of the layers as aquifers or as reservoirs for hydrocarbons.

Although older sediments occasionally outcrop on the seafloor, the history of their genesis and burial is more spectacularly displayed on reflection seismic lines. Today, we are witnessing some very exciting changes in stratigraphic methodology. In the past, stratigraphers carefully studied sections outcropping on land or sections penetrated by wells. Correlations between such sections were often ambiguous. Today, reflection seismic lines offer x-ray-like sections that display stratigraphy in considerable detail.

Our newly acquired ability to make fairly detailed paleooceanographic reconstructions further enhances the scope of stratigraphy. Reconstructions are based on detailed paleontologic correlations and the paleomagnetic studies that led to the development of plate tectonics. Much work, including drilling, is needed to reconcile seismic and stratigraphic evidence with postulated paleooceanographic configurations. The sedimentary record of the margins holds clues to past climatic and oceanographic conditions that are vital ingredients of modern climate-prediction models. These will help us to understand the effects of future climate changes.

Passive continental margins are now recognized to be the modern equivalent of some of the old geosynclines. During the last decade, we have learned that the deeper portion of passive margins conceals an early rifting history that preceded the actual opening of the ocean. Geologists infer much from seismic lines, but in fact, there are few places where the early rifting sequence has been documented in detail by seismic lines that are calibrated by drilling evidence. Such studies could greatly improve our understanding of the restricted early rifting environments visualized by geologists. These environments sometimes lead to the deposition of salt and organic-rich source beds, which can be critical factors in determining and explaining the presence or absence of hydrocarbons on passive margins.

Geophysical models portray the evolution by subsidence of passive margins as the consequence of crustal cooling combined with sediment loading. To confirm the validity of these models, we need reliable subsidence data based on wells not located on structural anomalies.

A vexing problem on both active and passive margins is the nature of the ocean–continent boundary and its associated gravity and magnetic anomalies. Multichannel seismic techniques combined with refraction seismic methods have improved the geophysical resolution of the continent–ocean boundary. The answers to some aspects of this problem await testing by the drill in locations that are carefully selected after much geophysical work. However, a basic understanding of this problem will require the application of geophysical methods capable of resolving lateral variations in upper-mantle structure extending to several hundred kilometers depth.

Major problems in geology remain to be solved on active margins. Plate boundaries were initially defined by the distribution of earthquakes. In some cases, two plates slip laterally past each other (San Andreas Fault), while other cases (e.g., island arcs of the West Pacific) suggest that a cold lithospheric slab is sinking under another lithospheric plate. First-motion studies of earthquakes suggest that some segments in the sinking slab are under compression and that others are under extension. Often, these studies lack detail and precision. The use of ocean-bottom seismometers (OBS) allows much more detailed studies of microearthquakes on a more local scale and should greatly help augment existing data. The same instruments, used for refraction profiling, allow us to study the nature of the lower crust and the upper mantle on active margins.

The study of earthquake dynamics is an especially important aspect of continental-margins research. Emphasis should be placed on investigating the tectonic stress regime of the margins. Digitally recording global networks of seismometers and OBS's will be useful for this purpose. This research is directly relevant to earthquake forecasting.

Seismic-reflection data suggest that great wedges of structurally deformed sediments exist on the inner side of deep-sea trenches. These are believed to be sediment scraped off the top of deep-sea trenches. Only a few wells have penetrated the top of these sequences. Here again, deep drilling based on carefully planned geophysical surveys will help in explaining the genesis of these frontal zones of island arcs.

As we move landward from the deep-sea trenches, we observe extensive volcanic arcs. Farther inland, we encounter their deep-seated equivalents, the igneous intrusions, which have been uplifted during complex, mountain-building events. Geoscientists now realize that active margins are the places where mountain building can be caught in the act. A large data gap separates the deep-earth information given by seismologists from the surface data gathered by surface geologists and geochemists.

To close this gap, we need crustal refraction and reflection studies both on the oceans and on land. Current progress in deep crustal reflection work on land is particularly encouraging. This work should be the foundation for crustal drilling on land that would complement the marine margin drilling planned for the International Phase of Ocean Drilling (IPOD).

The genesis of marginal basins behind island arcs is obscure. We need deeper insights into the origin of these basins, as they appear to represent an important stage in mountain-building (orogenic) processes. Such processes often seem to involve the collapse of marginal basins.

This brings us to the importance of understanding ancient continental margins, which often are deformed by mountain-building processes. Suffice it to say that mountain ranges are the locus of some of the most important metallic miner-

5

als. Also, the richest petroleum provinces occur in foredeeps intimately associated with orogenic processes resulting from continental collisions. Old folded belts allow us to see rocks formed by past processes analogous to those occurring at depth in today's continental margins. Study of old mountain ranges is essential if we are to understand contemporary mountain-building processes.

THE ROLES OF GOVERNMENT, INDUSTRY, AND ACADEME

We are a society fond of checks and balances for our institutions, a basic attitude that transcends politics. The same concept applies particularly to scientific research that aims to provide a solid background for important economic decisions. Three major active constituencies—government, industry, and academe—engage in such research. The tensions between these three groups are creative tensions; they have spurred much of our recent progress and new fundamental insights.

Some recent developments seem to be shifting research efforts away from academe and industry into the mission-oriented agencies of the federal government. Some of the more important trends in this direction that we perceive are as follows:

• Overall federal money for oceanographic research has increased substantially in recent years, but most new funds have gone to mission-oriented government agencies. Funding for basic oceanographic research done by scientists

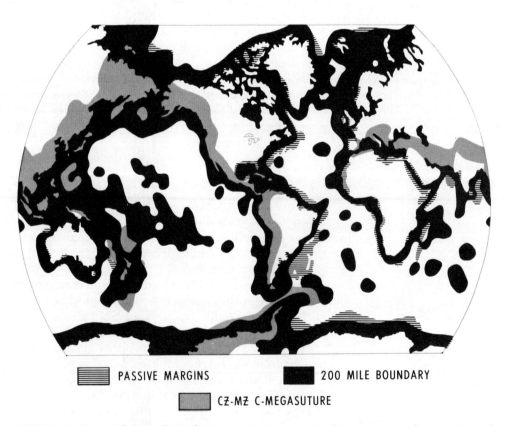

PASSIVE MARGINS   200 MILE BOUNDARY

CZ-MZ C-MEGASUTURE

FIGURE 3 Proposed 200-mile exclusive economic zones (EEZ's) superimposed on continental margins. Compare with Figure 2 and note that, in essence, continental margins and the 200-mile EEZ's coincide. For definition of Cenozoic–Mesozoic Megasuture, see Figure 8.1.

6

in academic institutions has barely kept up with inflation. Of course, this situation, as well as the next, noted below, is common to much of all government-supported academic science.

• To get the same support as in the past, ocean scientists must devote substantially more time to proposal writing and administration.

• Real environmental concerns that must be dealt with have introduced difficulties and delays in obtaining approvals, particularly for U.S. margins research. Penetration of the seafloor by drilling, jetting, or even, in some cases, long piston coring now requires permits from the U.S. Geological Survey. Seismic work using explosive sound sources requires special permits because of the danger of killing fish. Delays in securing these permits are often discouraging.

• The Law of the Sea Conference has led to increased unilateral claims of 200-mile exclusive economic zones, which effectively cover all continental margins of the world (Figure 3). Sanctioning and continuing this trend will make research on foreign continental margins the subject of lengthy international negotiations.

We are disturbed by these trends. Our views regarding the distribution of work on continental margins are simple. Government should govern and regulate the use of public lands, as whatever mineral wealth they contain is the proper heritage of the people. This involves setting policy and administration in connection with resource exploitation and management of the environment. Industries should explore for and exploit resources, competing with each other to provide needed services for a profit. Academic institutions should obtain and transmit knowledge and teach. All three constituencies do research appropriate to their roles.

The traditional distribution of effort between industry, academe, and government was quite effective. For example, academics gave birth to the theory of plate tectonics, one of the most spectacular scientific advances of this century. Industry developed a superb geophysical technology and geological expertise that assisted in the discovery and exploitation of significant hydrocarbon reserves in geologically complex and logistically hostile areas. Federal agencies (e.g., the U.S. Geological Survey) have provided much-needed general background and surveys in addition to their research contributions.

Gradually, the study of continental margins has progressed to a point where, for much of the world (although by no means for all), a generalized knowledge is now available for strategic resource planning. Simultaneous with this evolution, industry developed and now can provide strategic and tactical knowledge adequate for its own needs and far beyond the capability of academic institutions. This evolution suggests that future academic research on continental margins should emphasize studies of fundamentals rather than surveys or inventories and that they should be conducted in a climate of intense dialogue between scientists in academe, industry, and government.

RECOMMENDATIONS FOR CONTINENTAL-MARGINS GEOLOGICAL AND GEOPHYSICAL RESEARCH IN THE 1980'S

PREMISES

Our recommendations* are designed primarily to contribute to the solution of major geological problems on continental margins. Because, in our view,

*Refer to Part III for more details concerning the recommendations discussed in this section.

continental-margins research by government and industry is growing and healthy, our recommendations are aimed at increasing the vitality of academic research in this field.

• We propose that academic research on our domestic margins be increased—if necessary, at the expense of work on foreign continental margins. Such an emphasis will help to clarify many fundamental scientific questions and, at the same time, provide important conceptual background for the economic development of our own continental margins. A strong program on our own margins may help to set standards for research on foreign margins.

• Academic research on continental margins is aimed at developing principles and thus differs from prospecting by industry. It does not duplicate work done under the missions of either industry or government. Government agencies and industry should remain sensitive to the fact that, ultimately, they have to rely on universities to provide and train scientists for their future programs on continental margins. It is beneficial to mission-oriented agencies that they support research in academic institutions, and they should be encouraged to do so further.

• To carry out a geophysical program effectively, academic institutions should have access to the best available technology. It is wasteful to undertake geophysical surveys using substandard equipment.

• Work on continental margins should be designed to take advantage of all available information, if possible, including that from wells drilled by industry. Geological and geophysical reconnaissance should precede detailed surveys. Drilling should be undertaken only after thorough detailed studies of such surveys and careful selection among alternate drilling locations.

The geological and geophysical work preceding drilling constitutes independent research. Drilling for scientific purposes without adequate geophysical research constitutes seeking answers to questions not yet asked.

• Like industry and government, academe should also undertake large-scale multidisciplinary studies of continental margins. Advisory committees should involve representatives from government, academe, and industry. Such committees have successfully directed the Deep Sea Drilling Project (DSDP), the U.S. Geodynamics Project, and, more recently, the COCORP (Consortium for Continental Reflection Profiling) Project on land. Groups of this type have provided effective means through which industry and government scientists can participate in the planning of such projects and can provide pertinent information. Even so, the necessary and desirable use of these advisory structures imposes a major burden in the time taken from their participants to develop and get support for large-scale science efforts. Therefore, simpler advisory structures would help to conserve the efforts and time of our best researchers.

HIGH-PRIORITY RESEARCH FOR THE 1980'S

A Sediment Dynamics Program—High Priority

We recommend that a coordinated program be developed to study sediment dynamics on continental shelves, slopes, rises, and marginal basins. This program would study the entrainment, transport, and deposition of continental-margin sediments.

Relevance

Sediments and human particulate wastes are transported and deposited by the same processes that determine sediment distribution. Therefore, studies in sedimentation are deeply relevant to the solution of waste-disposal problems.

Studies of seafloor sediment stability are a legally imposed prerequisite to the design and installation of seafloor-mounted structures (e.g., petroleum production platforms, subsea safety and production systems, pipelines, and other structures). This is an area in which science and engineering are thoroughly entwined.

Areas for Concentrated Efforts

- Marine currents, the movement and settling of sediments in estuaries, continental shelves, slopes, rises, and marginal basins.
- Interdisciplinary research efforts on the benthic boundary layer with studies of solution–sediment, organism–sediment interactions, sediment–water interface metabolism, water turbulence, velocity gradients, and seabed consolidation and stability.
- Quaternary deposition and erosion in coastal areas and on the continental slope and rise.
- Evaluation of differences in sediment distribution and sedimentary processes on modern continental shelves as compared with pre-Quaternary continental shelves.
- Multibeam swath-mapping to determine the microbathymetry of continental slopes and rises.

Coordination and Communication

Existing federal efforts in this field need coordination. One single agency should be declared as lead agency. Basic research should be done by academic scientists with support from the mission-oriented agencies. This is especially necessary for studies of the benthic boundary layer and the Quaternary.

Estimated Level of Expenditure (in 1977 Dollars)

Conversion of multibeam arrays on two NOAA
  Class I vessels to deep-ocean swath-mapping
  capability $ 500,000/vessel

Operating costs   1,500,000/vessel/year

Sediment dynamics studies (combined effort
  of federal agencies and academic
  institutions)   18,000,000/year

A Program for Geotraverses on Domestic Continental Margins—High Priority

We recommend recording long, multisensor (reflection, refraction, potential, and electrical) geophysical and geological traverses across land and marine segments of U.S. continental margins. Traverses should extend across both active and passive margins into the continental interiors. General problems that are involved include the kinematic, dynamic, and thermal evolution of active and passive margins. Geotraverses over land and water are recommended for the

East Coast, Gulf Coast, and West Coast, with a large proportion of the whole geotraverse effort devoted to Alaskan continental margins.

Relevance

Our current understanding of the genesis of active and passive margins is in its infancy. Geological history and thermal evolution of passive margins influence the potential for oil and gas. Understanding the geochemical evolution of active margins is important as a conceptual basis for the exploration of mineral deposits. Earthquakes occur frequently on active margins, and they also occur, although less frequently, on passive margins. To improve forecasting, detailed earthquake studies and monitoring are necessary.

Areas for Concentrated Efforts

- Rifting on passive margins
- Stratigraphy, diagenesis, rates of subsidence, paleooceanography during drifting of passive margins
- Evolution of continental fragments
- Continent–ocean boundaries on all margins
- Thermal history of passive margins
- Formation of accretionary wedge and forearc basins on active margins
- Geochemical evolution of active margins
- Formation of marginal basins
- Effective monitoring of seismicity on all domestic margins

Approach and Methods

The sequence of activities should begin with a compilation of previous work, progress to regional geological and geophysical studies on land and sea, and narrow down to a transect band (typically some 100 miles wide and 400–600 miles long) that is half on land and half over water. Advantage should be taken of existing well control, but in many cases, only targeted drilling for scientific purposes may answer essential questions. Bottom sampling and drilling for deeper samples will help to achieve most objectives, but such drilling should follow appropriate geological and geophysical surveys and studies.

Methods employed include earthquake seismology; multibeam sounding; multichannel seismic-reflection and -refraction work; ocean-bottom seismometers; gravity, aeromagnetics, and electrical methods; geological sampling; drilling; submersibles; and other techniques.

Communication and Coordination

We recommend that a committee be established to set up four regional working groups to draft the details and costs for geotraverses and related work in the East Coast, Gulf of Mexico, West Coast, and Alaska. The project should enlist scientists from government, academe, and industry, selected for their competence in the field. The COCORP Project and the DSDP are excellent administrative models.

Estimated Level of Expenditure (in 1977 Dollars)

The costs of a transect obviously will vary with accessibility, length, and complexity of the geological problems involved. For one typical land–sea geotransect, a very rough estimate of the costs is in the following range.

Geology and Geophysics	$ Millions
Marine geophysics	3.0 to 5.0
Land geophysics	1.0 to 2.0
Marine geological and geochemical work	1.0
Land geological and geochemical work (including helicopter support as needed)	1.0 to 2.0
Total cost for one typical net geotraverse (without drilling)	6.0 to 9.0

Outfitting Modern Geophysical Research Vessels—High Priority

Cost studies for marine surveys show that the major expenditures are for the vessel, precise navigation, and scientific and technical personnel. Once committed to these expenditures, the incremental cost for the best geophysical and geochemical equipment is comparatively small and is desirable for cost-effective acquisition of geological–geophysical information. To work with inadequate equipment is a waste of money, scientific talent, and time.

The sediment cover in most of the deep oceanic areas is fairly thin. While it is true that there are significant piles of sediment in such deepwater areas as the Argentine Basin, the Somali Basin, and the deep Gulf of Mexico, the major sediment accumulations are associated with continental margins. Scientists will not be able to understand the deep structure of these margins without modern multisensor geophysical and geochemical instrumentation, including multichannel seismic reflection and refraction and onboard data processing. Moreover, scientists will likely have to use this equipment in unconventional ways. Since the equipment involved is bulky and complex, it must be permanently mounted on a ship. The nature of operations with this equipment suggests that the ship necessarily be dedicated to work with this equipment and would not, therefore, be available for other oceanographic experiments.

The fundamental problem, then, is to find a method by which the necessary data—particularly 48- or 96-channel seismic-reflection data—can be obtained for continental-margin studies. At present, the academic research community does not have 48- or 96-channel capability. The U.S. Geological Survey does not have in-house 48- or 96-channel capability either, and the Survey needs to assess whether such capability is or is not necessary to accomplish its assigned mission.

Given that the scientific problems of continental margins require 48- or 96-channel capability to solve them, how might this be obtained?

(a) One possibility is to purchase seismic lines from a commercial contractor. This is reasonable if conventional data are sought, but it is unlikely that data from other geophysical or geochemical sensors could be collected simultaneously. The opportunities for experiments of a nonconventional sort, or innovations, would be limited, and the exposure and training of research scientists and students to these techniques would be limited.

(b) A second possibility would be to arrange cooperative programs with the U.S. Geological Survey or industry. The Survey, as noted earlier, does not have 48- or 96-channel capability and uses contractors for a significant portion of its seismic-reflection work. Moreover, it has heavy responsibilities to assess the resources of the continental margins. Industry contracts most of its seismic-reflection work, and research vessels operated by the petroleum companies are kept very busy acquiring proprietary data for these companies. Industry's vessels are not available for cooperative research programs as they are largely dedi-

11

cated to hydrocarbon exploration. Moreover, the time schedule imposed by Outer Continental Shelf lease sales has created a new urgency in the allocation of vessel time for industrial research. Current lease sales schedules suggest that, during the next decade, industry will continue to be preoccupied with the immediate problems associated with Outer Continental Shelf leasing.

(c) A third possibility would be to convert ships in the present academic fleet and to equip them for comprehensive geological and geophysical continental-margins research. These ships would then be removed from the fleet and made unavailable for other oceanographic purposes. All indications are that pressure for ship time is very strong and will remain so in the near future. There is the question of whether existing vessels in the academic fleet are suitable for such a conversion as well as whether a ship would be available.

(d) Finally, there is the possibility of adding to the oceanographic fleet ships that have been constructed and equipped specifically for multisensor geophysical investigations appropriate for modern continental-margin studies. This alternative would provide vessels that are best suited for the purpose and opportunity for innovative experiments, and the academic scientific community would possess the potential to determine the deep structure of the margins. To construct such vessels from scratch and equip them for this work would be expensive, but in the view of this Panel, this approach promises to be the most fruitful.

Therefore, we recommend that funds be made available to outfit two such vessels to operate principally (1) on the East Coast and Gulf of Mexico margins and (2) on the West Coast and Alaska margins. The two ships should have reinforced hulls to permit summer work in Arctic waters. The *Hollis Hedberg* (described in Appendix C) is an example of such a vessel.

The vessels should be equipped with onboard digital processing capable of real-time monitoring of all sensors, including seismic deconvolution and stacking of sufficient channels (at least half of those recorded) to ascertain that the data recorded are qualitatively and quantitatively adequate to solve the scientific objectives. Outfitting vessels now with less than 48 to 96 seismic-reflection channels would fail to take advantage of available technology.

While a substantial amount of processing would take place onboard, it would be necessary to plan facilities and/or contracts for postcruise playback processing to support detailed interpretations.

From industry experience, it is estimated that to build (or purchase) and equip vessels as described above would cost $12 million to $14 million (in 1977 dollars) each. Total annual costs, including shipboard operations, onboard and postcruise processing, and interpretation are estimated at $9 million to $11 million for each vessel. The availability of two modern geophysical vessels would substantially reduce the demand on ships currently used for geophysical research and permit them to be used more heavily in other fields of oceanographic research or permit a reduction in the size of the existing fleet.

The geophysical vessels should be highly available to and shared by qualified scientists, primarily members of the academic community. The ships should be national facilities based at two selected academic oceanographic institutions. Disposition of the vessels and their long-term programming would be a major responsibility that should respond to national scientific goals, while component projects should be selected on the merit of the science proposed. Planning and advice could be structured in analogy to other "big science" efforts, such as IPOD-DSDP, IDOE, or COCORP.

The two ships proposed would not necessarily have to be built or acquired simultaneously. However, in weighing whether to recommend one or two ships, we reasoned that the geology of the Pacific coast is sufficiently different from that of the Atlantic to justify two vessels. Because each coast has a representative scientific community, it seems likely that it would be difficult for each community to have access to a single ship on alternate years. Furthermore, if scientists are to make significant progress in Arctic Ocean reconnaissance, and because the Arctic season open for work is short, research would proceed faster and more efficiently with two ships.

SECOND-PRIORITY RESEARCH FOR THE 1980'S

Geotraverses and Ocean–Continent Boundary Studies on Foreign Continental Margins—Second Priority

We recommend that foreign geological and geophysical traverses be carried out if the specific problem cannot be studied on U.S. margins or if the problem is complementary to the domestic work. In general, most problems outlined for the domestic geotraverse program can also be studied on other continental margins. The sequence of activities and the research methods would be similar, and the relevance is consonant. (See pp. 8–9.)

Areas of Concentrated Effort

Areas of particular importance (in order of priority) are the following:

• *The Arctic Ocean,* where the logistic difficulties and common interests suggest that an international (U.S., Canada, Scandinavia, and U.S.S.R.) program should be undertaken. Part of this program may overlap the traverse program in Alaska. The cost for a typical Arctic Ocean transect is about $5 million. To make five transects would require about five years and a total cost of about $25 million. (See Part I, Chapter 7.)
• *The Caribbean,* where the structural and stratigraphic evolution is not understood and is the subject of intensive scientific speculation.
• *Conjugate passive margins,* which appear to be symmetrical to a midocean ridge. Studying them helps to emphasize differences that may be due to (a) climatic control of sedimentation, (b) differing late structural histories, and, possibly, (c) discontinuities across spreading centers.
• *West Pacific margins and their marginal seas.* This type of margin is not well represented in the United States. Study should be continued because the formation of marginal seas is a principal, still unsolved plate-tectonics puzzle.

Drilling on Continental Margins—Second Priority

We fully and emphatically concur with the recommendation of the JOIDES Subcommittee on the Future of Scientific Ocean Drilling (FUSOD) that an ambitious drilling program be undertaken, **but only if adequate funding is assured for scientific studies** *that include (1) broad-scale problem definition, (2) small-scale site examination and preparation, (3) sample analysis and well logging, and (4) interpretation and synthesis.*

Because continental margins extend onto land, geophysics and land drilling should be part of the same program. Doing both the science and the drilling on the marine and landward portions of U.S. continental margins would provide the United States, for the first time, with a comprehensive approach to solid-earth research on domestic margins.

This Panel also agrees with FUSOD that the high total cost of the proposed plan

13

for continued *Glomar Explorer* drilling (about $450 million) dictates that the best possible geological and geophysical reconnaissance be undertaken to provide a selection from which to choose the best drilling sites.

The recommendations of the Panel on Continental Drilling of the (now disbanded) FCCSET Committee on Solid Earth Sciences do not address the problem of continental-margin drilling on land. Such drilling could be complementary to and provide continuity with the program envisaged in the JOIDES/FUSOD proposal. We recommend that a committee be established to design specific plans for geological and geophysical traverses on the landward extensions of continental margins (see p. 13) with a view toward continental-margin drilling on land. The approximate costs for such a program need to be established.

Relevance

Drilling gives scientists a major opportunity to test their scientific predictions. To quote from the JOIDES/FUSOD report:

It is readily apparent that a knowledge of the nature and origin of oceanic crustal rocks is of utmost importance for our understanding of the structural and lithologic history of both active and passive continental margins, and indeed, for an understanding of more ancient parts of continents which were generated during earlier episodes of ocean crust formation and destruction. . . . Through deep sea drilling and field geology using submersibles we are beginning to understand the structure of the ocean . . . the (drill) hole is not the experiment, it is the ground truth that translates geophysical parameters into geological reality.

Discussion and Estimated Level of Expenditure (in 1977 Dollars)

Drilling costs are very high. Our crude estimates for a typical transect such as that described on pp. 10–11 would cost on the order of

Drilling 3–4 deep holes on land and on the continental shelf	$ 45 million–60 million
Drilling 2–3 deep holes in deeper waters	$ 60 million–90 million
TOTAL	$105 million–150 million

A rough estimate for a total drilling program on continental margins for the decade of the 1980's would be

About one third of the proposed *Glomar Challenger* drilling program	$ 40 million
Most of the proposed *Glomar Explorer* program	$160 million
About 10 shelf tests for scientific purposes, using commercial drilling vessels	$150 million
TOTAL estimated drilling costs for the 1980's	$350 million

In view of the large costs for a drilling program, we emphasize

- That previous well control—if available—should be used in the planning of all work;
- That enough geophysical and geological work must be done to provide several geotraverses from which we can choose the most suitable drilling sites to solve the scientific problems—haste in the selection process could be very wasteful; and

14

- That any drilling on continental margins requires government approval and, therefore, it is of great importance that representatives of government agencies be consulted and involved early in planning.

It is difficult to assess the importance of drilling for scientific purposes. There is no doubt in our minds that drilling key research wells on continental margins will be a fitting ultimate test of the concepts and models developed from geophysical and geological studies. The prospect of having the necessary tools available in time to test new concepts is exciting. Nevertheless, the Panel gave the following reasons for giving a second priority to the drilling program*:

(1) The fact that we are so heavily insistent on adequate and intensive preparatory geological and geophysical work preceding any drilling suggests that, at this time, the ideas for a drilling program are not sufficiently specific.

(2) In our judgment, the high-priority efforts that we have recommended stand as valuable research targets quite independent from any drilling plans. Some of us are particularly concerned that a traverse program as well as other oceanographic research projects might be refashioned into a simple surveying process focused mainly on finding a sufficient number of drill sites in time, while other scientific objectives are downgraded.

(3) We believe that, while some drilling targets could be reached only by the proposed 12,000-foot riser technology, many objectives could probably be reached with current commercial drilling technology. We judge that, with proper planning, commercial drilling vessels could be made available for scientific drilling in the 1980's. Note also that the FUSOD report identifies some sites with 13 km of sediment over basement and deep crustal holes in 18,000 feet of water with 9000 feet of penetration. The proposed 12,000-foot riser technology will not be able to reach these deep targets.

(4) The overall logic inherent in a research program leading from a geological–geophysical reconnaissance to detailed surveys and then drilling suggests that drilling should have a second priority. In other words, first the high-quality geophysical research, then the drilling.

To sum up, we believe that it is far preferable that the basic science be healthy and adequately funded before more expensive drilling is planned.

RECOMMENDED LEVEL OF EXPENDITURES FOR BASIC EARTH SCIENCE RESEARCH ON CONTINENTAL MARGINS IN THE 1980'S

Looking at the high-priority components of the proposed program (which, as previously mentioned, concentrates on academic research) and allowing for work on certain research items that fit less easily into the main program, we recommend the following estimated overall level of expenditures (in 1977 dollars) over the next decade†:

---

*Note from the chairman: These comments concern an overall, large-scale, expensive drilling program. They do not address the question of whether there are some single locations that may be ready for drilling as separate objectives outside the context of a large-scale drilling program.
†Note from the chairman: The proposed level of expenditures does not address the question of drilling. Should there be sufficient funds available to permit an additional extensive drilling effort, then the level of expenditures may be increased by some $350 million, as mentioned on p. 14.

15

Ships

Conversion of two NOAA vessels to deep-ocean swath-mapping capability	$1 million
Construction and instrumentation of two modern geophysical research vessels	$24 million to 29 million

Research

A basic geological/geochemical/geophysical research program undertaken mainly by scientists in academe and complementary to and coordinated with research undertaken by scientists in government and industry:

	Annual Costs
Marine geophysical research*	$18 million to 22 million
Land geophysics†	$ 2 million to 4 million
Marine geology and geochemistry†	$ 1 million to 2 million
Land geology and geochemistry†	$ 2 million to 4 million
TOTAL estimated costs for geological/ geochemical/geophysical research operations	$23 million to 32 million
Sediment dynamics program (combined federal agencies and academic programs)	$18 million/year
RESEARCH TOTAL:	$41 million to 50 million/year

These figures are very rough estimates, and they are large and considerably higher than proposals being considered by other groups. In light of this, we re-emphasize the following points:

- The proposed program concentrates on domestic continental margins (two thirds or more of the total program).
- The proposal includes substantial expenditures for geophysical and geological work on land.
- A large part of the land effort would be spent in Alaska.
- A significant part of the domestic marine effort would be spent off Alaska and in the Arctic Ocean.
- The geological and geophysical work is a *sine qua non* condition for any drilling plans. That is, *if we cannot afford the geology and geophysics, we should not embark on the drilling.*

RECOMMENDATIONS CONCERNING PROPOSAL WRITING AND THE DISTRIBUTION OF GRANTS

Writing proposals for research, attending committee meetings, and particularly the ever-increasing budget monitoring and administrative work required to verify the time (in minutes!) spent on each project are fast becoming dominant and time-consuming activities for many of the nation's best researchers. The time many scientists spend on various planning committees with overlapping scopes is time that they cannot devote to their research.

---

*Two geophysical vessels working on transects and nontransect work, i.e., continent–ocean boundary, seismic stratigraphy, deep crustal refraction, Arctic research. Realistically, some vessel time will be devoted to deep-ocean research off continental margins.
†Work equivalent to two *net* transects.

We recommend that the National Science Foundation (NSF) take the lead to review and streamline current research funding procedures with a view to

(a) Designing a more standardized format for research funding procedures that NSF and most other federal agencies could use;

(b) Minimizing the length of research proposals;

(c) Limiting the required length of vital statistics and scientific credentials of the requesting researcher(s);

(d) Standardizing budget forms; reducing budget details to the minimum acceptable to federal auditors; and

(e) Streamlining committees and their procedures to avoid duplications and to limit the scope and size of such advisory groups.

We further recommend that NSF study the feasibility of "progressive grant status" for institutions engaged in well-circumscribed basic research fields, e.g., continental-margins research, as follows:

(1) Project support—for individual projects;

(2) Coherent—for groups of projects at an individual institution; and

(3) Institutional—for large programs at institutions that are generally acknowledged as having a broad base of competent activity in a given field.

A move in this direction could help to streamline procedures and, at the same time, provide more meaningful relations between grantors and grantees.

CONCLUSION

The essence of our thinking is that (a) the time is ripe to concentrate more research on domestic continental margins, (b) the best geological and geophysical technology should be used for such research, and (c) drilling for scientific purposes can be justified only if it is preceded by detailed geological and geophysical surveying.

We attempted to compare expenditure levels for solid-earth continental-margins research with past expenditures but were unable to isolate the relevant figures from the multitude of budgets from federal agencies and academic institutions. The levels of expenditure we propose are high. Our most expensive recommendations (i.e., first priority, two geophysical vessels; and, second priority, drilling) address primarily the problems of data quality and, to a lesser degree, increases in the pace of gathering or volume of data. Consequently, the number of people working on continental margins on land and at sea may not increase proportionately to the increased expenditures.

Science—and particularly oceanography—is international. However, we believe that an effort with increased emphasis on national concerns is needed. The scientific problems are there in abundance, and to remain leaders in the international field, we must couple the best technology and a strongly developed understanding of our own efforts with efforts undertaken by our North American neighbors. There, we must be mindful of the obvious: Our neighbors have their own style of work on their resources and their environment. We should not take them or their cooperation for granted. However, we do share the same resources and environmental concerns and, therefore, should also share the fruits of our mutual research.

We recommend that these goals for geological and geophysical research on continental margins be part of an overall research plan for the 1980's that addresses the evaluation of resources and the management of the environment of North America and its surrounding seas—in short, the North American natural heritage. We are recommending an increase in the momentum of solid-earth science research in our own backyard.

Our recommendations should be but a small part of a larger plan that places emphasis on cohesive research efforts on continental interiors and in the neighboring deep oceans. This overall plan should be paralleled by similar efforts in the biosphere, the hydrosphere, and the atmosphere. These research concerns need to be evaluated in the context of the whole.

To undertake such a plan, we must rely on the traditional bridge-builders, the professional societies and such programs as the U.S. Geodynamics Project and its successor, to reduce stress between the constituencies and to provide much-needed neutral ground. In the absence of good communication, it is not surprising that there is a tendency toward polarization between government, academe, and industry, and even perhaps questioning of motives. If the geological and geophysical problems are to be successfully attacked with proper strength and full perspective, scientists from industry, government agencies, and the academic community must gather in a neutral setting, unfettered by hangups, inhibitions, regulations, or jealousies, to identify the challenges and opportunities that lie ahead—and to bring these to the attention of the supporting public in clear language.

I | DESCRIPTION AND DEFINITION OF RESEARCH PROBLEMS

1 | Concepts and Terminology

As illustrated by the hypsographic curve, the surface of the earth has two dominant levels: the continental platform and the oceanic, or deep-sea, platform. The transition zones between the two levels are continental margins. Oceanographers see three major submarine provinces (proceeding from the shoreline to the deep plains of the oceans).

(1) Continental shelves occupy some 7 percent of the seafloor. They are usually 65–100 km wide but may extend offshore as far as 1200 km. Typical depths are around 130 m, but can range up to 550 m. Gradients rarely exceed 1°.

(2) Continental slopes occupy 8–9 percent of the ocean floor and are typically 15–100 km wide, as measured horizontally from the shelf edge to the bottom of the slope. From the edge of the shelf, the slope ranges to about 5000-m depth, with gradients between 2° and 6°.

(3) Continental rises occupy about 3 percent of the ocean floor. Rises are not always present, but when they are, they occupy as much area as the adjacent shelf and slope, and their widths range from 0 to 600 km, with depths of 1400–5000 m.

Geologists realize that submarine margins are extensions of provinces that occupy large areas on land and that have variable morphologies. On Atlantic-type margins, a wide coastal plain marks the landward continuation of the shelf. In sharp contrast, the margins surrounding the Pacific display mountains, island arcs with volcanoes, and limited local coastal plains. The striking morphologic difference is related to the structural evolution of the various margins. The following paragraphs express that evolution in a global and geological perspective. Figure 1.1 illustrates and explains several terms that are used in this text.

Geophysical measurements and inferences indicate that the earth has an outer rigid shell (the lithosphere), which overlies a hotter, weaker, and more viscous zone (the asthenosphere). The thickness of the lithosphere is not well established, because its bottom is not easily measured or defined. Depending on the point of departure, geophysicists make differing measurements leading to different inferences (Jordan and Fyfe, 1976). An example of a map that is based on the relation of the age of the crust to heat flow is shown in Figure 1.2.

For years, seismologists have observed a sudden increase in the propagation velocity of seismic waves from lower velocities to velocities greater than 8 km/sec. This discontinuity in velocity was named after its discoverer, Andrija Mohorovičić, and is colloquially referred to as "the Moho." Rocks overlying the Moho form the crust. Deeper rocks form the upper mantle. Together, the crust and upper mantle form the lithosphere. The base of the crust beneath the oceans is usually less than 10 km deep, while the crust beneath continents is 30–60 km thick. The map in Figure 1.3 gives the approximate thickness of the crust.

PLATE TECTONICS - CONTINENTAL MARGINS

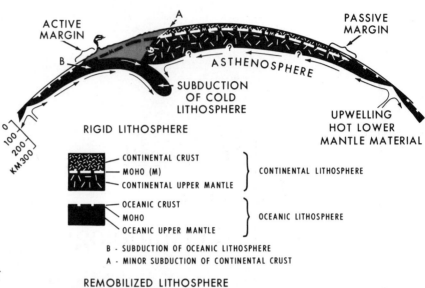

FIGURE 1.1 Schematic cross section of the lithosphere and pertinent nomenclature. No horizontal scale; vertical scale only approximate.

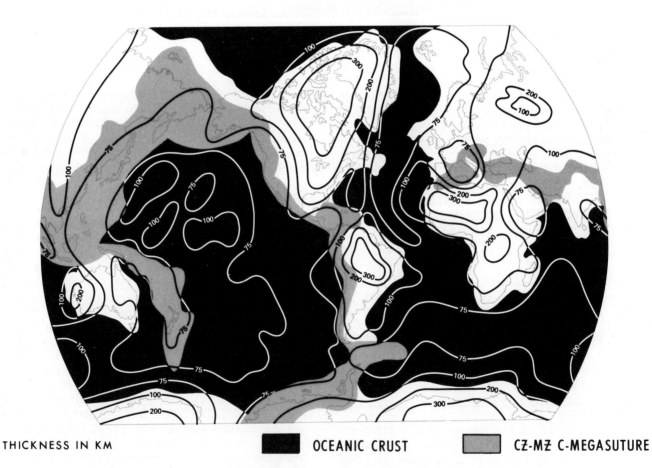

FIGURE 1.2 Thickness of the lithosphere derived from geothermal data. Surface heat-flow variations have been used to map the thickness of the lithosphere. Redrawn after Chapman and Pollack (1977) and modified by A. W. Bally.

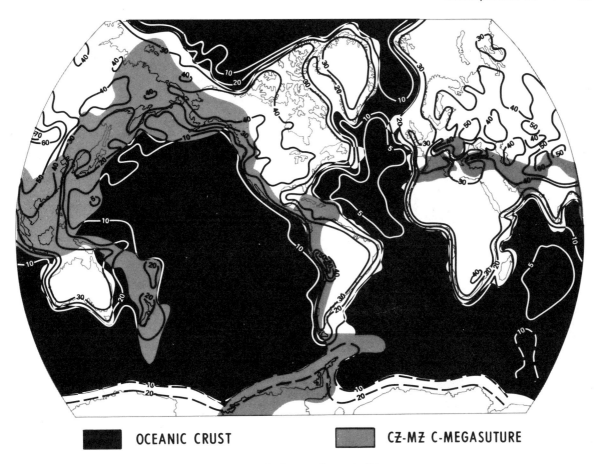

OCEANIC CRUST   **CZ-MZ C-MEGASUTURE**

FIGURE 1.3   Thickness of crust in kilometers.

The distribution of historic earthquakes suggests that the lithosphere is segmented in a number of rigid, shell-like plates. The boundaries of these plates are circumscribed by worldwide earthquake belts (see Figure 1.4). Based on the seismic motion of these earthquakes and on geological observations, three kinds of boundaries are commonly differentiated:

(1) Rifts associated with shallow-focus earthquakes (indicate extension of the crust);
(2) Transform faults with shallow-focus earthquakes (indicate a strike-slip motion along which crustal segments slide past each other); and
(3) Subduction boundaries associated with shallow- to intermediate- or deep-focus earthquakes. (Motions are complex, but compressional thrust faults dominate, while extensional and strike-slip faulting is less predominant. This zone—within which most earthquakes occur—is known as the Benioff Zone.)

Figures 1.4 and 1.5 show the distribution of the major lithospheric plates. Plate boundaries appear to be unrelated to the distribution of oceanic and continental crust. For instance, the extensional boundary of the midocean rift leaves the ocean and bisects continental crust in the area of the Red Sea and the East African Rift Zone. A similar situation is observed in the California–Great Basin area. Also, the earthquake belt surrounding the Pacific continues to the west into the Himalayas, Central Asia, and the Alpino–Mediterranean area. Thus, earthquake belts and the associated plate boundaries descend much deeper than the Moho, which underlies both continental and oceanic crust.

The plate-tectonics hypothesis is concerned with the motion of lithospheric plates. This concept visualizes the intrusion of hot igneous rocks derived from the mantle along the axes of midocean ridges. The magma cools to form new lithosphere emplaced between older lithosphere. As the molten mass cools, it registers the polarity of the earth's magnetic field. In the past, that polarity was reversed several times. Thus, series of igneous rocks of alternating polarity are successively emplaced parallel to midocean ridges. These reversal sequences allow us to follow and reconstruct the history of a growing crust beneath the oceans.

The concept of ocean-floor spreading was spectacularly confirmed by the Joint Oceanographic Institutions for Deep Earth Sampling (JOIDES) Deep Sea Drilling Project

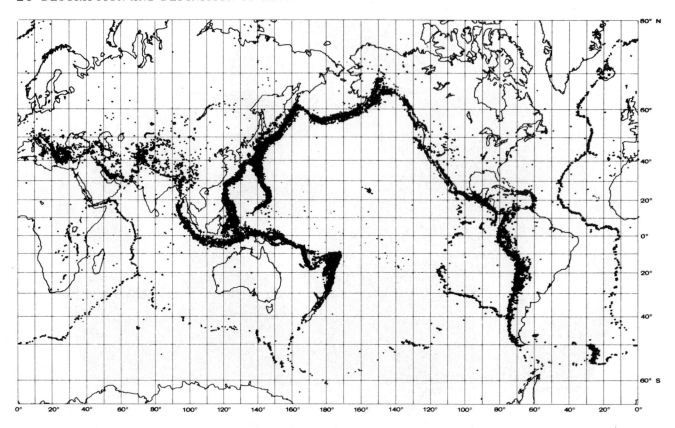

FIGURE 1.4 Worldwide earthquake epicenter distribution (1963–1977). The dark bands on the map consist of thousands of small dots, each representing the epicenter of an earthquake. The data plotted are for earthquakes with magnitude 4.5 and greater for which 10 or more stations were used to determine the epicenters. The earthquake distribution shown on Figure 2 is a highly schematic version of this distribution array. (Courtesy, Environmental Data Service, NOAA, 1978.)

(DSDP), which documented the ocean floor's history in considerable detail. Specifically, the age of sediments overlying what are believed to be the topmost (and therefore the youngest) basalt flows was predicted with remarkable accuracy. Occasionally, sediments were discovered underneath these basalts, but, so far, they have not been demonstrated to be substantially older than the basalts. Also, reflection seismic data suggest that layers exist below the basalts in some areas of the ocean. These layers need to be cored. The results of such drilling may suggest substantial revisions of the plate-tectonics concept.

Subduction is the process by which a lithospheric slab dips under a continent or an island arc deep into the asthenosphere. The most common case involves subduction of oceanic lithosphere with Benioff Zones (B-subduction). Another case involves more limited subduction of continental crust and is critical to geologists who want to know how mountain ranges are formed. This is A-subduction (A after Ampferer, who originated the concept early in this century).

In a nutshell, new lithosphere is created at midocean ridges and returns to the mantle at subduction zones.

Earthquake belts provide a fair outline of present plates, but it is more difficult to infer the outline and distribution of earlier plates. The ocean-spreading hypothesis documents the history of today's oceanic lithosphere backward into Jurassic time. Magnetic lineations and their offsets by transform faults record a relatively orderly accretion of igneous material in a dominantly extensional regime. In essence, the floor of the world oceans can be perceived as an immense extensional scar.

We can make reconstructions based on matching symmetrical linear anomalies and the presumed boundaries of continents. Such reconstructions show a steadily changing distribution of continents and ocean through time. These changes greatly influenced past ocean currents, climate, geography, and the distribution of past active and passive margins (Figures 4.7 through 4.11).

At this point, we must ask: Where are the subduction zones of the past? It turns out that the world-encircling fold belt that was formed during the Mesozoic and Cenozoic is the *counterpart* of the extensional scar. Many sedimentary basins are associated with the fold belt. On Figure 1.6 this zone is shown in white and referred to as the Cenozoic–Mesozoic Megasuture, a unit that represents the product of subduction-related processes that complemented the ocean spreading occurring since the breakup of a huge supercontinent (Pangea) in early Mesozoic times. In fact, many rocks that outcrop in today's mountain ranges are products of these deep-seated ancient processes.

Subduction may be responsible for converting oceanic crust into continental crust, although that is a simplistic view of a complex and barely perceived process. Possibly, large segments of pre-existing continental crust became heated and remobilized. Structures in metamorphic and igneous rocks suggest deformation by (ductile) flow and widespread intrusions. The rigid continental lithosphere was perhaps "softened."

Putting continental margins in perspective, we can subdivide the surface of the earth in the following way:

(1) The Cenozoic–Mesozoic extensional scar of the oceans, commonly called ocean crust.

(2) The Cenozoic–Mesozoic megasutures of the world.

(3) The combined Paleozoic fold belts represent the Palezoic Megasuture, which is bounded on both sides by sialic crust and shows Paleozoic A-subduction. Paleozoic B-subduction can only be surmised, because Paleozoic oceanic crust has not been preserved in its pristine form.

FIGURE 1.5 Distribution of lithospheric plates.

Only minor amounts of it can be found as ophiolitic sequences in folded belts. This suggests that the B-subduction process was so effective that virtually all Paleozoic oceanic crust vanished at depth during and after Paleozoic times.

(4) The Precambrian fold belts of the world represent several complex Precambrian megasutures. Again, no oceanic Precambrian crust is believed to have been preserved.

Figure 1.6 illustrates this summary and also indicates the approximate age of the economic basement for hydrocarbon accumulations. In other words, the map roughly predicts the age span of sediments that could yield hydrocarbons. For instance, basins on Precambrian crust contain Paleozoic and younger sediments, those on Paleozoic crust contain Mesozoic and younger sediments, and basins within the Cenozoic–Mesozoic Megasuture are filled with predominantly Tertiary sediments. This

illustrates how the new global tectonics—which is based on fundamental scientific research—helps to give economic geologists a simple and relevant overview of the age of sedimentary basins.

The basic types of continental margins are (see Figure 2) as follows:

(1) Passive margins: These mark the ocean–continent transition that is within a rigid lithospheric plate. Generally, they are associated with and facing a spreading midocean ridge. In our view, passive margins include the coastal plains on land and extend to the deep oceans.

(2) Cratonic margins: These are located on continental crust and, therefore, do not strictly qualify as continental margins. However, several cratonic basins are submarine and extend onto the shore. Others are entirely on land and would not concern people primarily interested in continental margins.

(3) Active margins: These are continent–ocean transi-

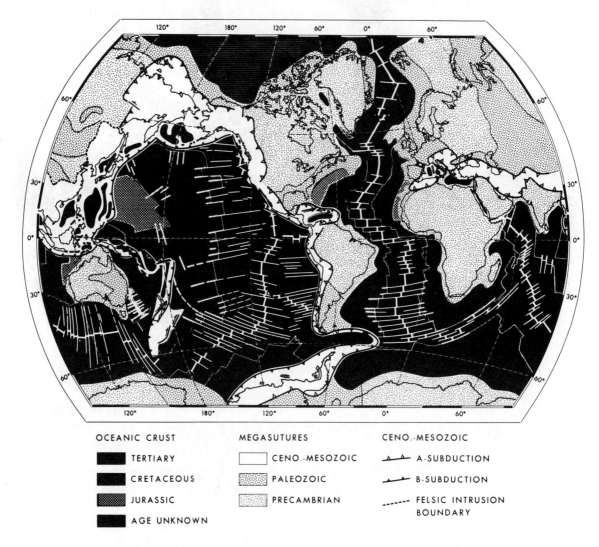

FIGURE 1.6 Economic basement and tectonics of the world.

tions associated with Cenozoic–Mesozoic subduction processes at plate boundaries. They coincide with the circum-Pacific to Mediterranean earthquake belt. In a very general manner, active margins coincide with the Cenozoic–Mesozoic Megasuture, but keep in mind that today's earthquakes indicate present subduction, while the megasuture is the product of past subduction.

It is useful to differentiate further between

(3a) Active margins associated with B-subduction of oceanic crust and

(3b) Active margins associated with A-subduction of continental crust.

(4) Transform margins: Transform-fault systems can intersect both passive and active margins. In this text, we discuss them as subdivisions of those two basic margin types.

This Panel's views are deeply rooted in the modern plate-tectonics theory. There are scientists who do not accept plate tectonics as a workable hypothesis (see Beloussov, 1975; Carey, 1977; and Meyerhoff's several publications, as cited at the end of this and other chapters). They judge that the evidence is not sufficiently compelling, or they require plausible mechanisms to explain the purported mechanics. Others believe in a dramatically expanding earth and concurrent vertical tectonics. We respect these views but believe that plate tectonics offers a fine reference frame for this report.

REFERENCES AND BIBLIOGRAPHY

American Association of Petroleum Geologists (1977). Geology of continental margins, *Continuing Educ. Course Notes, Series 5.*

Bally, A. W. (1975). A geodynamic scenario for hydrocarbon occurrences, *Proceedings of the Ninth World Petroleum Cong.,* Tokyo, Paper P.D. 1:33–44.

Beloussov, V. V. (1975). *Foundations of Geotectonics,* Nyedra, Moscow, 260 pp.

Boillot, G. (1979). *Géologie des Marges Continentales,* Masson, Paris, 139 pp.

Burk, C. A., and C. L. Drake, eds. (1974). *The Geology of Continental Margins,* Springer-Verlag, New York, 1009 pp.

Carey, W. S. (1977). *The Expanding Earth,* Developments in Geotectonics Series, No. 10, Elsevier, Amsterdam, 488 pp.

Chapman, D. S., and H. N. Pollack (1977). Regional geotherms and lithospheric thickness, *Geology* 5:265–268.

Continents Adrift and Continents Abroad, readings from *Scientific American* (1976), W. H. Freeman and Company, San Francisco, 230 pp.

Cox, A. (1973). *Plate Tectonics and Geomagnetic Reversals,* W. H. Freeman and Company, San Francisco, 702 pp.

Dennis, J. G., compiler and ed. (1967). International tectonic dictionary—English terminology, *Commission for the Geological Map of the World Memoir 7,* Am. Assoc. Phys. Geol., Tulsa, Okla.

Drake, C. L., J. Imbrie, J. A. Knauss, and K. K. Turekian (1978). *Oceanography,* Holt, Rinehart, and Winston, New York, 447 pp.

Jordan, T. H., and W. S. Fyfe (1976). Lithosphere–asthenosphere boundary, a Penrose Conference report, *Geology* 4:770–772.

Le Pichon, X., J. Francheteau, and J. Bonnin (1973). *Plate Tectonics,* Elsevier, New York, 300 pp.

Meyerhoff, A. A., M. A. Meyerhoff, and R. S. Briggs, Jr. (1972). Proposed hypothesis of earth tectonics, *J. Geol.* 80:663–692.

Pollack, H. N., and D. S. Chapman (1977). On the regional variation of heat flow, geotherms, and lithospheric thickness, *Tectonophysics* 38:279–296.

Press, F., and R. Siever (1978). *The Earth,* 2nd ed., W. H. Freeman and Company, San Francisco, 945 pp.

Rona, P. A. (1977). Plate tectonics, energy and mineral resources: Basic research leading to payoff, *EOS, Trans. Am. Geophys. Union* 58(8):629–639.

Skinner, B. J., and K. K. Turekian (1973). *Man and the Ocean,* Foundations of Earth Science Ser., Prentice-Hall, New York, 149 pp.

Smith, A. G., J. C. Briden, and G. E. Drewry (1973). Phanerozoic world maps, in *Organisms and Continents Through Time,* N. F. Hughes, ed., *Spec. Paper in Palaeontology No. 12,* The Palaeontological Assoc., London, pp. 1–42.

Steiner, J. (1977). An expanding earth on the basis of sea-floor spreading and subduction rates, *Geology* 5:313–318.

Strong, D. F., ed. (1976). Metallogeny and Plate Tectonics, Geological Association of Canada, Spec. Paper No. 14, 660 pp.

Verhoogen, J., F. J. Turner, L. P. Weiss, C. Wahrhaftig, and W. Fyfe (1970). *The Earth, An Introduction to Physical Geology,* Holt, Rinehart and Winston, New York, 748 pp.

Wright, J. B., ed. (1978). Mineral deposits, continental drift and plate tectonics, *Benchmark Papers in Geology 44,* Dowden, Hutchison and Ross, Inc., Stroudsburg, Pa.

Wyllie, P. J. (1971). *The Dynamic Earth: Textbook in Geosciences,* John Wiley & Sons, New York, 416 pp.

Yorath, C. J., E. R. Parker, and D. J. Glass (1975). *Canada's Continental Margins and Offshore Petroleum Exploration,* Canadian Soc. Petrol. Geol., Calgary, 898 pp.

2 | The Surface

SUBMARINE TOPOGRAPHY AND TOPOGRAPHIC SURVEYS

The classic passive-margin profile consisting of continental shelf, continental slope, and continental rise implies the dominance of sedimentation processes over tectonism. Sediments are derived primarily from the adjacent continent, then transported by a variety of processes over the shelf, progressively down the slope, and through submarine canyons to form the prism of sediment that comprises the continental rise.

In the Pacific, the typical active margin consists of a continental shelf and slope bounded on the seaward side by a trough, a trench, or a ridge. These topographic lows or highs often act as effective sediment traps to block further seaward transportation of terrigenous sediment.

Figures 2.1–2.3 illustrate some of the topographic variations of North American margins. Figure 2.1 shows the Atlantic shelf. It is dominated by sand ridges and channels, some of which change into deeply incised canyons on the continental slope. Figure 2.2 shows the trench-arc margin of the Aleutians and Kuriles—a classic case. Figure 2.3 reveals the unique California Continental Borderland that extends roughly 1000 km from California's Point Conception to Viscaino Bay in Baja California. The Borderland is a checkerboard of some 24 deep basins averaging 20–50 km in width and 30–110 km in length.

Complete description of a margin involves its topog-
raphy, the materials composing its features, and the processes operating in or on the mass. As a product of the last century's work by the National Marine Survey and its successors, much is known about the morphology of the surface of margins. For the shelves and coastal embayments, there are extensive records of changes in bathymetry over a century that permit scientists to investigate rates of change in the relative stability of relief forms. An adequate data base exists for describing the large-scale morphology (relief of tens of meters; horizontal scales of hundreds of meters or more), although all the data are not equally precise. Scientists are just entering the phase of investigation that focuses on smaller-scale relief forms (meters and less in relief; tens of meters or less in horizontal scale).

From a dynamic point of view, and over time scales of 100 years, appreciable changes occur in the smaller relief features at all depths over the margin. These changes result from sedimentation, erosion, or episodic movement of the substrate generated by surface and internal waves, tidal waves, surficial and deep currents, and gravity acting directly on masses on slopes. Biological agents also generate small-scale land forms; they are the critical agents affecting internal sedimentary structures in surficial deposits.

Topographic surveying is done by three government agencies and some academic institutions. They generally use 3.5- and 12-kHz echo-sounding techniques along with

28

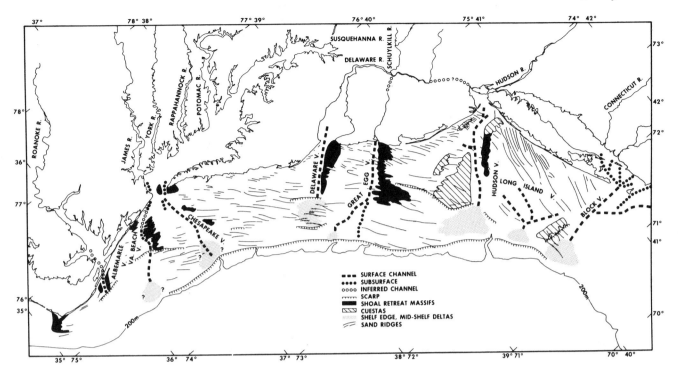

FIGURE 2.1 Morphologic elements of the Middle Atlantic Bight, North America. (From Swift, 1974.)

air-gun, sparker, and boomer sound sources. Nearly all research ships use satellite- and, along the U.S. seaboard, shore-based electronic and acoustic navigation resources. Industry performs detailed surveys over hydrocarbon lease areas and along the nearshore and shelf for mineral prospects both on the U.S. continental margin and elsewhere in the world. Deep-towed, sub-bottom-penetrating (500-Hz to 3.5-kHz), side-scanning instruments are now used by government and industry to determine the origins of shelf microbathymetry and the geological relation between surficial temporary features (e.g., sand waves) and the underlying consolidated strata (Figure 2.4). Surveys of this kind, which, in many cases, use bottom-transponder navigation, are extensive over offshore-platform and drill-rig sites and along submarine pipelines. (See Offshore Services, 1977, for a summary of the current capabilities of commercial subsea surveying.)

The principal problem, however, is still outstanding, i.e., the lack of detailed bathymetric information for the deeper parts of the shelf, the slope, and the upper rise regions of the world. The details of the tectonic fabric in these areas can best be studied with multibeam, swath-mapping arrays mounted on a research vessel (see Figure 2.5). One of the most exciting and productive techniques for detailed bathymetric and tectonic-fabric mapping of the ocean floor has been developed and used extensively by the U.S. Navy. This technique, which uses the Harris arrays, is the inertially controlled, multibeam, swath-mapping sonar. It uses a scanning mode to produce a de-

tailed contour map (along a 3- to 5-km swath of the ocean floor). The multibeam system recently used over the Mid-Atlantic Ridge showed ocean-floor features on the order of 10 m in wavelength. This made it possible to develop a finely detailed, tectonic-fabric map of the ridge. (See Ballard and van Andel, 1977.)

We recommend that the nonmilitary sector of the U.S. Government develop a multibeam, swath-mapping capability, particularly for studies on the continental slope, upper rise, and some of the adjacent deep-ocean regions. We recommend that this facility be used in cooperative studies with academic and other research organizations.

Such mapping must be accompanied by high-resolution navigation accuracy. In the past, the U.S. Navy has run, at irregular intervals, swath-mapping studies over Alvin dive sites, e.g., for Project FAMOUS. The Navy cannot devote its facilities on a scheduled basis to academic institutions. The program can best be carried out by NOAA scientists and ships, although scientists from academic institutions can augment these studies with deep-tow and submersible investigations, particularly in defining and sampling outcrops of older strata.

Swath-mapping instrumentation requires a sizable vessel, and since three of the NOAA ships—Researcher, Discoverer, and Surveyor—already have elements of the array mounted on their hulls, we recommend that NOAA install deep-ocean, swath-mapping capability on two of these vessels.

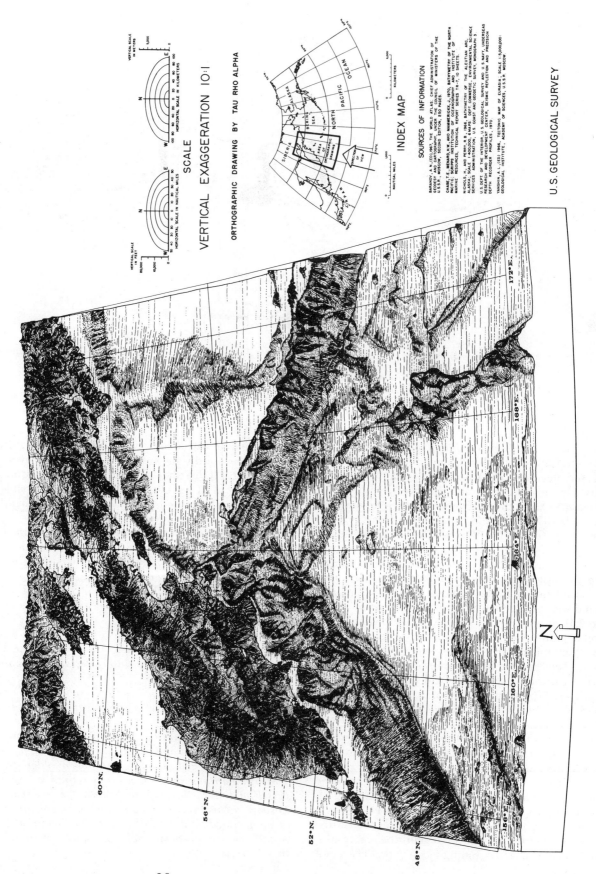

SCALE

VERTICAL EXAGGERATION 10:1

ORTHOGRAPHIC DRAWING BY TAU RHO ALPHA

INDEX MAP

SOURCES OF INFORMATION

BARANOV, A.N.,(ED),1967, THE WORLD ATLAS, CHIEF ADMINISTRATION OF GEODESY AND CARTOGRAPHY, UNDER THE COUNCIL OF MINISTERS OF THE U.S.S.R., MOSCOW, SECOND EDITION, 250 PAGES.

CHASE, T.E., MENARD, H.W. AND MAMMERICKX, J., 1970, BATHYMETRY OF THE NORTH PACIFIC, SCRIPPS INSTITUTION OF OCEANOGRAPHY AND INSTITUTE OF MARINE RESOURCES, TECHNICAL REPORT SERIES TR-7, 10 SHEETS.

NICHOLS, H., AND PERRY, R.B., 1966, BATHYMETRY OF THE ALEUTIAN ARC, ALASKA, SCALE 1:400,000, 6 MAPS DEPT. COMMERCE, ENVIRONMENTAL SCIENCE SERVICES ADMINISTRATION, U.S. COAST AND GEODETIC SURVEY, MONOGRAPH 3

U.S. DEPT. OF THE INTERIOR, U.S. GEOLOGICAL SURVEY AND U.S. NAVY, UNDERSEAS RESEARCH AND DEVELOPMENT CENTER, SEISMIC REFLECTION AND PRECISION DEPTH RECORDER PROFILES, 1970.

YANSHIN, A.L. (ED.) 1966, TECTONIC MAP OF EURASIA, SCALE 1:5,000,000: GEOLOGICAL INSTITUTE, ACADEMY OF SCIENCES, U.S.S.R. MOSCOW

U.S. GEOLOGICAL SURVEY

FIGURE 2.2 Physiographic diagram of the Aleutian–Kamchatka convergence. (Courtesy of the U.S. Geological Survey.)

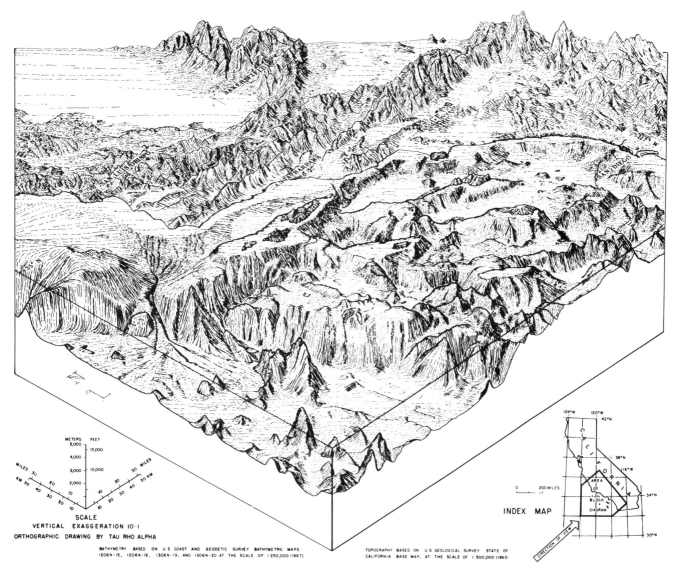

FIGURE 2.3 Physiographic diagram of the California Borderland. (Courtesy of the U.S. Geological Survey.)

SURFACE GEOLOGICAL MAPS

Geological maps of the land surface are commonplace, but they are not routine for continental margins. The main reason for this is that large areas of bedrock on continental margins are thoroughly blanketed by varying thicknesses of recent sediments. Such a sediment cover inhibits mapping bedrock distribution. Moreover, bedrock outcrops are considerably more difficult to sample at sea than they are on land.

Despite the difficulties, fairly detailed geological maps have been made for the California Borderland (Junger and Wagner, 1977; Vedder *et al.*, 1974), Canadian Atlantic continental margins (King and MacLean, 1976), and Hudson Bay (Figure 2.6); also for offshore areas of France, England, and Italy. Some maps show only the distribution of bedrock under the veneer of sediments. These maps are useful for engineering purposes, for constructing geological sections, and for stratigraphic maps.

A useful tool for collecting rock samples in water-

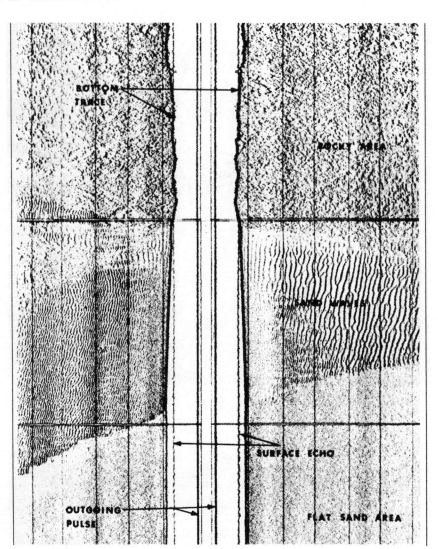

FIGURE 2.4 Typical side-scan record. Reading from the center out to the edges, differences in seabed topography are clearly shown. (Courtesy of Klein Associates, Inc.)

covered areas to construct bedrock maps is a small underwater drill, such as one developed by the Bedford Institute of Oceanography (BIO) (see Fowler and Kingston, 1975). Traditionally, hard-rock dredges were used for sampling bedrock, but it is often difficult to locate dredge-sample sites with precision or to determine whether representative bedrock is being collected. Small, electric, underwater drills avoid the high costs of surface-drilling facilities. The BIO drill obtains cores slightly longer than 5 m, and the locations of coring sites can be pinpointed with reasonable accuracy.

It is important to remember that submersibles offer unusual opportunities for surface mapping (Ballard and van Andel, 1977).

We recommend that the U.S. Geological Survey publish a series of seafloor-geology maps at appropriate scales for all the continental margins of the United States.

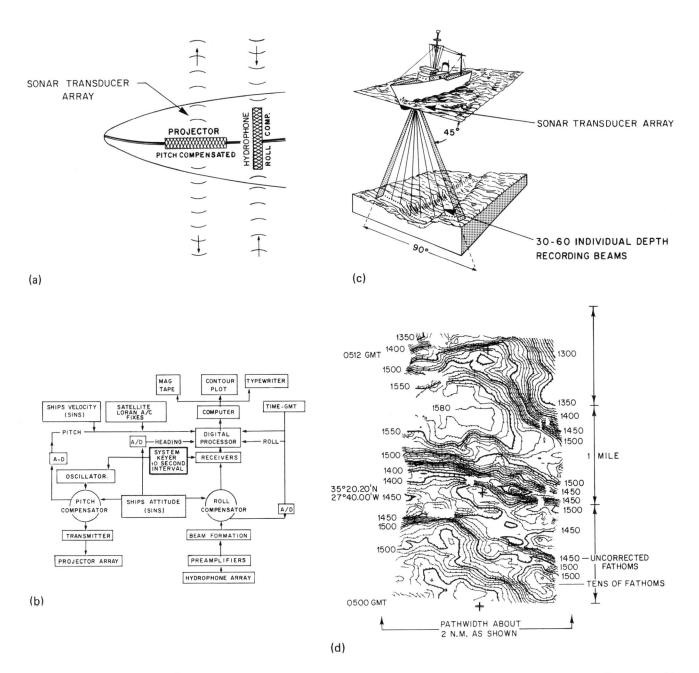

FIGURE 2.5 The SONARRAY bathymetric charting system: (a) Arrangement of the hull-mounted, crossed-fan sonar transducer array; (b) flow chart showing sonar signal processing; (c) configuration of the multibeam "fan" of sound used to depict ocean floor; (d) example of bathymetric contour chart drawn aboard ship while under way. (Courtesy of J. D. Phillips and H. S. Fleming.)

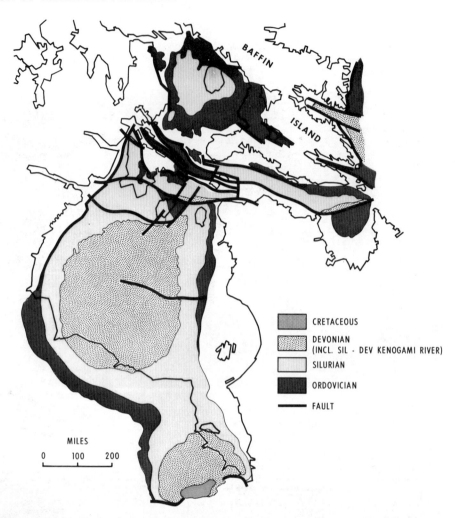

CRETACEOUS

DEVONIAN
(INCL. SIL - DEV KENOGAMI RIVER)

SILURIAN

ORDOVICIAN

—— FAULT

MILES
0   100   200

FIGURE 2.6   Geological map of Hudson Bay. The white areas are underlain by Precambrian. (Courtesy of B. V. Sanford, Geological Survey of Canada, simplified and redrawn by A. W. Bally.)

REFERENCES AND BIBLIOGRAPHY

Ballard, R. D., and T. H. van Andel (1977). Project FAMOUS: Operational techniques and American submersible operations, *Geol. Soc. Am. Bull.* 88:495–605.

Fowler, G. A., and P. F. Kingston (1975). An underwater drill for continental shelf exploration, *J. Soc. Underwater Technol.* 1(4):18–22.

Geological Society of America (1977). Spec. issues of the *Bulletin* on Project FAMOUS. Part I, 88(4):481–608; Part II 88(5):609–736.

Hutchins, R. W., D. L. McKeown, and L. H. King (1976). A deep tow high resolution seismic system for continental shelf mapping, *Geosci. Can.* 3(2):95–100.

Junger, A., and H. C. Wagner (1977). Geology of the Santa Monica and San Pedro basins, California Borderland, U.S. Geol. Survey, Miscellaneous Field Studies Map MF-820.

King, L. H., and B. MacLean (1976). Geology of the Scotian Shelf, Marine Sciences Paper 7, Geol. Survey of Can. Paper 74-31.

Offshore Services (1977). *Subsea Surveying: A Guide for Oilmen*, April.

Part I contains the following relevant articles:

Bailey, R. T. Position fixing; 1:Optical and shore based radio systems, *Offshore Services*, pp. 44–48.

Hayes, J. W. Echo sounding and seismic profiling, *Offshore Services*, pp. 55–63.

Jarvis, R. Position fixing; 2:Satellite navigation, *Offshore Services*, pp. 51–55.

Klein, M. Side scan sonar, *Offshore Services*, pp. 67–75.

Offshore Services (1977). *Subsea Surveying: A Guide for Oilmen*, May.

Part II contains the following relevant articles:

Burns, F. J. Deep-tow and digital techniques, *Offshore Services*, pp. 47–48.

Darling, G. How to select a ship for survey support, *Offshore Services*, pp. 56–60.

Krause, E. R. Soil investigations for jack-ups, pile design and fixed platforms, *Offshore Services*, pp. 50–55.

Swift, D. J. P. (1974). Continental shelf sedimentation, in *The Geology of Continental Margins*, C. A. Burk and C. L. Drake, eds., Springer-Verlag, New York, pp. 117–135.

Vedder, J. G., L. A. Beyer, A. Junger, G. W. Moore, A. E. Roberts, J. C. Taylor, and H. C. Wagner (1974). Preliminary report on the geology of the continental Borderland of Southern California, U.S. Geol. Survey, Miscellaneous Field Studies Map MF-624.

3 | Sediments and Sedimentary Rocks

SEDIMENTS AND THEIR ORIGIN

INTRODUCTION

As rocks weather, some parts go into chemical solution. Other parts degrade into sands and clays. Streams then carry these products to the continental margins and deep into the seas. Garrels and Mackenzie (1971) estimate that each year some 250×10^{14} g of material are carried into the sea. Streams carry about 85–90 percent of this matter, ice accounts for approximately 7 percent, groundwater for 1–2 percent, and wind, less than 1 percent. About 80 percent of the stream load is particulate matter, and the remainder is in dissolved form.

Nearly 80 percent of this enormous volume is trapped in continental margins. Sedimentologists track the fate, distribution, and redistribution of all this material. They began their work by mapping the topography and sediment distribution. Then they inferred systems of dispersal and transport across coastal plains, the continental shelf, the continental slope, and the deep sea.

Today, sedimentologists wish to directly measure transport rates and patterns and compare them with measurements of the fluid motions that move sediments. Sediment distribution studies were and still are important to petroleum geologists and hydrologists, who use them as analogs to map and predict subsurface reservoirs. To study and directly observe dispersal mechanisms helps to explain different distribution patterns. This work is criti-

cal, for without it, it is impossible to manage the environment of continental margins properly, where areas close to the shore are heavily polluted because of inadequate waste-disposal practices. These matters are under scrutiny by other groups, one of which is working with the Environmental Studies Board of the Commission on Natural Resources of the National Research Council (NRC). In particular, we call attention to the recent reports published by the National Academy of Sciences, *Disposal in the Marine Environment* (Ocean Disposal Study Steering Committee, 1976) and *Estuaries, Geophysics, and the Environment* (Geophysics of Estuaries Panel, 1977).

SEDIMENTATION ON ALLUVIAL PLAINS AND COASTS

Before they ever reach the marine realm, substantial amounts of sediments are deposited in coastal plains and the coastal zone. The coastal zone is the area in which terrestrial and marine processes mix. It includes the intertidal zones, large estuaries, lagoons, barrier islands, and other coastal features. Figure 3.1 illustrates the great variety of depositional environments in coastal zones.

Government surveys, academic institutions, and oil companies have been studying the coastal zone for years. Mission-oriented efforts will no doubt continue, but, fundamentally, we need a better understanding of sea-level changes as manifested in and as exhibiting the recent evolution of our coasts. As stated earlier, the present-day sea level is an ephemeral happenstance—a momentary stop in a sequence of numerous transgressions and regressions of the sea occurring over millions of years. Thus, it is in the coastal zone that sedimentary geometries have to be related to environment and where their evolution has to be placed in the frame of recent time (see Figure 3.2).

At this stage, we have no grand scheme that would allow broad-based predictions. We are still in a data-collecting phase. Many types of coastal zone need to be studied, each more or less separately. The importance of coastal monitoring has been stressed by Morisawa and King (1974). As an example, the glaciated coast of New England contrasts strongly with the transgressive

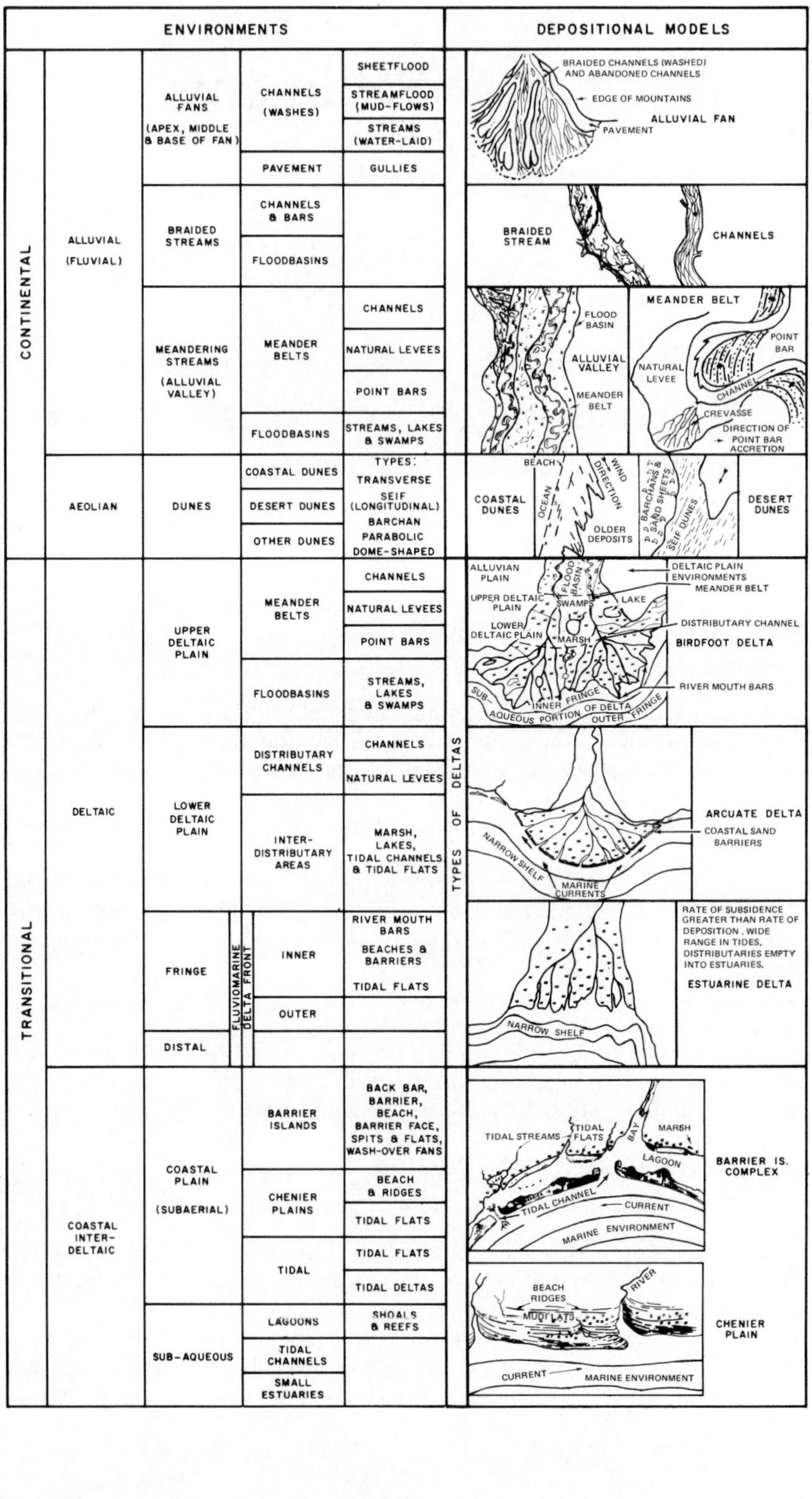

ENVIRONMENTS				DEPOSITIONAL MODELS
CONTINENTAL				
ALLUVIAL (FLUVIAL)	ALLUVIAL FANS (APEX, MIDDLE & BASE OF FAN)	CHANNELS (WASHES)	SHEETFLOOD	BRAIDED CHANNELS (WASHED) AND ABANDONED CHANNELS; EDGE OF MOUNTAINS; ALLUVIAL FAN; PAVEMENT
			STREAMFLOOD (MUD-FLOWS)	
			STREAMS (WATER-LAID)	
		PAVEMENT	GULLIES	
	BRAIDED STREAMS	CHANNELS & BARS		BRAIDED STREAM; CHANNELS
		FLOODBASINS		
	MEANDERING STREAMS (ALLUVIAL VALLEY)	MEANDER BELTS	CHANNELS	FLOOD BASIN; ALLUVIAL VALLEY; MEANDER BELT; NATURAL LEVEE; MEANDER BELT; POINT BAR; CHANNEL; CREVASSE; DIRECTION OF POINT BAR ACCRETION
			NATURAL LEVEES	
			POINT BARS	
		FLOODBASINS	STREAMS, LAKES & SWAMPS	
AEOLIAN	DUNES	COASTAL DUNES	TYPES: TRANSVERSE, SEIF (LONGITUDINAL), BARCHAN, PARABOLIC, DOME-SHAPED	BEACH; COASTAL DUNES; OCEAN; WIND DIRECTION; OLDER DEPOSITS; BARCHANS & SAND SHEETS; SEIF DUNES; DESERT DUNES
		DESERT DUNES		
		OTHER DUNES		
TRANSITIONAL				**TYPES OF DELTAS**
DELTAIC	UPPER DELTAIC PLAIN	MEANDER BELTS	CHANNELS	ALLUVIAN PLAIN; DELTAIC PLAIN ENVIRONMENTS MEANDER BELT; UPPER DELTAIC PLAIN; LOWER DELTAIC PLAIN; FLOOD BASIN; SWAMPS; LAKE; MARSH; DISTRIBUTARY CHANNEL; **BIRDFOOT DELTA**; RIVER MOUTH BARS; INNER FRINGE; OUTER FRINGE; SUB-AQUEOUS PORTION OF DELTA
			NATURAL LEVEES	
			POINT BARS	
		FLOODBASINS	STREAMS, LAKES & SWAMPS	
	LOWER DELTAIC PLAIN	DISTRIBUTARY CHANNELS	CHANNELS	**ARCUATE DELTA**; COASTAL SAND BARRIERS; NARROW SHELF; MARINE CURRENTS
			NATURAL LEVEES	
		INTER-DISTRIBUTARY AREAS	MARSH, LAKES, TIDAL CHANNELS & TIDAL FLATS	
	FRINGE (FLUVIOMARINE DELTA FRONT)	INNER	RIVER MOUTH BARS	RATE OF SUBSIDENCE GREATER THAN RATE OF DEPOSITION, WIDE RANGE IN TIDES, DISTRIBUTARIES EMPTY INTO ESTUARIES. **ESTUARINE DELTA**; NARROW SHELF
			BEACHES & BARRIERS	
			TIDAL FLATS	
		OUTER		
	DISTAL			
COASTAL INTER-DELTAIC	COASTAL PLAIN (SUBAERIAL)	BARRIER ISLANDS	BACK BAR, BARRIER, BEACH, BARRIER FACE, SPITS & FLATS, WASH-OVER FANS	TIDAL STREAMS; TIDAL FLATS; BAY; MARSH; LAGOON; TIDAL CHANNEL; CURRENT; MARINE ENVIRONMENT; **BARRIER IS. COMPLEX**
		CHENIER PLAINS	BEACH & RIDGES	
			TIDAL FLATS	
		TIDAL	TIDAL FLATS	BEACH RIDGES; RIVER; MUD FLATS; CURRENT; MARINE ENVIRONMENT; **CHENIER PLAIN**
			TIDAL DELTAS	
	SUB-AQUEOUS	LAGOONS	SHOALS & REEFS	
		TIDAL CHANNELS		
		SMALL ESTUARIES		

FIGURE 3.1 Coastal environments and models of clastic sedimentation. (After LeBlanc, 1972.)

TRANSGRESSIVE LAGOON-BARRIER COAST

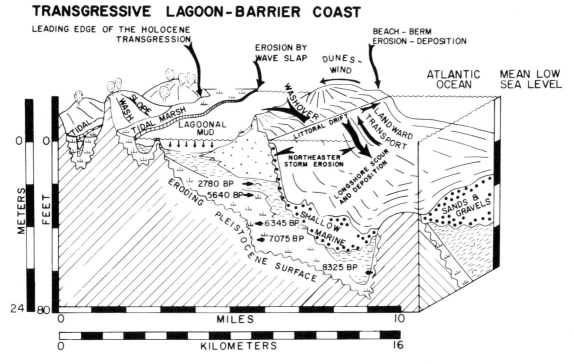

FIGURE 3.2 Three-dimensional representation of the Delaware Coastal Zone, showing both modern environments and those detected by subsurface analysis. (Slightly modified after Kraft *et al.*, 1973.)

barrier–lagoonal coast of the middle Atlantic states, the carbonate-dominated Florida Keys, the sandy lagoonal coasts of the Gulf states, the Mississippi Delta, the Gulf of California, the cliff-dominated Pacific coast, the fjords of British Columbia, and the Aleutian Archipelago. Processes occurring in one area may be entirely different from those in others. A fine example of a systematic approach to coastal geology may be found in the *Environmental Geologic Atlas of the Texas Coastal Zone* (Brown, 1972–1976).

Coastal zones constitute a field of study in which much detailed local work needs to be done and in which many disciplines should participate. The work should be tackled with the hope of evolving concepts that possess general validity. A great number of studies should be undertaken, with a premium placed on novel and differing approaches. Universities and state coastal laboratories may offer the best clusters of professionals to attack the problems. It is particularly important that local groups do these studies, as they usually can better explain the importance of their investigations to the public.

MARINE SEDIMENTS ON CONTINENTAL MARGINS

As in morphology, major regional differences exist in the broad aspects of surficial deposits. These sedimentologic provinces are not accidental. They reflect the relative influences of supply rates of material, hinterland geology, available dispersive energy, morphology, climate, and

oceanography. In some areas, provinces defined for shelves may not hold for slopes or rises.

Surficial sediments form the uppermost layers of continental margins. Typical thicknesses range from zero to a few tens of meters. General dimensions (at least of muds) can be determined in the field by low-frequency (3.5-kHz) echo sounding. Surficial sediments can be sampled using conventional grabs, dredges, box cores, piston cores, or vibracores. Usually, the sediments are Holocene or Pleistocene in age and, as such, may contain locally varying subunits that reflect the complex sea-level history of this latest portion of geological time (Figure 3.3).

At the Woods Hole Oceanographic Institution, a large (14 cm in diameter), long (20–40 m) piston corer has been developed and successfully tested. It obtained high quality cores of undisturbed sediments in water depths from 80 to 5500 m, with recovery ratios ranging between 0.76 and 0.87 (Hollister *et al.*, 1973). In our judgment, this tool will prove to be very valuable in studying details of surficial and Quaternary sediments. If scientists wish to understand paleooceanography and paleoclimatology, they must understand the history of sedimentation in the Quaternary period.

Quaternary sea-level indicators may be directly observed on shelves and can be deduced on slopes and rises. During low sea-level stands, most shelves are subaerially exposed, and large quantities of material are delivered directly to slopes and rises. At times of high sea level, the coastal zones and shelves may absorb sedi-

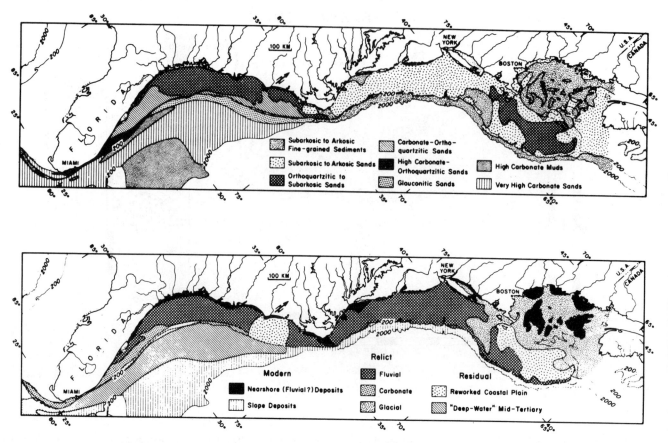

FIGURE 3.3 *Top*: Sediment types of the Atlantic continental margin off the eastern United States. *Bottom*: Source and age of the Atlantic continental-margin sediments off the eastern United States. [After Milliman *et al.* (1972). Reprinted from the *Bulletin* of the Geological Society of America.]

ments, and slopes and rises may be "starved." Thus, the surficial sediment prisms on margins may contain many diastems, disconformities, and unconformities.

Frequent eustatic and isostatic sea-level fluctuations due to the waxing and waning of large glacial ice shields have distorted the distribution of shallow-water and deep water sediment facies along the passive margins. Many continental margins have not yet returned to equilibrium since the rapid rise of sea level during the past 15,000 to 20,000 years. Therefore, the corresponding sediments are not in equilibrium with present environmental conditions. A substantial amount of modern river sediment is trapped within coastal estuaries and marshes. As a result, large portions of the continental shelves are still covered by glacial and relict sediments from times of lowered sea level. It is not clear when and how they will return to a facies distribution mode typical for nonglacial (basically pre-Quaternary) times.

Modern shelf muds generally occur only off rivers, in depressions, and in coastal areas. In many areas, modern shelf sands are predominantly biogenic in origin. Slope muds have been assumed to be modern, although locally they may represent either nondepositional or erosional conditions. Many mud deposits may be dynamic accumulations rather than static, and the rate at which material fluxes through these types of deposits should be studied.

ATLANTIC MARGIN SEDIMENTS

Along the Atlantic coast *shelf* provinces (Figure 3.3), Cape Hatteras is a prominent cusp that divides the shelf into a northern region dominated by detritus and a southern region dominated by carbonates. North of Cape Hatteras, shelf sediments are composed predominantly of terrigenous sands that reflect glacial sources off New England and Canada and fluvial input plus reworked coastal deposits from the central Atlantic states. Fine-sediment lenses are generally restricted to the nearshore. South of Cape Hatteras, the sediment cover is thin and may locally contain large amounts of reworked Miocene detritus. Skeletal sands and clean quartzitic sands are typical shelf sediments, with carbonate increasing south of Cape Kennedy in Florida and in the Bahamas. Fine, silty sands form a belt along the adjacent inner shelf. Carbonate increases offshore. Phosphorite and glauconite are common on both the outer shelf and the upper slope.

The *continental slope* north of Cape Hatteras is covered with fine-grained sediments: sandy silt on the upper slope and silty clay on the lower slope. The relative abundance of planktonic remains from coccolithophorids, diatoms, radiolarians, foraminifera, and pteropods increases offshore. Calcium carbonate content increases offshore and toward the south, with maximum values reaching 95–98 percent on the Florida Hatteras slope. Although modern Holocene muds are assumed to be accumulating over most of the slope, there are relicts of sediments that were deposited some 10,000 to 30,000 years ago. In addition to this, there are, of course, outcrops of much older sedimentary rocks. This is particularly true south of Cape Hatteras, where Miocene outcrops dominate the Florida Hatteras scarp. On the Blake Plateau, Miocene lime and phosphatic sands are prevalent.

Turbidity currents and slumps are partially responsible for the deposition of sands and sandy muds on the *continental rise*. But these clastics can also be carried by and settle out of geostrophic currents that parallel the topographic contours of the rise. The latter transport mechanism explains the dominance of high-latitude sands found on the rise off the southern United States, where clastic influx from the mainland is limited. The same geostrophic currents are also responsible for the formation of erosional furrows on the rise (Figure 3.4).

GULF OF MEXICO MARGIN SEDIMENTS

Carbonate sediments (primarily sands) dominate the shelves off both southernwest Florida and Yucatan. Moving north along both shelves, the sediment becomes increasingly terrigenous (primarily quartzose sands) and

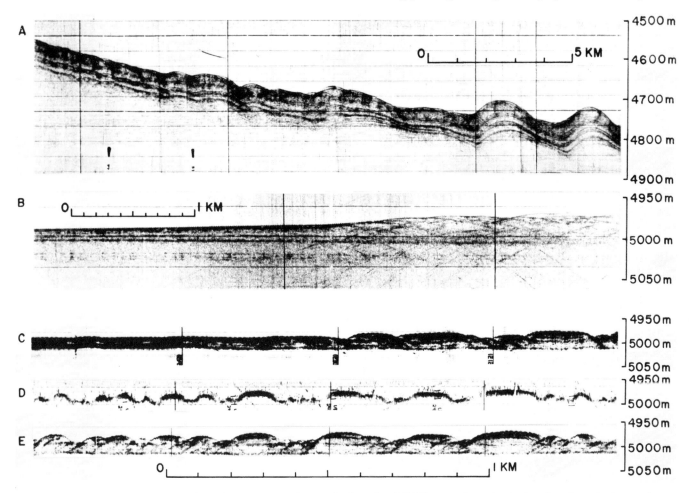

FIGURE 3.4 Abyssal furrows on the Bahama Outer Ridge. A, Surface-ship 3.5-kHz record starting in Area 1 and continuing upslope toward the crest of the Bahama Outer Ridge. The wave farthest to the right corresponds to the wave at 74°21.8′W. (East is on the left.) B, Surface-ship 3.5-kHz record of the Bahama Outer Ridge–Abyssal Plain contact. (West is on the left.) (Profile location is 28°35′N, 75°25′W.) C, Near-botton 4-kHz record of the same contact shown in B, demonstrating the abruptness of the contact. (Slight undulations in the record are caused by variations in fish elevation. The deeper reflectors are flat; see profile B.) (West is on the left.) D, Near-bottom, narrow-beam, echo-sounding (40-kHz) record of furrows responsible for hyperbolas seen on profile B in Area 2. (West is on the left.) E, Near-bottom 4-kHz record taken simultaneously with profile B in Area 2. (West is on the left.) [After Hollister *et al.* (1974). Reprinted from *Geology*, a publication of the Geological Society of America.]

often forms ridges that may be drowned longshore bars. The central northern Gulf margin is dominated by Mississippi River silts and clays, which form a thick lense that extends along the eastern Texas coast.

The carbonate province off Florida and Yucatan encompasses their steep slopes and adjacent rises. Mud-dominated sediments that display mass-wasting effects and that were deformed by salt diapir intrusions form the northern slope of the Gulf. This slope is a hummocky region of great width. It impinges upon the central Sigsbee Abyssal Plain. A large, fan-shaped deposit of turbidites forms at the mouth of the Mississippi Submarine Canyon at the eastern side of this broad slope.

PACIFIC MARGIN SEDIMENTS

North of Point Conception, California, the *shelf* is narrow, sandy, and covered with unconsolidated sediments that range from zero to a few meters thick. To the north are the broader and sedimentologically more complex Oregon and Washington shelves. These have been strongly influenced by effluent from the Columbia and several other rivers that drain the glaciated and, in the case of the Olympic Mountains, volcanic coastal ranges. Sediments are primarily sandy, and at least some of these sands are modern. Most of the shelf sediment appears capable of moving during major storms. The shelves off Alaska and Canada also are covered with reworked coarse glacial sediments. Broad, gentle, fine sand, and silt plains front the Bering Sea margins. These high-latitude shelves show an additional environmental factor—scour by sea ice (Figure 3.5).

The Southern California Borderland includes narrow (1–10 km wide) coastal shelves (Figure 2.3) with sandy to silty sediment cover. Shelf sediments between the basins contain important biogenic carbonate facies and bank-top silts, sands, and authigenic phosphatic coatings. Thick-

nesses vary but are generally thin. Many sills between the deep basins characteristic of this region are bare rock surfaces.

The Pacific *continental slope* is thinly covered by silts and clays along its entire length. Where sedimentation is high, mass wasting is an important process. In some of the Borderland basins, conditions become anoxic (or nearly so), thus allowing varved sediments to form.

The *well-developed rise* of coalesced submarine fans forms at the base of the slopes north of Point Conception, primarily in those areas in which offshore troughs or ridges are absent or where they have been breached by canyons. The Astoria Fan (Figure 3.6) off the Columbia River is a prime example.

Off Canada and Alaska, trenches, troughs, and ridges trap or block sediment flux and, therefore, have prevented fans from developing a significant rise. Similarly, the Borderland off southern California has trapped much of the sand-derived sediment.

PROBLEMS IN SEDIMENTATION

A detailed overview of problems related to shelf sediment dynamics is given in Gorsline and Swift (1977). Only selected key problems are noted in the following paragraphs.

UNSETTLING QUESTIONS REGARDING SEDIMENT SETTLING

Continental shelf sediments are dominated by sands. These relatively shallow-water materials are exposed to a large number of oceanographic influences—tidal currents, longshore transport, storms, internal waves, and so forth. Considering these forces, it is not surprising to see many different bedforms on the shelf. The problem, however, is in delineating the exact transporting agents, the

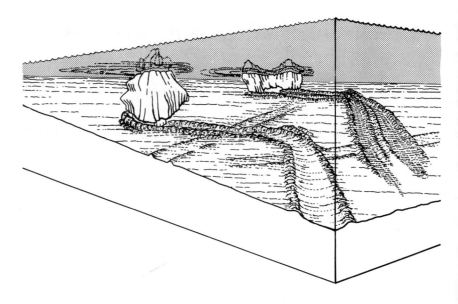

FIGURE 3.5 Schematic drawing of ice as a scouring agent; scour can occur in the Arctic at water depths as great as 50 m. (Courtesy of the U.S. Geological Survey. See also Reimitz *et al.*, 1973.)

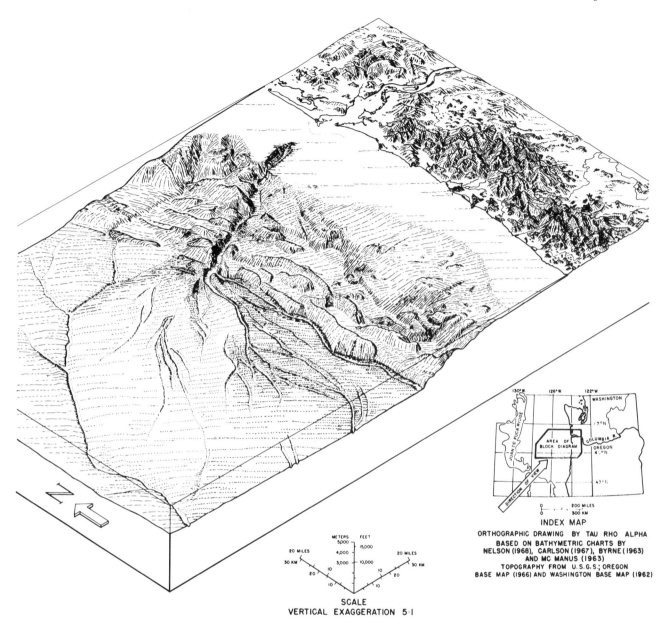

FIGURE 3.6   Physiographic diagram of the Astoria Canyon and Fan. (Courtesy of the U.S. Geological Survey. See also Nelson *et al.*, 1970.)

depth to which erosion and/or transport is effective, and the periodicity of such events. These problems have to be understood before offshore structures can be installed on shelf sands.

Therefore, people working on ocean dynamics problems quite rightly devote much time to questions of transport and to the distribution of suspended sediments and chemical elements or compounds. Their work leads naturally into a primary interest of sedimentologists: How does sedimentary material eventually settle to become sediments?

At present, rivers annually carry more than 8 billion tons of sediment into the world oceans. Most of this material settles out quickly in estuaries or the immediate nearshore area. This is particularly true in the eastern United States, where rivers tend to carry relatively small loads and empty into large estuaries. In fact, these estuaries not only trap land-derived material, they also collect shelf sands that are carried landward by shelf and nearshore ocean circulation. Only during major floods or storms can significant amounts of fluvial sediment escape their estuaries. As a result, most shelf sediments appear to be relicts of sands that were deposited during the last low stand of sea level.

How suspended fluvial material settles out within an estuary is not well documented. Presumably, sand-size material settles from suspension as a result of decreased flow velocities. Finer material, particularly clays, may flocculate as river waters mix with estuarine waters; salinities of 2–5 parts per thousand are thought to be sufficiently saline to facilitate the physiochemical flocculation of fluvial clays. Ingestion of suspended material, and excretion (fecal pellets) by both pelagic and benthonic animals may also account for considerable decrease in suspended loads in many nearshore waters.

As a result, the terrigenous fraction of the suspended load markedly decreases seaward from more than 10 mg/liter in most rivers and upper estuaries to less than 0.1 mg/liter in most middle- and outer-shelf waters. Organic constituents also decrease in absolute concentration, but far more gradually than do terrigenous constituents, with the result that suspended particulates are increasingly organic-rich offshore. Organic percentages of 10–20 percent in estuaries and 85–95 percent in outer-shelf waters are not uncommon (Figure 3.7).

Scientific interest in the distribution, composition, and fate of suspended particulates has increased in the past ten years. This is because there is more knowledge about geological and biological processes, but also because it has become necessary to know about the movement and dates of anthropogenic pollutants in the ocean. The standard method of measuring the suspended load of oceanic waters is to filter a known volume of water through a preweighed filter of constant pore size—normally, 0.45 μm. Water samples can be collected in water bottles, although settlement of particles within the bottles prior to filtering has been shown to be a potential

source of serious error. The vertical distribution of particulates is best defined by measuring the light transmission (or scattering character) of the water column by lowering a light transmissometer or nephelometer. This is often done in conjunction with the collection and filtration of several water samples to relate the light curves to actual particulate concentration. In turn, horizontal distribution of suspended solids can be defined by high-frequency (200-kHz) echo sounding, a technique that is gaining favor.

One measure of the fate of suspended particulates is the rate at which they settle through the water column. Measurement of this flux has biological and chemical implications as well as geological significance, since it represents a measure of the biological turnover and a pathway for deposition of trace metals (including pollutants) as well as normal downward sediment flux. Usually, measurement is made by placing a sediment trap into the water column or seafloor. Care is needed to ensure that the trap's configuration does not alter the hydrodynamic regime around its mouth.

Particles suspended in continental-margin and deeper oceanic waters probably do not experience physiochemical flocculation. More likely, they settle slowly through the water column, particle by particle, or more rapidly when smaller particles biologically aggregate into larger ones. Differentiating between these two types of settling is important, for the particle-by-particle settling implies a process sufficiently slow that most changes of the organic material would occur within the water column. Available data, however, show that particles settle through the water column at far greater rates, primarily as the result of the expulsion of ingested detritus by floating organisms. New research on the mode, rate, and implications of these fecal activities is in progress under the direction of Andrew Soutar at Scripps Institution of Oceanography and R. G. Douglas at the University of Southern California (personal communication).

It is recommended that a research program—as part of the DOE Program on Shelf and Nearshore Dynamics of Sediments (SANDS)—be launched to examine the dynamics of contemporary sediment transport from rivers to the continental rise and to examine processes in the benthic boundary layer and the water column.

The rates of these transport processes at various energy levels, particularly under storm conditions, need to be determined. Mass balances must be defined for the major transport systems, bed load, and suspended load. Mathematical models must be developed and field-tested.

How particles fall through the water column and move on, in, and through the sediment column once they have settled from suspension needs to be studied, especially through sediment-trap experiments. Although several such experiments are currently being conducted, few (if any) emphasize sediment settlement on marine continental margins.

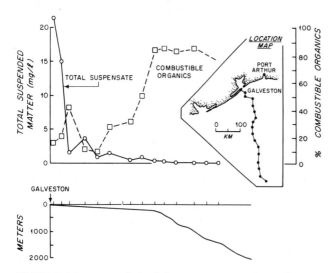

FIGURE 3.7   Suspended solids composition in the surface waters on the shelf and slope off Texas. (From Manheim et al., 1972.)

THE MYSTERY OF THE BENTHIC BOUNDARY LAYER

In 1974, under the auspices of the NATO Science Committee, approximately a hundred scientists met in France to study the benthic boundary layer (BBL). Two years and 323 pages later (McCave, 1976), they had yet to define the BBL. Biologists, chemists, and geotechnologists consider the uppermost sediments to be the BBL, while oceanographers, sedimentologists, and engineers think of it in terms of the water closest to the seafloor. Because all agree that some very important processes occur very close to the marine–sediment interface, the question of definition loses significance. The NATO meeting inaugurated a new area of interdisciplinary research in this region.

As sediment reaches the bottom of an estuary, a shelf, or the deep sea, it continues to interact with its surrounding environment. Of concern is the effect of water-mass movement very close to the water–sediment interface on sedimentary structures, for example, the creation of furrows on the abyssal floor. Engineers need to understand better the physical properties and the depositional fate of floccules formed out of fine-grained suspensates in estuaries. Temperature gradients and turbulence on the seafloor are only poorly known. The distribution of chemical elements dissolved in the interstitial waters of oxic and anoxic marine depositional environments needs to be evaluated. For instance, chemical diagenesis results in the production of gases that can weaken sediment cohesiveness. Biological activity can strengthen the substrate against erosion (e.g., worm tubes), weaken it (burrowing activity), or actually erode it (deposit feeders).

As part of the IDOE Program on Shelf and Nearshore Dynamics of Sediments (SANDS), we recommend a concentrated interdisciplinary research effort on the BBL.

Study of the BBL involves a large number of disciplines (physical, biological, and chemical oceanography, as well as geology) and will entail long-term, bottom-mounted stations to make and record such measurements. Some of the instruments (e.g., nephelometers, current meters) are available; others (e.g., recording respirometers, devices to measure sediment mixing and early diagenesis) are being tested in laboratories for ultimate use in the environment (Tamatmat, 1971; Goldhaber et al., 1977; and Guinasso and Schink, 1975). The magnitude of BBL problems requires a large-scale involvement of many scientists and correspondingly high levels of financial support.

We agree with the conclusions reached at the 1974 NATO conference that high-resolution data are necessary to understand the following problems:

- Solution–sediment chemical interaction;
- Metabolism at the sediment–water interface;
- Organism–sediment relationships;
- Variation in water turbulence and stability;
- Velocity gradients, other physical and chemical parameters; and

- Deposition, consolidation, and stability of the seabed.

Within these general problem areas, the following were defined as needing special attention:

- Long-term (1 day $< X <$ 1 year) measurements of turbulence characteristics and mean flow velocities within the BBL with different bottom-roughness elements to obtain reliable values of bed shear-stress and diffusion coefficients;
- Interchange processes between interstitial water and the viscous sublayer of the overlying water;
- Determination of chemical, biological, heat, and momentum fluxes and gradients within the BBL, the determinations of rates and the mechanisms that produce them, and their interrelationships throughout the BBL;
- Chemical, microbiological, and mineralogical studies of the suspended particles in the nepheloid layers, their comparison with the substrate and the rate of particular interchanges across the viscous sublayer;
- The effects of varying degrees of flocculation on rates of deposition and erosion under varying shear stresses and different structures on the BBL, and the causes of flocculation variations;
- Determination of factors that control the downward rate of sediment input from the BBL to the bottom;
- Determination of the *in situ* metabolism rates of the different size groups, taxa, and metabolic types of various benthic communities in both aerobic and anaerobic, high-pressure and low-pressure conditions;
- Determination of the relationship between chemical gradients and depth of bioturbation, transport vectors, and impact of pore-water exchange compared to particle advection;
- Long-term (time-lapse), *in situ* observations on macrofaunal behavior in order to understand their physical effects on the bottom and their importance in resuspending sediment;
- Identification of organic and metal-organic compounds in the sediment, their physical-chemical properties, and modes of formation of authigenic phases; and
- Determination of the critical factors that determine the basic geotechnical properties of sediment, e.g., shear strength, compressibility, elastic-wave speed, and attenuation.

This still gives us no single definition, but the problem areas and the activities suggest that the BBL involves the upper 5–20 cm of sediment on the seafloor that is particularly mobile, burrowed, and reworked by benthic organisms and that undergoes early diagenetic changes. Summarizing, the BBL deals with the soft top layer of the seafloor and the immediate water column overlying it. Why, then, so much interest?

1. Many, if not most, of the materials that ultimately become resources of one kind or another pass at some

time during their evolution through this layer and are modified in a major way in the process.

2. All resources of the sedimentary column, whether old or new, deep-seated or surficial, participate in the diagenesis of the sediment. The stage for diagenesis is set in the BBL. Diagenesis cannot be understood without first knowing considerably more about the processes that go on in the BBL.

3. The chemistry of ocean water itself may be influenced in a major way by fluxes across the BBL (Kaplan, 1974).

4. Especially in the deep sea, the importance of depositional and erosional processes has been grossly underestimated.

5. In extracting inorganic marine resources, it is essential to understand the effects of extraction on the seabed.

THE CONTINENTAL SLOPE: A KEY PROBLEM AREA

Morphology predicates most geological studies. In terms of morphology, we probably know less about the continental slope than any other physiographic region in the oceans. We need to know the morphology of the slope before we can begin to understand the nature of downslope sediment movement, stability of the slope sediments, and the late Quaternary history of the slope. Such studies require narrow-beam, swath-mapping echosounding capabilities, which, until recently, have been available only on a few Navy ships. The National Ocean Survey (NOS) has begun to develop such techniques for

their vessels. Also, more detailed, highly accurate navigation is required. Investigations from submersibles are necessary for closer mapping of both topography and outcrops.

Rates of accumulation on the North Atlantic slope and rise appear to have decreased markedly during the past 8000 years, again reflecting the relatively small amount of terrigenous material escaping the inner margin. If a major source of deep-sea sediment is the modern erosion of slope and rise sediments, a far greater portion of the outer margin than commonly believed may be nondepositional or erosional. Documenting this has significance for applied research as well as for basic research. For example, it is essential to know the long-term slope stability before an offshore structure, such as a drilling tower, is installed on the margin floor.

Quaternary glacial and interglacial sedimentation appears to be anomalous to the depositional regime throughout most of geological history. Hence, we must put aside uniformitarian ideals and study, compare, and contrast earlier, more typical, sequences with those deposited during a glacial eustatic regime.

We recommend an intensive study of morphology and the deposition and erosion of surficial sediments on the slopes and upper rises of both passive and active continental margins.

We would like to know whether sediments currently accumulate on the slope and rise or if large portions are being eroded. While the U.S. Geological Survey (USGS) has undertaken such studies on the shelf, both the Na-

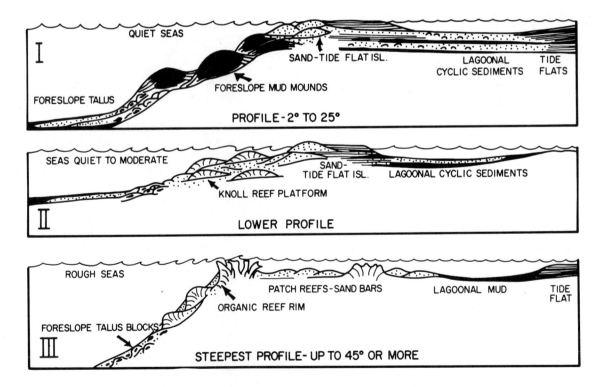

FIGURE 3.8 Three types of carbonate-shelf margins. (After Wilson, 1974.)

tional Science Foundation and the Office of Naval Research have less actively funded such studies on either the shelf or slope. Documenting the late Quaternary sequences and rates of accumulation will necessitate coring the upper few to tens of meters of the sediment column in close conjunction with detailed topographic studies.

CARBONATE GENESIS

Over the past two decades, there have been major advances in our knowledge of the processes of deposition and subsequent diagenesis that lead to the formation of carbonate rocks. This knowledge has made it easier to understand ancient carbonate rocks, an understanding necessary to solve exploration and production problems encountered in both mineral and hydrocarbon exploitation. (Three different carbonate models are depicted in Figure 3.8.) Most of this success has been realized in studies of ancient carbonates deposited over broad areas of the continent or in associated epicontinental basins. Relatively little is known about carbonate depositional processes at continental margins. The current state of knowledge is reviewed in Taylor (1977).

Carbonate platforms are essentially sediment factories that use biological and mineral nutrients from the sea to produce large volumes of carbonate particles. The potential for carbonate production must be high at continental margins because of their immediate access to the immense oceanic nutrient reservoir. Two questions that need answering are (1) what are the limiting factors that control carbonate sediment production rates, and (2) what is the disposition of the large volumes of sediments produced? Both of these questions are being pursued in the realm of general carbonate research, but they need to be studied in the specific context of continental-margin environments, where some factors (e.g., upwelling, complex surface-currents, relative sea-level changes related to tectonic activity) are certain to differ from the more stable epicontinental-shelf environment. Of particular interest are the distribution patterns of sediments generated in shallow water that find their way into deeper waters and interfinger with pelagic sediments.

GEOCHEMISTRY AND DIAGENESIS OF SEDIMENTS

So far, most of the discussion has centered on the distribution and deposition of surficial sediments. Of equal or perhaps even greater importance is the geochemical regime in both surface and subsurface sediments. We have less knowledge about the geochemical distributions and processes in continental-margin sediments than about deep-sea sediments. In discussing the various problems and needs, we have separated the inorganic from the organic aspects of the problem, realizing that in many instances, the problems involve complex interactions of

both. The following comments on inorganic geochemistry partially overlap and complement what was said in the section on the benthic boundary layer (see pp. 43–44).

INORGANIC GEOCHEMISTRY

A still-outstanding basic research task is to formulate a geochemical budget of the world oceans. The amount of solid and dissolved constituents entering the ocean by weathering and other processes can be estimated with a fair degree of confidence. Past estimates of the flux of these constituents into and out of margin sediments have depended mostly on analysis of deep-sea sediments, primarily because they have generally high concentrations of minor and trace elements. There are several problems with this approach. First, these earlier estimates have been outdated by new data gathered from ocean drilling, by new developments in analytical chemistry that enable a level of analytical precision previously unattainable, and by changing concepts of global tectonics. Second, deep-sea sediments constitute less than 20 percent of the total oceanic sediment volume, whereas continental-margin sediments comprise nearly 80 percent. Thus, in order to understand a global geochemical budget, scientists must take continental margins into account.

In particular, researchers need detailed trace-metal evaluation for those elements that are difficult to analyze (and thus frequently neglected), such as tin, tungsten, germanium, molybdenum, arsenic, antimony, bismuth, silver, gold, tantalum, niobium, selenium, and uranium.

Manganese nodules may play a particularly important role in controlling the budget of some trace metals in seawater. These nodules, like tree rings, record metal concentrations in their environment of deposition. They differ from tree rings in the extraordinary length of time that a single nodule can represent—as much as 4 million years. However, it is difficult to incorporate ferromanganese oxide phases into ocean-balance equations (as well as into paleoenvironmental considerations) because there are serious deficiencies in both our information and our theories. (1) The total distribution of nodules and their spatial concentrations on the seafloor are debatable. (2) The rate of nodule accumulation in the deep Pacific has been measured, but in other areas it is totally unknown. (3) The actual sources of metal that enrich nodules are not known. Most chemical resource evaluations quoted today are based on data more than 15 years old.

Continental-margin sediments offer significant potential as raw materials. A wide range of such materials is created by separation and sorting, by organic deposition, and by chemical precipitation. Some examples of known potential resources in addition to manganese deposits are sand and gravel, titanium placers, and phosphorite deposits. Further technological development may help to identify other potentially valuable mineral resources, such as tin and tungsten, that could be useful by-products in the refining of titanium minerals. Also, novel processes have been patented for using contaminated or naturally occurring, organic-rich, offshore muds in ceramic manufacturing.

To reiterate, we do not understand the flux of dissolved and particulate solids into and out of the seafloor. In some oceanic areas, nutrients and other dissolved constituents are supplied from bottom sediments in greater quantities than from organic decomposition either at the surface or within the water column. Until very recently, these fluxes have been almost completely ignored in ocean-balance calculations. They may be particularly critical for the nutrient budget in coastal waters (such as bays and estuaries) and in areas of high surface-productivity or in semi-enclosed basins (such as those in the California Continental Borderland). In shallow waters, excess nutrient supply may continue to promote secondary organic growth where pollutant influx via surface water has been stopped or greatly diminished.

A further complication in understanding the flux of dissolved constituents is that nutrient flux gradients are the highest in the upper 5–20 cm of the seabed—an interval sufficiently thin to have been sampled only lightly by previous workers (see Figure 3.9). Further study of more closely spaced samples is required.

HYDROLOGY AND HYDROCHEMISTRY OF SUBSURFACE WATERS

Fresh water is probably the most important resource we have. Subsurface reservoir systems extend from land areas into the continental margins. Although the hydrology and hydrochemistry of these water systems on land is relatively well known, their offshore continuation is only poorly understood.

Recent USGS drillings on the Atlantic margin have shown that virtually the entire Atlantic shelf is underlain by lenses of fresh water and brackish water. Presumably, some of these lenses represent relict groundwaters formed during former low stands of sea level; others may result from offshore flow of modern groundwaters. These lenses can form critical barriers against saline intrusion into coastal freshwater reservoirs (Figure 3.10). Because such phenomena have been largely ignored by land hydrologists, we need to explore the chemistry and migration of both onshore and offshore subsurface waters and how they relate to each other.

Another reason for pore-water studies is to get a clearer understanding of the sedimentary, chemical, and hydrochemical properties of shallow sediments in shelf areas. Moreover, we need information on fluid and gas distribution and on the permeability, sealing capacity, and hydrodynamic (migrational) continuity of waters in deep strata. Unfortunately, many of these data have not been collected, because drillers are reluctant to sample pore waters during drilling operations. However, sampling for pore waters takes little or no additional effort, and the insights we might have with these data could

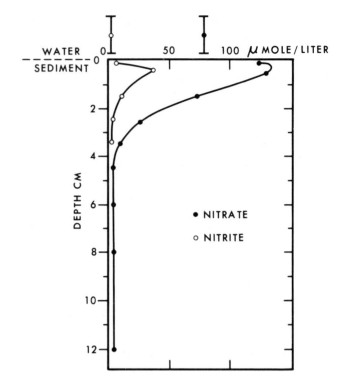

FIGURE 3.9 Regeneration of nitrate and nitrite in the pore waters within the upper centimeters of a core from the Scheldt Estuary, Belgium. [After Billen (1975); cf., Berner (1976).]

prove invaluable both for predictive purposes and for understanding the character and migration of subsurface waters.

We recommend that strong efforts be made to adequately sample and analyze subsurface waters on continental margins. This kind of data can be obtained in the course of drilling operations, and, therefore, the cooperation of the parties concerned with such operations is required.

CARBONATE DIAGENESIS

Included in carbonate diagenesis are most of the processes that transform a loose, porous carbonate sediment into a coherent rock that is usually hard and generally not very porous. The factors directly controlling diagenesis are temperature, pressure, and chemistry of the pore fluids. These direct factors are related to indirect factors such as relative sea-level fluctuations, geothermal gradients, burial rates, compaction, and subsurface fluid movement, all of which have a specific range of expression in the continental-margin environment. Tying these relationships together is necessary to predict the properties of carbonate rocks in the continental margins.

Essential to carrying out these studies are measurements of heat flow and the movement and chemical variations of subsurface waters, as mentioned elsewhere in this report. Geochemical and mineralogical studies of geologically well-defined rock groupings should also be made to determine if it is possible to sort out the history of diagenetic changes and how they relate to the history of the margin. Much could be gained from diagenetic studies on continuously cored deep wells on the Florida–Bahamas platform.

ORGANIC MATTER IN SEDIMENTS

Organic matter in recent and ancient sediments of continental margins comes from both terrestrial, transported, plant debris and indigenous marine life that grows and flourishes in the water column over the shelf. Areas of most abundant algal and microplankton growth, and thus the richest areas in terms of available carbon, are along the outer edges of the continental shelf. The abundance of marine life in this outer-shelf environment is often the result of nutrient-rich ocean currents that upwell at the shelf edge.

Another favorable environment rich in marine life and land-derived materials occurs in waters where major rivers discharge into the ocean. In a seaward direction, organic constituents decrease in absolute concentration, but far more gradually than terrigenous constituents. The result is that suspended particulates are increasingly organic-rich offshore. (See also Figure 3.7.) Particulates containing 10–20 percent organics in estuaries and 85–95 percent in outer-shelf waters are not uncommon.

The presence and diagenesis of organic matter in sediments can influence sediment formation and properties. Such physical properties as the degree of flocculation and cementation are influenced by organic materials and their transformation in pore waters. Organic matter coating mineral grains may change the surface properties of the minerals and thus affect their physical properties. These kinds of changes influence the susceptibility of continental-margin sediments to movement by such mechanisms as resuspension and slumping.

The most intense zone of biological activity, and therefore of chemical and biochemical transformation of organic matter, takes place in the upper few tens of centimeters of the surface sediment. Below this depth, biological and biochemical activity falls off rapidly. The organic matter available in this upper layer determines the numbers and types of organisms living in this zone. Their activity produces profound changes in the chemical environment, altering such parameters as Eh, pH, and metal equilibria between sediment and interstitial water. Bottom-feeding animals can stir up benthic sediments, causing them again to go into suspension and to move laterally.

Alternatively, in anoxic sediments, microorganisms can convert organic matter and CO_2 into methane gas. If the amount of methane present exceeds saturation at *in situ* bottom temperatures and if clathrate formation (see pp. 50–51) is not possible, methane might form gas bubbles in the sediments that may destabilize them enough to cause massive slumping. This mechanism could be of particular importance in continental margins, where massive slumping is common and where the rapid depositional rates that are required for microbial methane production are found.

At present, the source of outer-margin organics is not clear. How much is planktonic; how much is land-derived? Also, sediments serve as a sink for organic pol-

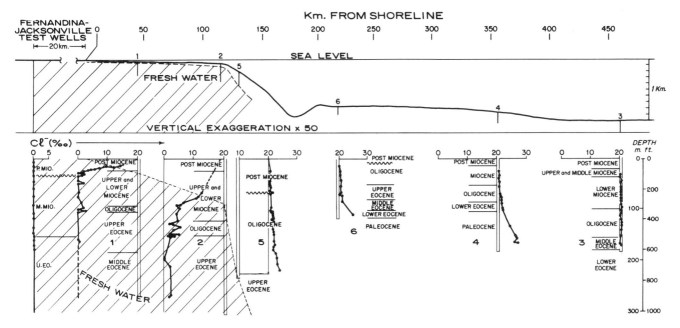

FIGURE 3.10 Distribution of interstitial chloride in JOIDES drill cores off northern Florida. "Fresh water" includes some brackish interstitial fluids. (From Manheim and Sayles, 1974.)

lutants, such as polynuclear aromatic hydrocarbons and polychlorobiphenyls (PCB's). As with naturally occurring organic molecules, the conversion, fate, and influence of these compounds on benthic communities is only poorly known. Conversions of these compounds may serve as useful tracers for geochemical processes, since, as with radionuclides, their introduction into the margin system has been fairly recent.

THE ORIGIN OF PETROLEUM

Much progress has been made in this field during the last few decades. (For summaries, see Tissot *et al.*, 1974; Hood and Castaño, 1974; Dow, 1977a; and Tissot and Welte, 1978.) Commonly, several stages of petroleum formation are differentiated.

Formation, Deposition, and Preservation of Organic Matter

Abundant organic life is usually restricted to the upper water layers. In the aerobic zone, most of the remnant organic matter oxidizes into CO_2 and H_2O by aerobic bacteria. The process is completed in the top few centimeters of the sediments, where bacteria abound, and where the high oxygen consumption causes anaerobic conditions to occur below a thin, superficial aerobic layer. There, anaerobic bacteria take over. They reduce nitrates and sulfates and may produce methane from much of the remaining organic matter. Microbially resistant particles such as woody particles, pollen, spores, waxes, and resin remain preserved.

In stagnant water bodies, the interface between aerobic and anaerobic realms moves from within the sediment into the water body. The same degradation of organic matter occurs, and typically, only minor amounts of organic matter remain preserved.

Only exceptional conditions, such as the abundance of nutrients from upwelling waters, lead to high photosynthetic activity and accumulation of organic matter. In such circumstances, bacteria and algae consume unusually large amounts of oxygen, causing a reduction of the aerobic zone. The dead remnants of these algae and bacteria sink to the bottom, where anaerobic bacteria start working on them. The end product of the reworking process consists of bacterial bodies mixed with microbially resistant parts of planktonics, pollen, spores, waxes, resins, etc. (Lijmbach, 1975). The incomplete degradation

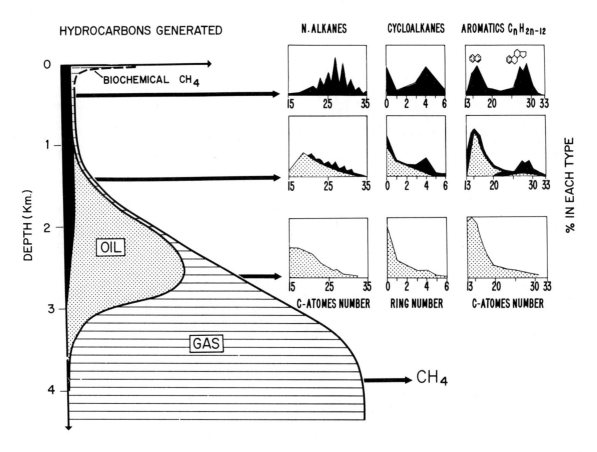

FIGURE 3.11 General scheme of hydrocarbon generation. Depth scale represented is based on examples of Mesozoic and Paleozoic source rocks. It is only indicative and may vary according to nature of original organic matter, burial history, and geothermal gradient. (Tissot *et al.*, 1974.)

of organic matter occurring under these exceptional circumstances leads to the accumulation of organic-rich sediments.

Soon after burial, the organic matter forms large molecules, or kerogen, which is characterized by its insolubility in organic solvents such as benzene, chloroform, or carbon disulfide.

Generation of Hydrocarbons

Lithified, organic-rich sediments form source rocks. These are dark, fine-grained, often laminated sediments that are believed to be currently or formerly capable of expelling petroleum. During burial, these source beds are heated and may generate commercial quantities of hydrocarbons. A commonly accepted scheme for hydrocarbon generation with depth is shown in Figure 3.11.

In recent sediments, the hydrocarbon content is low and originates from microorganisms contained in the sediment. An exception is bacterially formed methane (marsh gas). With increasing depth, the oxygen content of the original kerogen decreases by removal of CO_2 and chemically bound water. Then, at some critical level of organic metamorphism, new hydrocarbons are generated by thermal cracking of kerogen. This is the main stage of petroleum generation. At greater depths and with increased temperatures, cracking becomes excessive; the kerogen and the retained oil break down into light hydrocarbons and eventually into methane. This becomes the zone of gas generation.

In addition to the chemical measurements shown in Figure 3.11, other tools allow us to measure the progress of the maturation process of kerogen, i.e., of the translucency of pollen, spores, and the kerogen itself, for example, the vitrinite reflectance method. The transmitted-light kerogen method ranks kerogen color on a 1–5 scale, with Stage 1 representing thermally unaltered material and Stage 5 representing blackened carbonized kerogen.

Vitrinite is a coal material. Its reflectance change is measured as a function of diagenetic alteration. It places a rock unit within the coalification range of peat to anthracite. Further refinement of vitrinite reflectance permits inference on the stages of organic maturation and "deadlines" for oil and gas occurrence (Castaño and Sparks, 1974).

Three basic types of kerogen can be recognized on the basis of their alteration path with increased burial. In Figure 3.12, path I includes algal debris and excellent Middle East source rocks, path II includes good source rocks from North Africa and other basins, and path III corresponds to less productive organic matter and may include gas source rocks. Pertinent to remember is that different types of source rocks yield different mixes of oil and gas.

Expulsion and Migration of Oil and Gas

Following their generation, hydrocarbons have to be expelled from fine-grained source beds into porous and permeable carrier beds. These either allow the hydrocarbons to escape to the surface or lead them to traps where the hydrocarbons accumulate in commercial quantities. Structural traps (e.g., anticlines or fault traps) are differentiated from stratigraphic or permeability traps (e.g., reefs, sand pinchouts, or paleomorphologic traps on unconformities).

Physical, Thermal, and/or Biological Alteration of Crude Oil

Such alteration is common in reservoirs and leads to the formation of heavy oils and tar sands.

Characterization of Microscopic-Size Particulate Organic Matter (Kerogen) Along North American Continental Shelves—An Illustration of the State of the Art

A very limited quantity of geochemical information has been published from the continental shelf through the Deep Sea Drilling Project (DSDP). Most of the geochemical information consists of analyses of organic and carbonate carbon percentage. Hydrocarbon gas data are very rare. No observations of organic-matter (kerogen) types have been made in the DSDP geochemical program along the North American continental shelf, so existing conclusions were derived from literature sources.

Paleozoic Kerogen

Samples of Paleozoic rocks are limited to areas of the Grand Banks. Here, mainly continental Devonian and Carboniferous strata yield herbaceous, woody, and coaly material. Only trace quantities of the oil-generating, amorphous type of kerogen have been identified. No information on Paleozoic kerogen from the U.S. West Coast or Alaskan continental margins has been reported.

Organic Matter from Mesozoic Rocks

On the Scotian Shelf and Grand Banks, Upper Triassic and Lower and Middle Jurassic sediments, again, yield mainly terrestrially derived herbaceous, woody, and coaly material, with a paucity of amorphous kerogen. The same kerogen types prevail as far south as the Baltimore Canyon area, where the COST B-2 test well was drilled. In the upper Middle and Upper Jurassic, increasing recovery of amorphous kerogen probably reflects marine influences, yet amorphous kerogen never predominates. Marginally marine to nonmarine Lower Cretaceous strata again yield woody–coaly types of kerogen. The Albian–Cenomanian marine transgression is marked by increased percentages of amorphous kerogen and, in many areas, is the dominant type recovered. This same situation prevails throughout much of the Upper Cretaceous along the continental margins north of the Baltimore Canyon area. No information on organic-matter types is available for the West Coast or Alaskan continental margins.

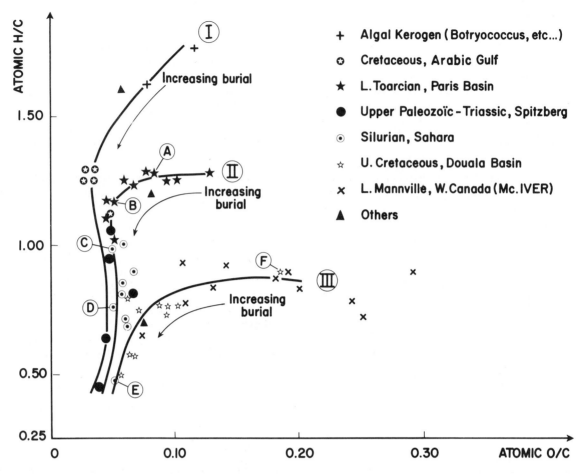

FIGURE 3.12   Examples of kerogen evolution paths. Evolution of kerogen composition is marked by arrow along each linear path. (Tissot *et al.*, 1974.)

Organic Matter from Tertiary Rocks

Amorphous organic matter is locally present in Tertiary sediments along the U.S. East Coast offshore. It becomes the dominant organic constituent in residues from the Labrador Shelf. This reflects the gradual return to marine sedimentation in the Tertiary. The same situation prevails on the U.S. West Coast offshore, where, based on limited data, amorphous organic matter and marine foraminifera appear to be common constituents in Tertiary sediments.

Kerogen Alteration Levels as a Reflection of Past Thermal History

Vitrinite reflectance of Paleozoic sediments on the Scotian Shelf are generally within the main generating phase for liquid and gaseous hydrocarbons. East-northeast on the Grand Banks, Newfoundland reflectance values in Mesozoic sediments tend to be lower than in the Paleozoic, and thermal requirements for liquid hydrocarbon generation are not met until well depths of 2400 m

are attained (see Cassou *et al.*, 1977). Transmitted-light kerogen observations confirm this depth figure in Mesozoic sediments and in the Baltimore Canyon area. Tertiary alteration values along both the East and West Coasts show the sediments to be immature and display preliquid hydrocarbon-generation characteristics of yellow kerogen colors and less than 0.5 R_0 percent vitrinite reflectance.

In conclusion, *we encourage specific projects relating to the formation, deposition, and preservation of organic matter in differing physiographic, hydrographic, and climatic settings. Projects in these fields would improve our understanding of the formation of hydrocarbon source beds and will need the cooperation of physical and chemical oceanographers, biologists, and sedimentologists.*

CLATHRATE FORMATION

At low temperatures and high pressures, many low-molecular-weight gases (including methane, ethane, carbon dioxide, and H₂S) combine with water to form icelike crystalline solids (Figure 3.13). The conditions in which these (so-called) clathrates are formed are well known from laboratory studies on clathrate formation. (See Figure 3.14 for conditions of hydrate formation in the methane–freshwater system.) These studies are related to gas-transmission pipelines in cold climates. Natural clathrates are found on land in the permafrost oil fields of Siberia and in the North American Arctic. Because the appropriate conditions are found throughout deep-sea sedimentary environments (Figure 3.15), it has been postulated that clathrates may exist there. This has been confirmed by the discovery of reflecting horizons that parallel the seafloor and cut across bedding planes. Such horizons have been observed on the Blake Outer Ridge off the southeastern United States (Figure 3.16) and beneath the upper continental rise off New Jersey and Delaware, as well as off the U.S. West Coast. Preliminary drilling by the Deep Sea Drilling Project (DSDP) has shown much higher-than-normal methane concentrations associated with the Blake Outer Ridge reflecting horizon. Similarly, high methane concentrations have been correlated with other high-amplitude reflecting horizons in the deep sea.

The presence of clathrates in deep-ocean sediments could be important in blocking the escape of hydrocarbons, often capping petroleum or natural gas accumulations. Conversely, failure to recognize clathrates during drilling or platform siting could cause some hazards, because they profoundly alter the physical properties of sediments in which they occur. Continental margins encourage conditions that favor the formation of clathrates. Biological productivity tends to be greater there than in the open ocean, and rapid deposition favors the presence of anaerobic, methane-producing bacteria. During geologic time, clathrates may form and then be destroyed. The change may be an important sedimentary process resulting in large-scale slumping of unstable sediments on the upper continental slopes.

Since clathrates can influence the geology of areas in which they occur as well as serve to indicate potential sources of hydrocarbons, a concentrated effort should be undertaken to investigate where they exist in continental-margin sediments. The only effective way to prove their existence is to core in an area that has the proper high-amplitude reflecting horizons and to recover the core under *in situ* conditions of pressure, temperature, and gas concentration. Several attempts to do this were made by the DSDP, but in all cases, a considerable amount of gas escaped during core recovery and prior to sampling. Nevertheless, methane concentrations of at least 20 mmole/liter were indirectly estimated to be present in interstitial waters in a Cariaco Trench DSDP core before it was brought to the surface. While this is a minimal value, it does show that methane concentrations approaching saturation levels are possible in the deep sea.

We recommend intensified studies of the distribution and physical and chemical properties of seafloor clath- *rates. These studies should include the development of adequate and safe sampling methods and an evaluation of clathrate formation and destruction as a significant sedimentary process.*

MARINE HYDROCARBON DETECTION

A major application of marine geochemistry to the continental margins is "sniffing." This is a technique for locating oil or natural gas seeps by measuring the concentra-

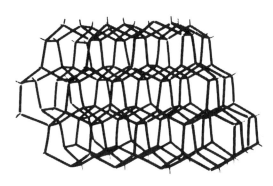

NORMAL ICE

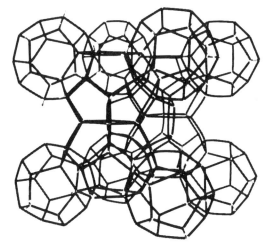

HYDRATE

TETRAKAIDECAHEDRON

DODECAHEDRON

FIGURE 3.13 Photographs of models of "cages" and crystal lattice of Type I hydrate and ice. (Courtesy of R. D. McIver, Exxon Production Research, 1977.)

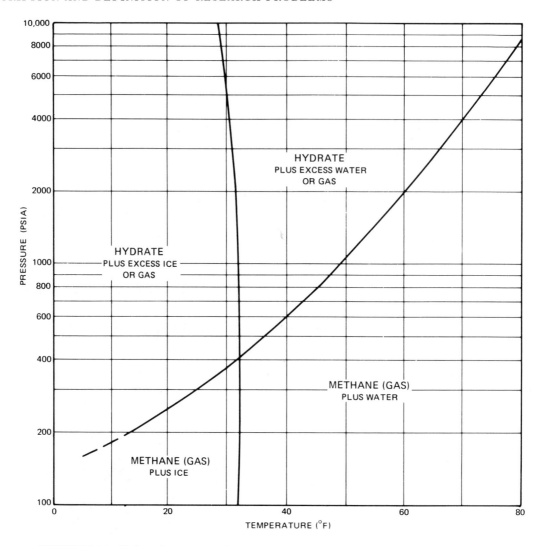

FIGURE 3.14 Hydrate formation in the methane–freshwater system. (After Katz *et al.*, 1959.)

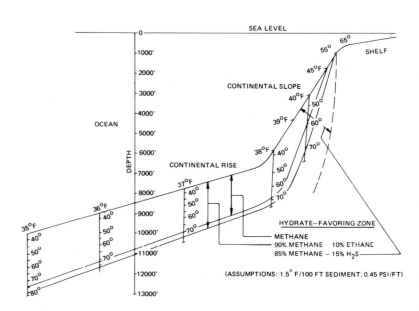

FIGURE 3.15 Hydrate-favoring zone under the ocean. (Courtesy of R. D. McIver, Exxon Production Research.)

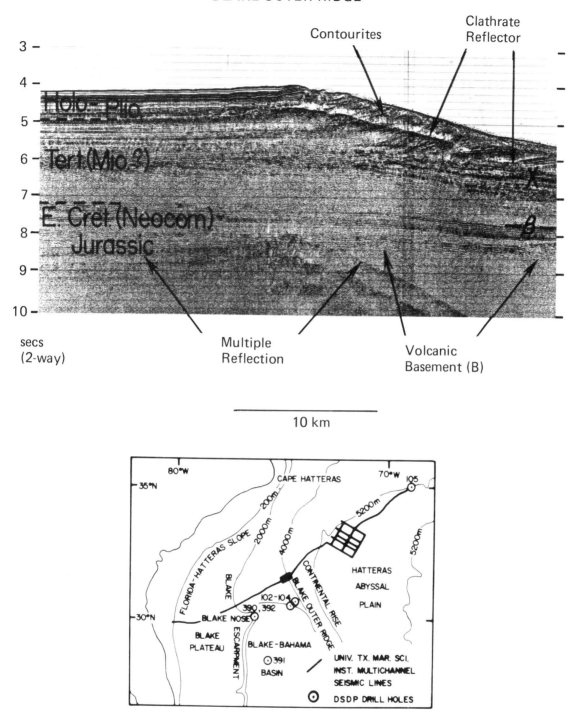

FIGURE 3.16 Strong clathrate (gas–hydrate) reflector beneath Blake Outer Ridge, U.S. Atlantic continental margin. Reflector follows contour of seafloor and cuts across dipping contourite buildups that form Blake Outer Ridge. Thick Late Tertiary sedimentary section overlies unconformably much thinner Early Cretaceous (Neocomian) through Jurassic sedimentary section. Top of Neocomian correlated with horizon beta (B) from DSDP 105. Middle to possibly Early Jurassic (?) strata onlay and fill volcanic basement (horizon B) (Layer 2). [Modified from Shipley *et al.* (in press) and Buffler *et al.* (in press).]

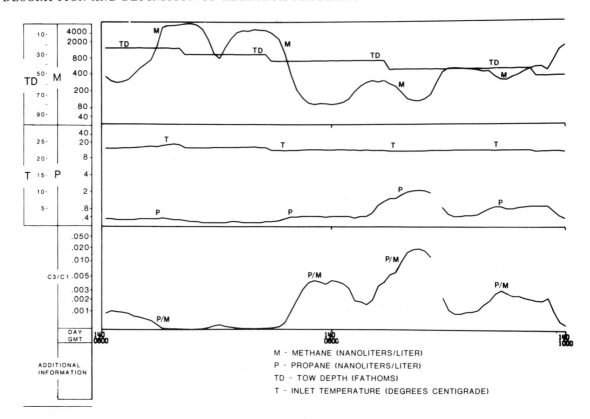

FIGURE 3.17 Sniffer data in profile form. Methane anomaly (on left) may be due to seep from known gas field near this location. Propane anomaly (on right) may indicate an undiscovered oil field. (Courtesy of J. D. Burgess and R. J. Mousseau, Gulf Science and Technology Company.)

tion of gaseous hydrocarbons in the sea. The underlying assumption is that areas with anomalously high hydrocarbon concentration can be traced to seeps. Other compounds from petroleum could also be used as seep indicators, but for practical reasons (primarily, the difficulty in analyzing those compounds), sniffing is usually restricted to the measurement of the gaseous hydrocarbons.

Practicing this detection technique, petroleum and natural-gas companies have acquired systems that can rapidly collect data on the amount of these compounds present in the ocean. These companies are understandably discreet about their exploration activities, but it seems quite certain that several of them employ such systems for this type of study. Prough (1976) has described a system that InterOcean Systems, San Diego, offered to industry. In addition to the commercially supplied systems, there are companies, e.g., Gulf, that have developed and employ their own systems.

Gulf's system on the R/V *Hollis Hedberg* employs three separate water inlets that continuously supply sample streams from the near surface, the intermediate depths, and the deeps. The deep-towed sample inlet can operate down to nearly 180 m while the ship is moving at normal seismic-survey speeds. Each sample stream is analyzed for seven hydrocarbon gases once every 3 min with a sensitivity that depends on the hydrocarbon but that is, for

example, for propane, about 5×10^{-11} liter of propane at STP per liter of seawater.

Compared with the techniques by which most gas in seawater concentrations has been determined, and which involve stripping a discrete sample for each analysis, it is obvious that the industrial marine hydrocarbon analyzer can generate larger volumes of data much faster (up to 1440 samples in a 24-hour day when all three inlets are employed). This amount of data requires automated handling methods. The Gulf system records the gas data along with ancillary data, such as inlet temperature and sampling depths, on digital magnetic tape. Routine processing provides listings, profiles, and data maps. Figures 3.17 and 3.18 show examples of such processing.

Interpretation of these data for exploration is built on the assumption that local areas of enhanced hydrocarbon concentration are caused by seepages from oil or natural-gas accumulations. Figure 3.17 illustrates one form in which "sniffer" data can be used as an exploration tool. It shows geochemical data from a deep-tow inlet in profile form. The strong methane anomaly on the left of the figure probably discloses seepage from a known gas field at this location. On the right, a propane anomaly may indicate seepage from an undiscovered oil field.

Figure 3.18, another data display, shows methane data

from the deep-tow inlet as a contour map. The area of high methane concentration on the right of the map may reflect a gas seep at the seafloor. Note, incidentally, the detail in the methane distribution when it is sampled by a "sniffer" system. Such detail would not appear in the usual station-type survey.

In order to use sniffing tools effectively and to interpret the data adequately, it is necessary to have information about phenomena other than hydrocarbon seeps that contribute to hydrocarbon gases in the sea. Two features in the published data make it increasingly evident that methane, the primary component of natural gas, is actually formed in the ocean within the water column itself. First, studies on baseline concentration show that the methane concentration in surface ocean water slightly exceeds what would be present if the surface water were in equilibrium with the atmosphere. In other words, the

ocean is acting as a source of methane to the atmosphere (see report of the Ocean Sciences Board, 1978, p. 147).

Second, several investigators have observed that the methane concentration often demonstrates a maximum value near the depth of the thermocline (Figure 3.19). Scranton and Brewer (1977) have modeled their observations to rule out the possibility that this maximum may be due to the advection of high concentrations from nearshore sources. They conclude, in concurrence with Lamontagne *et al.* (1973), that methane is being generated, probably biogenically, near and above the level of the observed maximum concentrations. Obviously, such a phenomenon cautions against blind acceptance of "methane anomalies" as indicators for oil or gas seeps.

Fortunately, there is evidence that other gaseous hydrocarbons that are directly related to petroleum do not

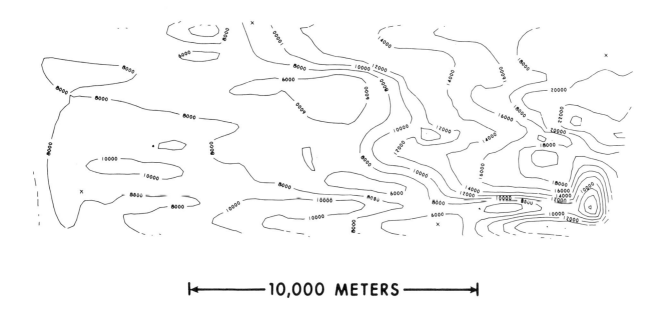

FIGURE 3.18   Methane concentration from a deep inlet displayed in contour form. Numbers on contours are relative methane concentrations. (Courtesy of the Gulf Oil Corporation.)

have a biogenic source in the sea. Note in Figure 3.19 that, at the same depth at which methane exhibits a maximum concentration, ethane and propane do not increase in concentration. There is an extensive literature on biological production of methane, which is summarized by Mechalas (1974). Neither of the two major pathways contending for metabolic production of methane, CO_2 reduction or acetic acid fermentation, produces higher molecular-weight homologs. Hunt (1974) states, "... none of the alkanes in the butane–heptane range are formed biologically as far as is known at present." These findings support the belief that saturated hydrocarbons heavier than methane are not produced in the water column at sea.

Still, the possibility remains that methane and heavier homologs are introduced into the sea by contemporaneous generation of gases in the shallow sediments. Obviously, gases forming in yesterday's sediments may have no relationship to deeper petroleum deposits. There is ample evidence for methane generation in shallow (i.e., a few tens of meters) sediments. However, it is not clear how much enters the sea. Reeburgh (1969) published profiles of methane concentration in the upper couple meters of Chesapeake Bay sediments (Figure 3.20) that show methane concentrations reaching the gas ebullition point at depths of a meter or two below the seafloor. These profiles also show very low methane concentrations within the upper 25 cm. Reeburgh (1969), Reeburgh and Heggie (1974), and Martins and Berner (1974) considered

alternative mechanisms that might lead to the observed profiles. The preferred explanations are that sulfate reduction occurs in the upper sediments and methane generation occurs in the lower sediments. The small amount of methane that was observed in the upper sediments was attributed to diffusion from below. Reeburgh concludes that the upward concavity of the upper profile was due to physical or biological mixing. This interpretation implies a significant methane flux to the seafloor.

Barnes and Goldberg (1976) believe that their data from the Santa Barbara Basin (showing a similar profile) indicate a mechanism whereby methane is generated throughout the sediment column but is simultaneously consumed biochemically (probably by *Desulfovibrio*— bacteria) within the upper section. Reeburgh also advances this theory to explain methane behavior for the Cariaco trench. Such an interpretation eliminates (or, at least, greatly reduces) the inferred rate at which methane is purportedly introduced to the seafloor. An answer to these questions would be relevant to petroleum exploration.

The presence in recent shallow sediments of small amounts of hydrocarbons in the ethane-through-heptane range has been reported by several investigators. Hunt (1975) found that the total of these gasoline-range hydrocarbons typically amounts to 10–500 ppb by weight. Since the amount of these hydrocarbons correlates with the organic carbon content of the sediments, he believes that the diagenetic reactions forming them are occurring much

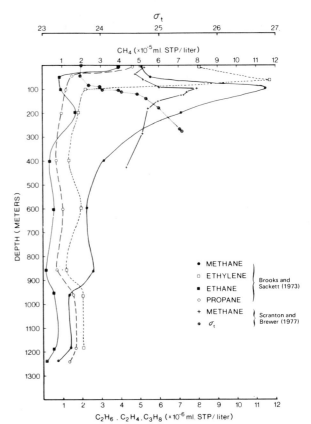

FIGURE 3.19 Depth profiles of methane and other gases in open ocean waters. [After Brooks and Sackett (1973) and Scranton and Brewer (1977).]

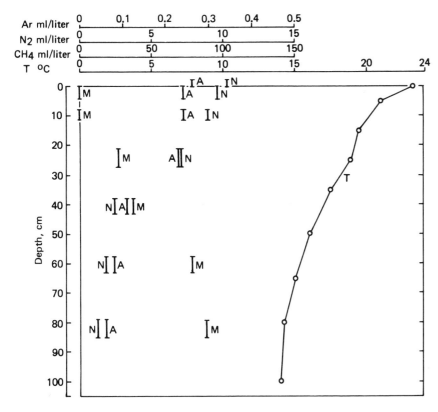

FIGURE 3.20   Methane, argon, nitrogen, and temperature with depth distributions in shallow Chesapeake Bay sediments. (From Reeburgh, 1969.)

earlier and at much lower temperatures than most geochemists had expected. Their low concentration, dilution by the other organic material, adsorption on the sediment surfaces, and possibly other effects may lower the escaping tendency of these hydrocarbons to the point where they do not appear to enter the water column in sufficient quantities to interfere with this exploration technique. Further research is needed in this area to establish the limits of possible interference. Pore-water measurements for these heavier hydrocarbons would be especially useful.

Hunt's finding may be more indicative of possible interference in the exploration technique described by Horvitz (1968). In this technique, the shallow sediment sample itself is analyzed for the low-molecular-weight hydrocarbons it contains. Allegedly, contours of hydrocarbon concentration that are in the same range of values that Hunt reports define deep oil or gas accumulations. Normal fluctuations in the indigenously produced gas may be the primary control for the contours.

We recommend that environmental and sediment-dynamics studies make full use of commercially available sniffing techniques. These techniques should be further developed to trace and map the distribution of nonhydrocarbon chemical components in waters overlying the seafloor.

REFERENCES AND BIBLIOGRAPHY

SEDIMENTS AND THEIR ORIGIN

Berner, R. A. (1976). The benthic boundary layer from the viewpoint of a geochemist, in *The Benthic Boundary Layer*, I. N. McCave, ed., Plenum Press, New York, pp. 33–35.

Blatt, H., G. Middleton, and R. Murray (1972). *Origin of Sedimentary Rocks*, Prentice-Hall, Englewood Cliffs, N.J., 634 pp.

Braunstein, J., ed. (1974). Facies and the reconstruction of environments, *Am. Assoc. Petrol. Geol. Reprint Series #10*, Tulsa, Okla., 224 pp.

Brown, L. F., Jr., Project Coordinator (1972–1976). *Environmental Geologic Atlas of the Texas Coastal Zone*, a folio of 63 maps accompanied by text in 6 vols., Bur. Econ. Geol., University of Texas, Austin, Tex.

Busch, D. A. (1974). Stratigraphic traps in sandstones: Exploration techniques, *Am. Assoc. Petrol. Geol. Mem. #21*, Tulsa, Okla., 174 pp.

Commission on National Resources and the Steering Committee for Analytical Studies, National Research Council (1977). *Perspectives on Technical Information for Environmental Protection, Vol. 1*, National Academy of Sciences, Washington, D.C.

Davis, R. A., and R. L. Ethington, eds. (1976). *Beach and nearshore sedimentation*, Soc. Econ. Paleontol. Mineral. Spec. Publ. #24, Tulsa, Okla., 175 pp.

Garrels, R. M., and R. T. Mackenzie (1971). *Evolution of*

Sedimentary Rocks. W. W. Norton & Company, New York, 397 pp.

Geophysics of Estuaries Panel (1977). *Estuaries, Geophysics, and the Environment*, Studies in Geophysics Ser., Geophysics Research Board, Assembly of Mathematical and Physical Sciences, National Research Council, National Academy of Sciences, Washington, D.C., 127 pp.

Ginsburg, R. N. (1974). Introduction to comparative sedimentology of carbonates, *Am. Assoc. Petrol. Geol. Bull. 58*(4).

Goldhaber, M. B., R. C. Aller, J. K. Cochran, J. K. Rosenfeld, C. S. Martens, and R. A. Berner (1977). Sulfate reduction, diffusion, and bioturbation in Long Island Sound sediments: Report of the FOAM Group, *Am. J. Sci. 277*(3):193–237.

Gorsline, D. S., and D. J. P. Swift, eds. (1977). *Continental Shelf Sediment Dynamics: A National Overview*, A report of a workshop held in Vail, Colo., Nov. 2–6, 1976, sponsored by the Office of the International Decade of Ocean Exploration of the National Science Foundation, the Atlantic Oceanographic and Meteorological Laboratories of the National Oceanic and Atmospheric Administration, the Energy Research and Development Agency, and the Office of Marine Geology of the U.S. Geological Survey, 134 pp.

Guinasso, N. L., Jr., and D. R. Schink (1975). Quantitative estimates of biological mixing rates in abyssal sediments, *J. Geophys. Res. 80*(21):3032–3043.

Hollister, C. D., A. J. Sylva, and A. Driscoll (1973). A giant piston-corer, *Ocean Eng. 2*:159–168.

Hollister, C. D., R. D. Flood, D. A. Johnson, P. Lonsdale, and J. B. Southard (1974). Abyssal furrows and hyperbolic echo traces on the Bahama Outer Ridge, *Geology 2*(8):395–400.

Inman, D. L., and B. M. Brush (1973). The coastal challenge, *Science 181*:20–32.

Kaplan, I. R., ed. (1974). *Natural Gases in Marine Sediments*, Plenum Press, New York, 324 pp.

Kraft, J. C., and J. J. Chako (1976). The Geological Structure of the Shorelines of Delaware, DEL-S6-14-76, College of Marine Studies, University of Delaware, 106 pp.

Kraft, J. C., R. B. Briggs, and S. D. Halsey (1973). *Coastal Geomorphology*, Publ. State U. of New York, pp. 321–354.

LeBlanc, R. J. (1972). *Geometry of Sandstone Reservoir Bodies, from Underground Waste Management and Environmental Implications*, Am. Assoc. Petrol. Geol. Mem. No. 18, Tulsa, Okla., pp. 133–189.

Manheim, F. T., and F. L. Sayles (1974). Composition and origin of interstitial waters of marine sediments, based on deep sea drill cores, in *The Sea*, Vol. 5, E. D. Goldberg, ed., John Wiley and Sons, New York, pp. 527–568.

Manheim, F. T., J. C. Hathaway, and E. Uchupi (1972). Suspended matter in surface waters of the northern Gulf of Mexico, *Limnol. Oceanog. 17*(1):17–27.

Marcus, P. A., ed. (1976). *Directory to U.S. Geological Survey Program Activities in Coastal Areas, 1974–76, Geol. Surv. Bull. 1428*, U.S. Government Printing Office, Washington, D.C.

McCave, I. N., ed. (1976). *The Benthic Boundary Layer: Proceedings of a Conference*, Les Arcs, France, Nov. 1974, Plenum Press, New York, 324 pp.

Middleton, G. V., ed. (1965). *Primary Sedimentary Structures and Their Hydrodynamic Interpretation*, Soc. Econ. Paleontol. Mineral. Spec. Publ. #12, Tulsa, Okla., 265 pp.

Milliman, J. D. (1974). *Recent Sedimentary Carbonates—Part 1: Marine Carbonates*, Springer-Verlag, New York, 375 pp.

Milliman, J. D., O. N. Pilney, and D. A. Ross (1972). Sediments of the continental margin off the eastern United States, *Geol. Soc. Am. Bull. 83*:1315–1334.

Morgan, J. P., and R. H. Shaver, eds. (1970). *Deltaic Sedimentation, Modern and Recent*, Soc. Econ. Paleontol. Mineral. Spec. Publ. #15, Tulsa, Okla., 315 pp.

Morisawa, M., and C. A. M. King (1974). Monitoring the coastal environment, *Geology 2*:385–388.

Nelson, C. H., P. R. Carlson, J. V. Byrne, and T. R. Alpha (1970). Development of the Astoria Canyon—fan physiography and comparison with similar systems, *Marine Geol. 8*(3/4):259–291. Physiographic diagrams from Open File, Dec. 5, 1972, U.S. Geol. Survey.

Ocean Affairs Board (1975). *Petroleum in the Marine Environment*, A report of a workshop on Inputs, Fates, and the Effects of Petroleum in the Marine Environment, May 21–25, 1973, Airlie House, Airlie, Va., Commission on Natural Resources, National Research Council, National Academy of Sciences, Washington, D.C.

Ocean Disposal Study Steering Committee (1976). *Disposal in the Marine Environment: An Oceanographic Assessment*, Ocean Affairs Board's Ocean Science Committee in the Commission on Natural Resources of the National Research Council, National Academy of Sciences, Washington, D.C.

Ocean Sciences Board (1978). *The Tropospheric Transport of Pollutants and Other Substances to the Oceans*, a report prepared by the Steering Committee for the Workshop on Tropospheric Transport of Pollutants to the Ocean, Ocean Sciences Board, Assembly of Mathematical and Physical Sciences, National Research Council, National Academy of Sciences, Washington, D.C., 243 pp.

Panel on Operation Safety in Marine Mining (1975). *Mining in the Outer Continental Shelf and in the Deep Ocean*, Marine Board, Assembly of Engineering, National Research Council, National Academy of Sciences, Washington, D.C.

Reimetz, E., P. W. Barnes, and T. R. Alpha (1973). Bottom features and processes related to drifting ice, Miscellaneous Field Studies Map MF-532, U.S. Geological Survey, Menlo Park, Calif.

Rigby, J. K., and W. K. Hamblin, eds. (1972). *Recognition of Ancient Sedimentary Environments*, Soc. Econ. Paleontol. Mineral. Spec. Publ. #16, Tulsa, Okla., 300 pp.

Scranton, M. I., and P. G. Brewer (1977). Occurrence of methane in the near-surface waters of the western subtropical North Atlantic, *Deep-Sea Res. 24*(2):127–138.

Selley, R. C. (1970). Ancient Sedimentary Environments, Cornell Press, Ithaca, New York, 237 pp.

Shepard, F. P., F. B. Phleger, and T. H. van Andel, eds. (1960). *Recent Sediments, Northwest Gulf of Mexico*, Am. Assoc. Petrol. Geol., Tulsa, Okla., 394 pp.

Stanley, D. J., and D. J. P. Swift, eds. (1976). *Marine Sediment Transport and Environmental Management*, papers from a short course, Key Biscayne, Fla., Nov. 1974, Wiley-Interscience, New York, 602 pp.

Tamatmat, M. M. (1971). Oxygen consumption by the sea bed. IV: Shipboard and laboratory experiments. *Limnol. Oceanog. 16*:536–550.

Taylor, D. L. (1977). *Proceedings of the Third International Coral Reef Symposium, Vol. 2—Geology*, U. of Miami, Miami, Fla., 623 pp.

Wilson, J. L. (1974). Characteristics of carbonate platform margins, *Amer. Assoc. Phys. Geol. Bull. 58*(4):810–824.

GEOCHEMISTRY AND DIAGENESIS OF SEDIMENTS

Barnes, R. O., and E. D. Goldberg (1976). Methane production and consumption in anoxic marine sediments, *Geology 4*:297–300.

Brooks, J. M., and W. M. Sackett (1973). Sources, sinks, and concentrations of light hydrocarbons in the Gulf of Mexico, *J. Geophys. Res.* 78:5248–5258.

Buffler, R. T., T. H. Shipley, and J. S. Watkins (1978). Blake continental margin seismic section, *Am. Assoc. Petrol. Geol. Seismic Section No. 2.* (in press).

Bujak, J. P., M. S. Barss, and G. L. Williams (1975). Kerogen type and thermal alteration index in Scotian Shelf, Grand Banks and Labrador Shelf wells, *Am. Assoc. Stratig. Palynologists Proc.*, 9th Ann. Mtg., Halifax, Nova Scotia (in press).

Cassou, A. M., J. Connan, and B. Porthault (1977). Relations between maturation of organic matter and geothermal effect as exemplified in Canadian East Coast offshore wells, *Bull. Can. Petrol. Geol.* 25(1):174–194.

Castaño, J. R., and D. M. Sparks (1974). Interpretation of vitrinite reflectance measurements in sedimentary rocks and determination of burial history using vitrinite reflectance and authigenic minerals, *Geol. Soc. Am. Special Paper* 153:31–52.

Claypool, G. E., B. J. Presley, and I. R. Kaplan (1973). Gas analysis in sediment samples from legs 10, 11, 13, 14, 15, 18, and 19, in *Initial Reports of the Deep Sea Drilling Project*, *XIX*:879–884, U.S. Government Printing Office, Washington, D.C.

Dow, W. G. (1977a). Petroleum source beds on continental slopes and rises, in *Am. Assoc. Petrol. Geol. Cont. Educ. Course Notes Ser. #5*, Washington, D.C.

Dow, W. G. (1977b). Kerogen studies and geological interpretations, *J. Geochem. Expl.* 7:79–99.

Drewer, J. I., ed. (1978). Sea water, cycles of major elements, *Benchmark Papers in Geology 45*, Dowden, Hutchison and Ross, Stroudsburg, Pa.

Emery, K. O., and D. Hoggan (1958). Gases in marine sediments, *Bull. Am. Assoc. Petrol. Geol.* 42:2174–2188.

Feux, A. N. (1977). The use of stable carbon isotopes in hydrocarbon exploration, *J. Geochem. Expl.* 7:155–188.

Hitchon, B., ed. (1977). Application of geochemistry to the search for crude oil and natural gas, spec. issue of *J. Geochem. Expl.* 7(2), 216 pp.

Hood, A., and J. R. Castaño (1974). Organic metamorphism: its relationship to petroleum generation and application to studies of authigenic minerals, reprinted from U.N. ESCAP, *CCOP Tech. Bull.* 8:85–118.

Hood, A., C. C. M. Gutjahr, and R. L. Heacock (1975). Organic metamorphism and the generation of petroleum, *Am. Assoc. Petrol. Geol. Bull.* 59(6):986–996.

Horvitz, L. (1968). Hydrocarbon geochemical prospecting after thirty years, paper presented at the Symposium on Unconventional Methods in Exploration for Petroleum and Natural Gas, Southern Methodist U., Dallas, Tex., Sept. 12–13.

Hunt, J. M. (1974). Organic geochemistry of the marine environment, in *Advances in Organic Geochemistry*, 1973, Editions Technip., Paris, pp. 593–605.

Hunt, J. M. (1975). Hydrocarbon studies (DSDP leg 29), in *Initial Reports of the Deep Sea Drilling Project*, *XXXI*:901–903, U.S. Government Printing Office, Washington, D.C.

Katz, D. L., D. Cornell, R. Kobayashi, F. H. Poetman, J. R. Elenbas, and C. F. Weinand (1959). *Handbook of Natural Gas Engineering*, McGraw-Hill, New York, pp. 182–221.

Lamontagne, R. A., J. W. Swinnerton, and V. J. Linnenbom (1971). Nonequilibrium of CO and CH_4 at the air–sea interface, *J. Geophys. Res.* 76:5117–5121.

Lamontagne, R. A., J. W. Swinnerton, V. J. Linnenbom, and W. D. Smith (1973). Methane concentrations in various marine environments, *J. Geophys. Res.* 78:5317–5324.

Lijmbach, G. W. M. (1975). On the origin of petroleum, *Proc. Ninth World Petroleum Congress 2*, Applied Science Pub. Ltd., London, pp. 357–369.

Manheim, F. T. (1976). Interstitial waters of marine sediments, in *Treatise on Chemical Oceanography, Vol. 6*, 2nd ed., J. P. Riley and R. Chester, eds., Academic Press, London and New York, pp. 115–186.

Manheim, F. T., and F. L. Sayles (1974). Composition and origin of interstitial waters of marine sediments, based on deep sea drill cores, in *The Sea, Vol. 5 (Marine Chemistry)*, E. D. Goldberg, ed., Wiley-Interscience, New York, pp. 527–568.

Martins, C. S., and R. A. Berner (1974). Methane production in the interstitial waters of sulfate-depleted marine sediments, *Science 185*:1167–1169.

McIver, R. D. (1973). Hydrocarbons in canned muds from Sites 185, 186, 189, and 191—leg 19, in *Initial Reports of the Deep Sea Drilling Project, XIX*, U.S. Government Printing Office, Washington, D.C., pp. 875–877.

Mechalas, B. J. (1974). Pathways and environmental requirements for biogenic gas production in the ocean, in *Natural Gases in Marine Sediments*, I. R. Kaplan, ed., Plenum Press, New York, pp. 11–25.

Myers, E. P., and C. G. Gunnerson (1976). *Hydrocarbons in the Ocean*, MESA *Special Report*, U.S. Dept. of Commerce, Washington, D.C.

Ocean Sciences Board (1978). *The Tropospheric Transport of Pollutants and Other Substances to the Oceans*, a report prepared by the Steering Committee for the Workshop on Tropospheric Transport of Pollutants to the Ocean, Ocean Sciences Board, Assembly of Mathematical and Physical Sciences, National Research Council, National Academy of Sciences, Washington, D.C., 243 pp.

Prough, P. G. (1976). Sniffing for hydrocarbons in the sea, *Ocean Ind.* 6:111–112.

Reeburgh, W. S. (1969). Observations of gases in Chesapeake Bay sediments, *Limnol. Oceanog.* 14(3):368–375.

Reeburgh, W. S. (1972). Processes affecting gas distributions in estuarine sediments, in *Environmental Framework of Coastal Plain Estuaries*, P. W. Nelson, ed., *Geol. Soc. Am. Mem. 133*:383–389.

Reeburgh, W. S. (1976). Methane consumption in Cariaco trench waters and sediments, *Earth Planet. Sci. Lett.* 28(3):337–344.

Reeburgh, W. S., and D. T. Heggie (1974). Depth distribution of gases in shallow water sediments, in *Natural Gases in Marine Sediments*, I. R. Kaplan, ed., Plenum Press, New York, pp. 27–45.

Sackett, W. M. (1977). Use of hydrocarbon sniffing in offshore exploration, *J. Geochem. Expl.* 7:243–254.

Scranton, M. I., and P. G. Brewer (1977). Occurrence of methane in the near-surface waters of the western subtropical North-Atlantic, *Deep-Sea Res.* 24(2):127–138.

Shipley, T. H., R. T. Buffler, and J. S. Watkins (in press). Seismic stratigraphy and geologic history of the Blake continental margin, western Atlantic, in *Geophysical Investigations of Continental Slopes and Rises*, J. S. Watkins, L. Montadert, and P. Dickerson, eds., Am. Assoc. Petrol. Geol., Tulsa, Okla.

Sigalove, J. J., and M. D. Pearlman (1975). Geochemical seep detection for offshore oil and gas exploration, Paper No. OTC 2344, presented at the Seventh Annual Offshore Technology Conference, Houston, Tex., May 5–8.

Swinnerton, J. W., and R. A. Lamontagne (1974). Oceanic distribution of low-molecular-weight hydrocarbons: Baseline measurements, *Environ. Sci. Technol.* 8:657–663.

Swinnerton, J. W., and V. J. Linnenbom (1967). Determination of the C$_1$ to C$_4$ hydrocarbons in seawater by gas chromatography, *J. Chromatog. Sci.* 5:570–573.

Swinnerton, J. W., V. J. Linnenbom, and C. H. Cheek (1962a). Determination of dissolved gases in aqueous solutions by gas chromatography, *Anal. Chem.* 34:483–485.

Swinnerton, J. W., V. J. Linnenbom, and C. H. Cheek (1962b). Revised sampling procedure for determination of dissolved gases in solution by gas chromatography, *Anal. Chem.* 34:1509.

Tissot, B. (1977). The application of the results of organic geochemical studies in oil and gas exploration, in *Developments in Petroleum Geology, Vol. 1*, G. D. Hobson, ed., Applied Science Publishers, London.

Tissot, B., and F. Bienner, eds. (1973). Advances in organic geochemistry, *Actes du 6e Congrès International de Géochimie Organique*, Editions Technip., Paris.

Tissot, B., and D. H. Welte (1978). *Petroleum Formation and Occurrence, a New Approach to Oil and Gas Exploration*, Springer-Verlag, New York, 537 pp.

Tissot, B., B. Durand, J. Espitalie, and A. Combaz (1974). Influence of nature and diagenesis of organic matter in formation of petroleum, *Am. Assoc. Petrol. Geol. Bull.* 58(3):499–506.

Weber, V. V., and S. P. Maximov (1976). Early diagenetic generation of hydrocarbon gases and their variations dependent on initial organic composition, *Bull. Am. Assoc. Petrol. Geol.* 60:287–293.

4 | Stratigraphy

A. STRATIGRAPHY AND FACIES ANALYSIS

Traditionally, stratigraphy was based on careful field mapping, subsurface work, detailed measurements of sections, and elaborate lithologic and paleontologic descriptions. Construction and correlation of stratigraphic sections came next, rock sequences were interpreted in the light of recent sedimentation, and, finally, the grand synthesis showed one or more paleogeographic maps with vistas of ever-changing landscapes and seascapes.

In not much more than the last decade, two major developments revolutionized stratigraphy:

1. Seismic stratigraphy—high-resolution reflection techniques provided stratigraphers with x-ray-like pictures that revealed the fine anatomy of stratigraphic sequences (Figures 4.1 and 4.2).
2. Paleooceanography—via reconstructions based on the plate-tectonics concept of continents that shift through time, scientists became more acutely aware that there had been many major changes in ocean circulation and paleoclimatology.

Most of the traditional observations and studies remain valid and will be continued, but many of today's stratigraphers are reviewing them in an effort to reconcile them with the latest concepts and methodology. The entire field is in a splendid state of flux.

More than any other single group, geologists from the Exxon Production Research Company have worked out the methodology of *seismic stratigraphy*. The techniques and examples are discussed in detail by Vail *et al.* (1977). Seismic stratigraphy is a geological approach to the stratigraphic interpretation of seismic data. The unique properties of seismic reflections allow the direct application of geological concepts based on physical stratigraphy. Primary seismic reflections are caused by impedance contrasts across physical surfaces in the rocks, namely, stratal (bedding) surfaces and unconformities with velocity–density contrasts. Therefore, these reflections parallel stratal surfaces and unconformities. Since the rocks above such surfaces are younger than those below it, the resulting seismic section is preferably interpreted to be a record of the chronostratigraphic (time-stratigraphic) depositional and structural patterns, not a record of the time-transgressive lithostratigraphy (rock stratigraphy).

The resolving power of reflection seismic data varies with scale. With high-energy sources, deep penetration with relatively lower resolutions is obtained, whereas high-frequency sources permit high resolution (in meters and fractions of meters) of shallow layers. For a detailed account, see Des Vallieres (in press) and Sieck and Self (1977). Also, the chronostratigraphic record is more detailed in stratigraphic sequences deposited in a high-rate-of-subsidence regime. Low rates of subsidence lead

61

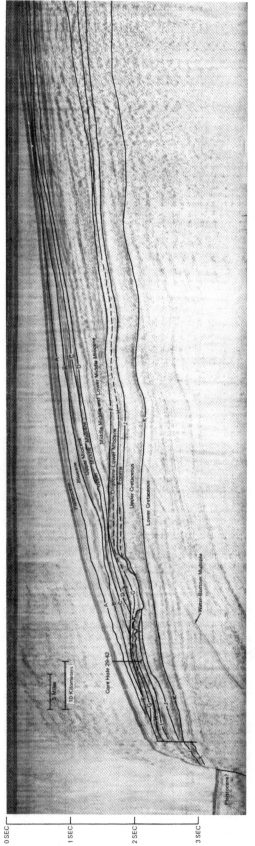

FIGURE 4.1 Seismic profile 126, reflection-time section, West Florida shelf. Marked change in sedimentary regime occurs at horizon F, with strong offlap of younger beds. Channel to right of core-hole 29-42 is part of ancestral DeSoto Canyon system. (After Mitchum, 1976.)

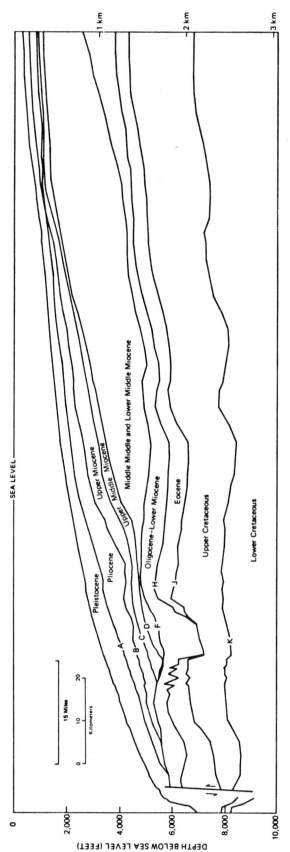

FIGURE 4.2 Seismic profile 126 converted to depth scale in feet. Vertical exaggeration is about 20:1. (After Mitchum, 1976.)

62

to more condensed sequences with fewer reflecting interfaces. In any event, a continuous seismic reflection constitutes an excellent *observable* chronostratigraphic correlation line. There are also exceptions where reflections are from hydrocarbon–water interfaces, from hydrates, or from within the crystalline basement. When these occur, they are easily recognized.

Because seismic reflections follow chronostratigraphic correlations, scientists can interpret postdepositional structural deformation and make the following types of stratigraphic interpretations from the geometry of seismic-reflection correlation patterns: (1) geologic time correlations; (2) definition of genetic depositional units; (3) thickness and depositional environment of genetic units; (4) paleobathymetry; (5) burial history; (6) relief, topography, and areal extent of unconformities; (7) eustatic and relative changes of sea level; and (8) paleogeography and geological history. A limiting factor is that rock type *cannot* be determined directly from the geometry of reflection correlation patterns. For lithologic information, true amplitude sections are becoming increasingly more important (see the papers in Section 3 of Payton, 1977).

To accomplish the geological objectives just listed, the following three-step interpretational procedure is used to analyze: (1) seismic sequences, (2) seismic facies, and (3) relative changes of sea level.

1. *Analysis of seismic sequences* is based on the identification of stratigraphic units composed of a relatively conformable succession of genetically related strata, i.e., *depositional sequences* (Figure 4.3). The upper and lower boundaries of these sequences are unconformities or their correlative conformities. The time interval represented by strata of a given sequence may differ from place to place, but the range is confined to synchronous limits that are marked by ages of the sequence boundaries where they become conformities. Depositional-sequence boundaries are recognized by reflections from lateral terminations of strata, termed *onlap, downlap, toplap,* and *truncation.* Depositional sequences provide an ideal framework for stratigraphic analysis, because they consist of genetically related strata that have chronostratigraphic significance.

2. *Analysis of seismic facies* is the delineation and interpretation of reflection geometry, continuity, amplitude, frequency, interval velocity, and the external form and associations of seismic-facies units within the depositional sequences. Once the parameters of seismic facies are described and mapped, we can predict the lithologic potential of the seismic facies based on an interpretation of the sedimentary processes and environmental settings.

3. *Analysis of relative changes of sea level* consists of constructing chronostratigraphic correlation charts and cycles of relative changes of sea-level charts on a regional basis and comparing them with global data. Similarities of the regional cycles to the global cycles (Figures 4.4 and 4.5) are significant because they introduce a dimension of

predictability into stratigraphy, allowing more accurate prediction of ages, times of unconformities, paleoenvironments, and lithofacies. Differences between regional and global curves indicate times of local structural deformation.

Numerous unconformities seen on seismic data can be correlated globally and appear to be related to eustatic changes of sea level (Figure 4.6). Thirteen appear to be related to major falls of sea level and are, therefore, considered major interregional unconformities. We define them as unconformities below which the youngest marine and coastal strata were widespread on the continental shelves at the time of deposition and above which the oldest strata were usually restricted to slopes or basins. Many minor interregional unconformities related to minor falls of sea level are present between the major interregional unconformities.

These conclusions are based on studies of regional grids of deep-penetration seismic data tied into well control with paleontologic and facies data. Much remains to be done to document these conclusions and understand the processes that created interregional unconformities. DSDP holes have provided significant data that support the present conclusions (Rona, 1973), however, deep-sea drilling, tied into a grid of deep-penetration seismic data in a variety of tectonic settings, is critical to understand these concepts.

The methodology is straightforward, but some of Vail's interpretations regarding worldwide simultaneous sea-level changes are controversial. (Figure 4.4). The intensity of the debate can be gauged by a quote from Dott and Batten (1971, pp. 69–70), who said: "Until the middle 20th century there still lingered a faith among geologists that the stratigraphic record was naturally divided by worldwide rhythms of mountain building reflected as very long-period, worldwide transgressive-regressive cycles, assumed to conform neatly with the system boundaries. This concept reflects a century-old influence of Hutton's and Lyell's cyclic view of the earth, and it provided a convenient rationale for a universal time scale. Modern stratigraphic studies have shown this scheme to be a fraud in its simple form; mountain building and un-

**SEISMIC STRATIGRAPHIC
REFLECTION TERMINATIONS**

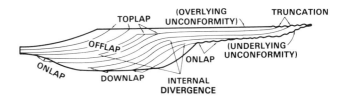

FIGURE 4.3 Schematic seismic sequence. Depositional sequence—stratigraphic unit composed of a relatively conformable succession of genetically related strata and bounded at its top and base by unconformities or their correlative conformities.

conformities have not been so perfectly uniform either in age or magnitude over large regions. Reaction set in and any widespread synchroneity was, for a time, denied categorically, but recently there has appeared evidence of certain long-term events which may, after all, prove to be more or less universal."

We recommend that some future deep-sea drilling on continental margins be aimed at drilling, coring, and testing well-described seismostratigraphic sequences. With such drilling data, scientists would be able to relate the lithofacies distribution to the geometries perceived on reflection seismic lines.

The problem is, clearly, to separate local from global (or interregional) effects. The Exxon concepts are based on reflection seismic lines that were calibrated by well locations aimed at testing potential hydrocarbon-bearing structures. These locations are not optimum to reveal the specific details of seismo-stratigraphic sequences. To understand these details, cores have to be obtained from scientifically selected targets, e.g., onlaps and offlaps.

Correlations between stratigraphic sections from neighboring segments of the coastal plain and shelf of any passive margin have often proved to be quite difficult and rarely have been done in detail. Only the most coarse correlations have been made for widely separated margins. The correlativity of the patterns of sedimentary onlap and offlap (regardless of their origin) suggests that multichannel seismic data, complimented by data from judiciously placed boreholes, may be used to develop a worldwide stratigraphy for passive margins of Mesozoic and Cenozoic ages.

Vail *et al.* (1977) already have demonstrated that developing a worldwide Mesozoic–Cenozoic stratigraphy for passive margins is a reasonable possibility. What is needed now are considerably more multichannel seismic data for margins of different ages. The experiments should be designed to obtain both shallow and deep sections. Most lines should be shot perpendicular to the margin, but some that are parallel to the margins are also needed. Both shallow and deep biostratigraphic boreholes will be needed to calibrate the seismic sections. These should be tied to the seismic lines. Lithofacies may be studied by analyzing the character of the seismic reflections and the borehole data.

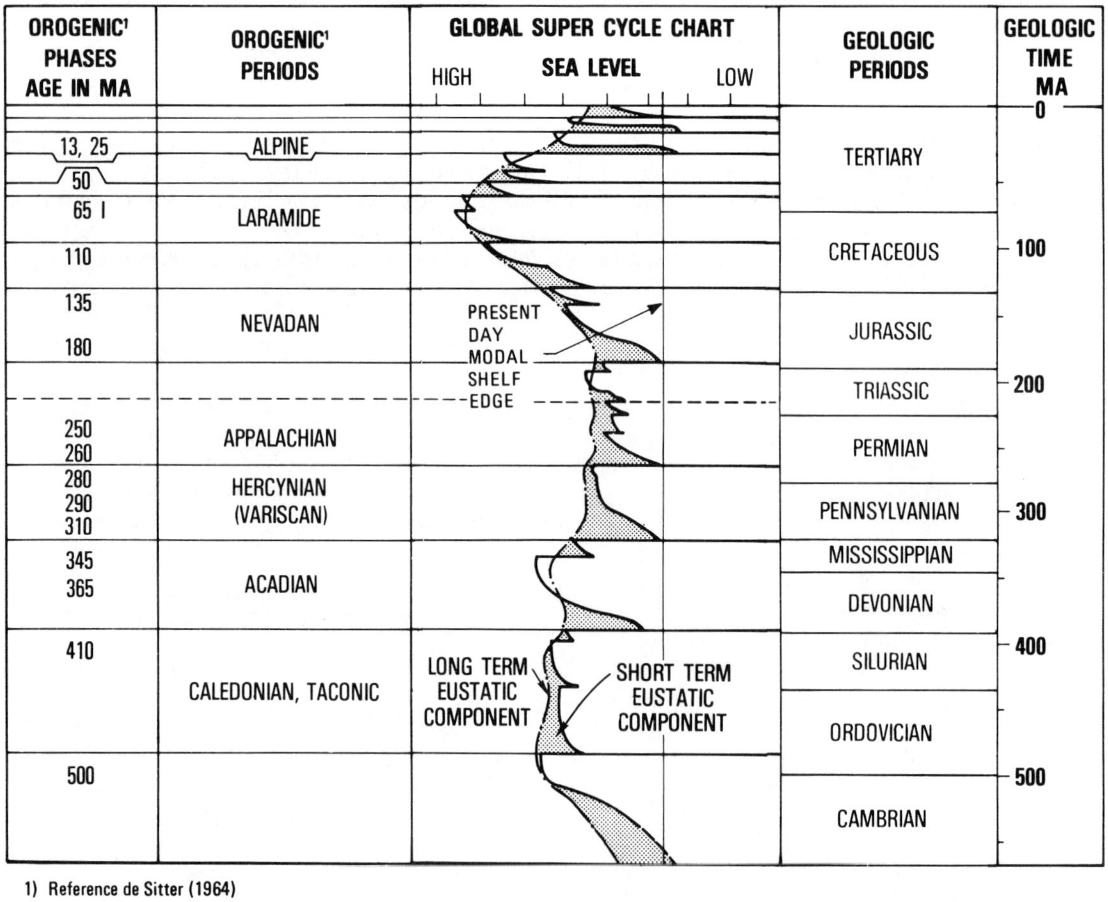

1) Reference de Sitter (1964)

FIGURE 4.4 Global supercycle chart and correlation with orogenic activity. (Compare and contrast also with Figure 9.2.) (From Vail *et al.*, 1977b.)

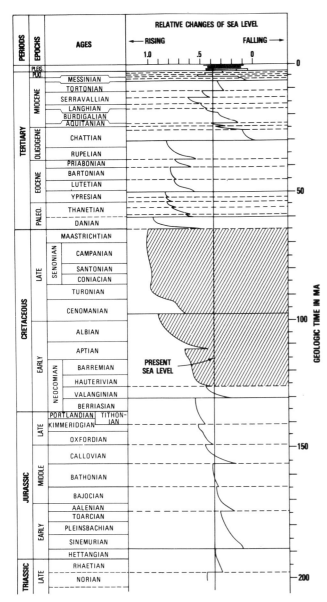

FIGURE 4.5 Jurassic–Tertiary global cycles of relative changes of sea level. (From Vail *et al.*, 1977a.)

B. PALEOOCEANOGRAPHY

As a consequence of study of DSDP cores and the development of plate tectonics, paleooceanography has made great advances. In the past, "paleogeographic" reconstructions were based on mostly land stratigraphic data displayed on base maps that showed the present distribution of continents or else on crude reconstructions based on Wegener's concept of continental drift. Much of this is summarized in the white paper of the JOIDES Advisory Panel on Ocean Paleoenvironment (1977), parts of which are reported in this report. In particular, the following

developments have changed the course of paleooceanographic work.

1. The distribution of present-day planktonic assemblages is related to salinities and densities of surface waters and to seasonal temperatures. That distribution can be compared and correlated with the distribution of Quaternary and, by analogy, older assemblages.

2. The study of the pelagic microfaunas obtained from DSDP cores now permits worldwide time–stratigraphic correlations of Cretaceous and Tertiary pelagic sequences with an accuracy of ± 1 million years.

3. Worldwide correlations based on microfaunas can be supplemented with paleomagnetic stratigraphic correlations to delineate precise stratigraphic horizons.

4. Worldwide transgressions and regressions can be traced and correlated by seismic stratigraphy (see preceding section). Long-term trends in sea level can be related to fluctuations in the spreading rate of midocean ridges.

5. Reconstructions based on matching continental outlines at deeper isobaths, continental geology, magnetic stripe patterns of the oceans, and fracture-zone trends provide detail on the distribution of continents and oceans during the Mesozoic.

6. Reconstructions using the determination of paleomagnetic poles of oriented rock samples supplement Mesozoic reconstructions and establish paleolatitudes. This technique also permits useful reconstructions to be made for the Paleozoic.

QUATERNARY AND LATE TERTIARY PALEOOCEANOGRAPHY

The CLIMAP project, sponsored by the International Decade of Ocean Exploration (IDOE), studied climatic changes over the last million years. This work has led to a reconstruction of the sea-surface temperatures, ice extent, ice elevation, and continental albedo for the earth some 18,000 years ago (the time of maximum extent of continental glaciation in the last ice age). A similar map for the penultimate interglacial maximum (20,000 years B.P.) is being made. Details of this work are reported in Cline and Hays (1976). Ruddiman and McIntyre (1976) report that seven climatic cycles occurred in the past 600,000 years, and within these cycles, at least 11 separate advances of polar water have occurred.

It appears that during the late Miocene–early Pliocene, the glacial history of the Antarctic, the Atlantic–Pacific interconnection across Middle America, and the flow in and out of the Mediterranean have greatly influenced climatic evolution.

During middle Miocene, the Antarctic ice cap grew to its present size (Kennett *et al.*, 1975), while the Calcite Compensation Level rose about 1 km (van Andel *et al.*, 1975). An even more fundamental Tertiary climatic change is thought to have occurred near the Eocene–Oligocene boundary (Shackleton and Kennett, 1975a, 1975b), resulting in a substantial increase in bottom-water current activity that produced widespread deep-sea unconformities and that profoundly modified continental slopes.

Much of this work has been based on interpretation of data from oceanic pelagic sediments. Our knowledge would be enhanced by studying long cores from the continental rises, where the rapid rates of sedimentation allow much greater stratigraphic resolution.

MESOZOIC–CENOZOIC RECONSTRUCTIONS

Much of the history of plate motion and evolution of ocean basins over the past 80 million years can be derived from the pattern of marine magnetic anomalies. The process of accretion that creates new ocean basins forms magnetic lineations parallel to and bilaterally symmetric with the midoceanic ridge axis. Each lineation represents a former ridge axis and has a magnetic signature that allows identification according to age. Paleogeographic reconstructions of an ocean basin for a particular age may be made by fitting together magnetic lineations of that age from opposite sides of the ridge axis, i.e., in a manner analogous to running the accretionary process backwards. In the Atlantic, where the surrounding continents have been passively rafted as spreading has occurred, the former relative positions of the surrounding continents (Figure 4.7) may thus be obtained (Pitman and Talwani, 1972).

A major problem lies in constructing paleogeographic maps so that the entire developmental history of an ocean basin can be deciphered. The oldest rifted margins that are known are the U.S. East Coast and Northwest Africa margins. The rift-to-drift transition occurred here about 170 m.y. ago (Early Jurassic). The rifts themselves may contain sediments as old as Late Triassic. These early rifts initiated the breakup of Pangea, which resulted in the changing shape and distribution of continental margins (Figure 5.1). The evolution of passive margins as related to that breakup is discussed on pp. 78–85.

The DSDP provided information on the oceans only as far back as the uppermost Jurassic. There is ample seismic evidence that older Early Mesozoic sections occur in the oceans and on continental margins. However, to reach

SEA LEVEL HIGHSTANDS	MAJOR GLOBAL SEA LEVEL FALLS	SEA LEVEL LOWSTANDS
	PRE-LATE PLIOCENE & PRE-PLEISTOCENE (3.8 & 2.8 MA)	LATE PLIOCENE - EARLY PLEISTOCENE
EARLY & MIDDLE PLIOCENE		
	PRE-LATE MIOCENE & PRE-MESSINIAN (10.8 & 6.6 MA)	LATE MIOCENE
MIDDLE MIOCENE		
	PRE-MIDDLE LATE OLIGOCENE (30 MA)	MIDDLE LATE OLIGOCENE
LATE MIDDLE EOCENE & EARLY OLIGOCENE		
	PRE-MIDDLE EOCENE (49 MA)	EARLY MIDDLE EOCENE
LATE PALEOCENE - EARLY EOCENE		
	PRE-LATE PALEOCENE (60 MA)	MID-PALEOCENE
CAMPANIAN & TURONIAN		
	PRE-MIDDLE CENOMANIAN (98 MA)	MID-CENOMANIAN
ALBIAN - EARLIEST CENOMANIAN		
	PRE-VALANGINIAN (132 MA)	VALANGINIAN
EARLY KIMMERIDGIAN		
	PRE-SINEMURIAN (190 MA)	SINEMURIAN
NORIAN & MIDDLE GUADALUPIAN		
	PRE-MIDDLE LEONARDIAN (270 MA)	MID-LEONARDIAN
WOLFCAMPIAN & EARLIEST LEONARDIAN		
	PRE-PENNSYLVANIAN (324 MA)	EARLY PENNSYLVANIAN
OSAGIAN & EARLIEST MERAMECIAN		
	PRE-DEVONIAN (406 MA)	EARLY DEVONIAN
MIDDLE SILURIAN		
	PRE-MIDDLE ORDOVICIAN (490 MA)	EARLY MIDDLE ORDOVICIAN
LATE CAMBRIAN & EARLY ORDOVICIAN		
		EARLY CAMBRIAN & LATEST PRE-CAMBRIAN

FIGURE 4.6 Table of major sea-level falls that cause major interregional unconformities. (From Vail *et al.*, 1977.)

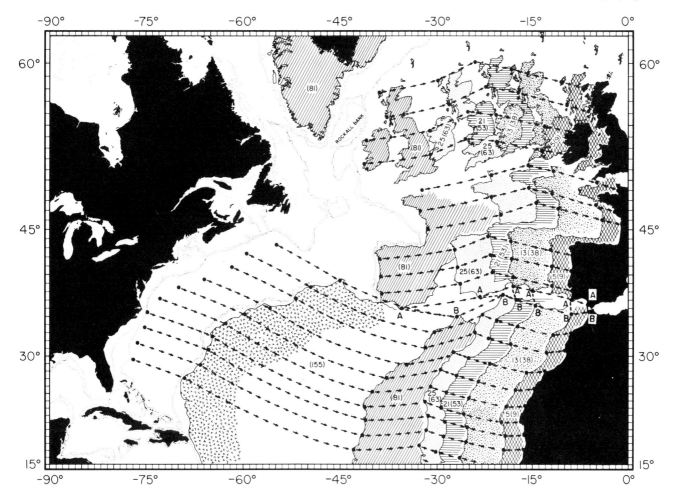

FIGURE 4.7 The relative position of Europe and Africa with respect to North America for specific times. Blacked-in continents represent present-day positions. Dates for earlier positions are indicated on the map. For Greenland only, the fully "closed" position [as deduced by Bullard *et al.* (1965) and presumed to be at 81 m.y.] and the present position are shown. The 200-m and 1000-m isobaths are shown for the relative positions at 81 m.y. Also shown are the positions of the Canary Islands and the New England Seamount Chain at this time. The arrows show the flow of Africa and Europe away from North America. The line *AB*, which joins a point in Spain and a point in Africa, is shown as a guide to the relative motion of Eurasia and Africa. [From Pitman and Talwani (1972). Reprinted from the *Bulletin* of the Geological Society of America.]

them by the drill requires capabilities beyond those of the *Glomar Challenger*. Because of the narrowness of a new ocean basin and the possible existence of inhibiting barriers at its ends (e.g., the present-day Red Sea), connection with the world ocean system may be severely restricted during its early development. This appears to have been the case for the South Atlantic north of the Walvis Ridge from about 130–106 m.y. B.P (LaBrecque, 1976).

Significant changes in oceanographic parameters may also take place during the evolutionary history of an ocean basin. For example, the Labrador Sea began to open about 81 m.y. ago, and the Norwegian Sea about 65 m.y. ago. In both instances, deepwater access between these young oceans and the Arctic was blocked by continent-bounded transform faults. There may have been

shallow-water communication, because sea level was unusually high at that time (Upper Cretaceous, Lower Paleocene; Hays and Pitman, 1973). A potential deepwater connection with the Arctic did not start to form until about 45 m.y. ago and may be one of the factors resulting in the formation of the Atlantic Eocene cherts. Today, connection with the Arctic is still restricted, but if present-day directions and rates of relative motion persist, a broad and deep connection will gradually be established.

Because of progressive plate motion, the geometry of an ocean basin, its connective links with other ocean basins, and its latitude may be radically changed within a few millions or tens of millions of years. For example, during the Pliocene, the Isthmus of Panama was uplifted, cutting off circulation between the central Atlantic and the central Pacific. Connection between the Mediterranean and the central Atlantic is also being cut off as Africa progressively moves northward into Europe (Minster *et al.*, 1974).

Although the timing of major changes in the geometry of the ocean basins may be calculated from models, the resultant effects must be determined from the sedimentary record. It is evident that changing the geometry, paleogeography, and interconnections of the ocean basins will affect sedimentation at margins as well as in the basins, but the precise details of these effects are not known. Analysis of the sedimentary record at the margins and in the deep sea would test the hypotheses about the nature of these changes. It is also possible that significant oceanographic changes might first be detected in the sediment record, indicating previously undetected changes in plate geometry or sea level. Passive margins serve as monitors of the oceanic surface-water circulation, of the vertical stratifications and characteristics of the oceanic

water-column, and of the climatically controlled paleoenvironment over the continental hinterland.

Transects of drill sites from continental margins would provide insight into the vertical structure of the oceanic water column because it is here that intermediate-depth and deepwater masses impinge upon the rise and slope of the continents. Because renewal of these water masses is much slower than surface- or bottom-water masses, the chances for the preservation of these oceanic paleoenvironments are considerably enhanced. The correlation of the Late Mesozoic and Cenozoic fluctuations of the Calcite Compensation Depth with the sequence of eustatic transgressions and regressions observed over the inner part of the continental margins proves the close linkage of the deepwater and shallow-water paleoenvironment. In short, transects across passive continental margins provide us with tools to study the horizontal and vertical gradients of the oceanic paleoenvironment.

Continental margins act as pathways to the seas for par-

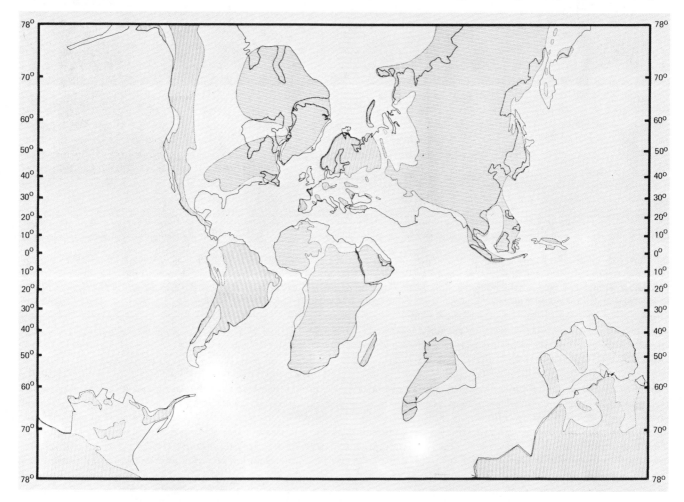

FIGURE 4.8   The arrangement of the various continental landmasses and the distribution of the seas for Coniacian time. This is a Mercator map. The pole of projection is the presumed rotational pole for the Coniacian as determined from paleomagnetic data. The position of the continents with respect to each other is based mainly on marine magnetic data. The distribution of the epeiric seas was determined from geologic data. (Courtesy of G. Mountain and W. C. Pitman, Lamont-Doherty Geological Observatory of Columbia University.)

ticulate or dissolved terrigenous material from the continental hinterland. The quantitative and qualitative composition of this input is controlled by the geology of the hinterland, by climatic processes, and by the area and relief of the drainage basins that supply the terrigenous material to the continental edge. The following climatic systems should be studied: (1) polar–subpolar, (2) transitional–temperate, (3) dry–subtropical, and (4) humid–tropical.

The global climatic zonation largely controls the amount and type of matter supplied from the continent to the ocean, causes the formation of different clay minerals, and encourages different sedimentation rates of terrigenous fines that result in a varying overburden, different physical properties, and different diagenetic processes of the organic and inorganic sediment components. In general, the benthic and planktonic communities of continental margins are highly diverse in tropical and subtropical regions, with diversities dramatically decreasing poleward. The variable proportions of organisms can change the flux of biogenous components from dominantly calcareous to dominantly siliceous, and vice versa.

Tropical continental margins trap large volumes of carbonate (either as reefal material or as platform carbonates) that markedly affect the ocean's carbonate budget.

Erosion along the continental slope generates hiatuses in the historic record. Slumping and mass-wasting seem to occur particularly frequently along temperate and subtropical continental margins, probably because of the specific physical properties of the deposits on these margins due to their high organic-carbon contents.

More work on paleogeographic reconstructions is needed. Areas such as the Central and North Pacific and Indian Ocean are particularly puzzling because of the obscure nature of the magnetic-anomaly (and fracture-zone) patterns in these areas.

Long-term variations in sea level may be dominantly controlled by the volume of the midoceanic ridge system, although other factors, such as continent–continent collision, sediment flux, and, of course, glacial fluctuation may produce significant effects on other time scales. Oceans are one of the most important factors controlling climate because of their ability to absorb, store, transport, and release heat.

The paleogeographic arrangement of the various continents for the mid-upper Cretaceous (Coniacian) is shown in a Mercator projection in Figure 4.8. Continental fragments have been arranged with respect to each other by fitting appropriate marine-magnetic lineations together. The positions of the continents are calculated with respect to the rotational axis, using paleomagnetic data. The distribution of the epeiric seas is determined from the geologic record. During the Coniacian, sea level was perhaps as much as 325 m above present sea level (Sleep, 1976). Climate at that time was significantly warmer at the poles than it is today.

Important changes in the size, geographic arrangement, and network of interconnections of the ocean basins have occurred since the Coniacian, brought about by the rifting and drifting of continents. Also, since the Coniacian, sea level has dropped by several hundred meters, changing the distribution of moderating epeiric seas, reducing the depth of interconnecting seaways, and probably affecting sources of oceanic bottom water.

In terms of geological time, the Coniacian paleogeography shown in Figure 4.8 represents but an instant in a continuously evolving mosaic of ocean basins generated by episodes of rift, drift, collision, and accompanying sea-level change.

The global change from a poorly oxygenated ocean to a well-oxygenated ocean is also of interest (Thiede and van Andel, 1977). Anaerobic conditions during mid-Cretaceous led to the deposition of black organic-rich shales in the Atlantic, the Tethys, and parts of the North Pacific and Indian oceans. The deposition and preservation of these sediments are related to changes of the oxygen minimum through geological time and are greatly influenced by transgressive and regressive cycles.

PALEOZOIC RECONSTRUCTIONS

The distribution of continents and oceans during Paleozoic and earlier times is far from clear. Here again, reconstructions allow us to pursue the fate of continental margins farther back in time. The main problems in this work involve (1) paleomagnetic reconstruction of continental cratons; (2) structural and paleomagnetic reconstruction of severely deformed folded belts; and (3) paleoclimatic interpretation of faunas, floras, and sedimentary rock sequences.

All three factors are fraught with uncertainties. Paleoclimatologic reconstructions are strengthened if they are based on the climatic compatibility of various observations. An interesting example of such a study has been published recently by Ziegler *et al.* (1977a, 1977b). Figures 4.9–4.11 show climatic and biogeographic maps for the Early Silurian.

C. THE NEED FOR BIOSTRATIGRAPHIC STUDIES

Stratigraphic and paleooceanographic studies are critically dependent on biostratigraphy. It is fair to state that the success of the DSDP would have been inconceivable without the help of many dedicated paleontologists and palynologists. Biostratigraphy provides the basis for time correlations; fossils contribute to an understanding of the paleoenvironments in which sediments accumulated and provide insights into the past conditions of ocean basins and their neighboring continents.

Studies of the areal and stratigraphic distribution of fossil taxa in Mesozoic and Cenozoic sedimentary sequences on continental margins are of fundamental importance in (a) providing a biostratigraphic and chronological framework for interpreting rates of geological processes

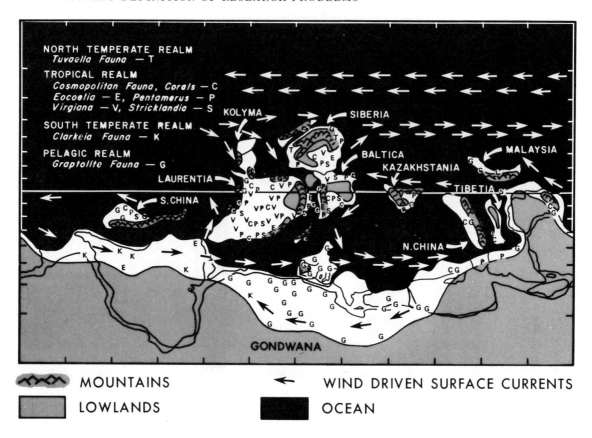

FIGURE 4.9 Biogeography and wind-driven surface currents (annual mean) in the Early Silurian in white shallow-water marine areas. (Redrawn from Ziegler *et al.*, 1977.)

associated with the evolution of continental margins and (b) interpreting the nature and extent of the temporal changes in sedimentary facies.

At present, most of our knowledge of oceanic stratigraphy has been derived from the study of zonations of calcareous and siliceous planktonic organisms. Extending these oceanic zones to continental margins and then to stratigraphic sections exposed on land is a fundamental problem. In some instances, this necessitates the integration of classical micropaleontology and palynology, as has been effectively demonstrated in Southern California (Figure 4.12). The information must come primarily from borehole cores, although many present data were gleaned from dredged rocks. In the future, submarine rock outcrops will be sampled more frequently by manned submersibles. In any event, sampling must be based on information from seismic profiles.

Of particular importance is the current work of the IPOD Stratigraphic Correlation Panel. This group is attempting to correlate the time stratigraphic zonation of benthic foraminiferal assemblages with the successful zonation based on pelagic foraminifera. Because pelagic as-

semblages are not common in shallow-water formations of continental margins, proper age determinations on margins have to be based on benthic assemblages. This work has to be done before scientific drilling in the shallow-water portions of continental margins begins.

Assuming that the sedimentary record will be adequately sampled, one area of particular interest is the depth zonation and morphologic gradients of fossil assemblages, particularly benthonic foraminifera. So far, attempts to link the characteristics of faunal assemblages from the abyss and shelf have been few and relatively unsuccessful. The relationship between environment and skeletal morphology is significant, because changes in phenotype morphology have been postulated to occur as a function of water depth. Others believe, however, that this is not an important factor and that morphologic differences with depth result from independent evolutionary lineages exhibiting trends of parallelism and convergence.

Many recent micropaleontological studies have demonstrated that the history of surface and abyssal ocean circulation can be determined from studies of marine

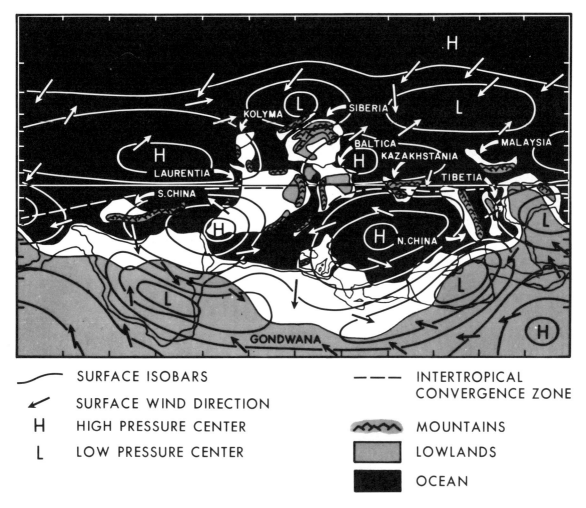

SURFACE ISOBARS

SURFACE WIND DIRECTION

H HIGH PRESSURE CENTER

L LOW PRESSURE CENTER

--- INTERTROPICAL CONVERGENCE ZONE

MOUNTAINS

LOWLANDS

OCEAN

FIGURE 4.10 Atmospheric pressure distribution and circulation for northern hemisphere winter in the Early Silurian in white shallow-water marine areas. (Redrawn from Ziegler *et al.*, 1977.)

microfossils. The more spectacular successes from this approach have been in reconstructing the surface hydrographic environment of the world oceans during global ice ages. These reconstructions were the fundamental input to (and historical tests of) several numerical models of climatic change. The primary data for these surface reconstructions are the observed distributions of marine planktonic organisms, a feature that restricts their application to the upper 500 m of the water column. The history of deeper-ocean circulation can be reconstructed from analogous studies of abyssal benthonic foraminifera, although the average minimum depth to seafloor in the open ocean generally restricts these reconstructions to depths greater than 3 km. Therefore, even though it is necessary to know the history of the intermediate and deep waters if we are to understand the dynamic link between past changes in surface and abyssal ocean circu-

lation, the paleocirculation of the depth interval between 0.5 and 3 km is generally unavailable for studies.

Studies of the present and past distribution of shelf and slope benthonic foraminifera can provide this knowledge of intermediate and deep waters. Such studies would require interpreting the ecological response of modern shelf and slope benthonic foraminifera to the local hydrography of the intermediate and deep waters and documenting changes in benthonic foraminiferal distributions through time.

We recommend that greater emphasis be given to studies relating biofacies to the environment of deposition on the slope and upper rise. Such studies should be based mainly on the distribution of benthonic foraminifera, because these organisms are the key to understanding the paleooceanographic conditions in intermediate and deep waters.

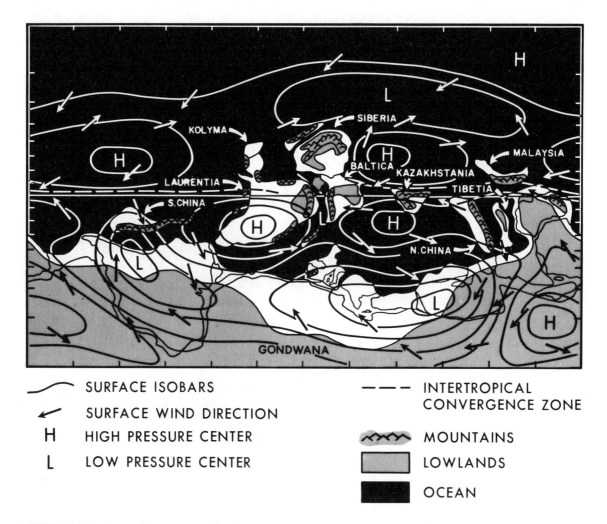

FIGURE 4.11   Atmospheric pressure distribution and atmospheric circulation for the northern hemisphere summer in the Early Silurian in white shallow-water marine areas. (Redrawn from Ziegler *et al.*, 1977.)

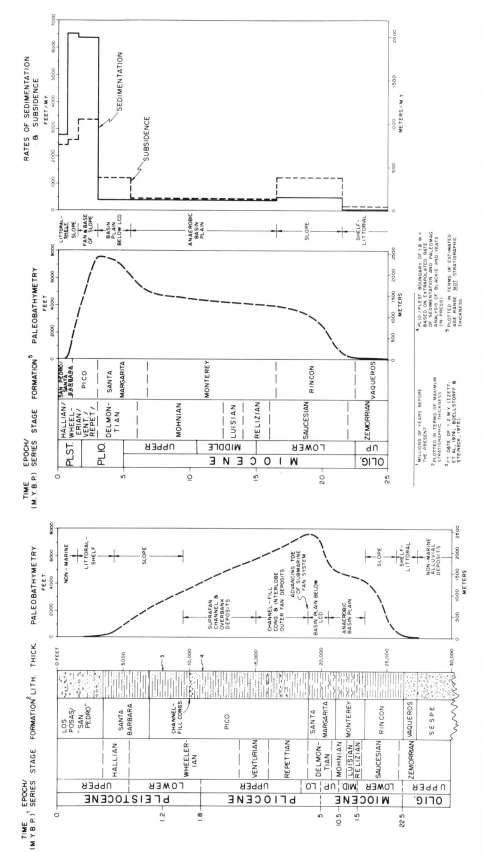

FIGURE 4.12 Paleobathymetry and marine paleoenvironments, and estimated rates of subsidence and sediment accumulation within Neogene deposits of the western Ventura Basin, Southern California Continental Borderland; figure from Ingle (in press). Note that the paleobathymetry is plotted both in terms of maximum stratigraphic thickness of each formation and alternately in terms of the estimated duration of each unit in time. Estimated paleobathymetry is based on biofacies analysis of benthonic foraminifera reported from each unit along with lithofacies analysis as discussed by Ingle (in press).

REFERENCES AND BIBLIOGRAPHY

Berggren, W. A., and C. D. Hollister (1974). Paleogeography, paleobiogeography and the history of circulation in the Atlantic Ocean, in *Studies in Paleo-oceanography*, W. W. Hay, ed., *Soc. Econ. Paleontol. Mineral. Spec. Pub. 20*, pp. 126–186.

Berggren, W. A., and C. D. Hollister (1977). Plate tectonics and paleocirculation—Commotion in the ocean, *Tectonophysics* 38:11–48.

Berner, R. A. (1974). Physical chemistry of carbonates in the oceans, in *Studies in Paleo-oceanography*, W. W. Hay, ed., *Soc. Econ. Paleontol. Mineral. Mem. 20*, pp. 37–45.

Briden, J. C., G. E. Drewry, and A. G. Smith (1974). Phanerozoic equal-area world maps, *J. Geol. 82*:555–574.

Bullard, E., J. E. Everett, and A. G. Smith (1965). The fit of the continents around the Atlantic, in A Symposium on Continental Drift, P. M. S. Blackett, E. Bullard, and S. K. Runcorn, eds., *Phil. Trans. R. Soc. London Ser. A 258*:41–51.

Cline, R. M., and J. D. Hays, eds. (1976). *Investigation of Late Quaternary Paleooceanography and Paleoclimatology, Geol. Soc. Am. Mem. 145*, 464 pp.

Des Vallieres, T. (in press). Test of various high resolution seismic devices in hard bottom areas, *Proceedings*, 10th Annual Offshore Technology Conference, May 8–11, Houston, Tex., 12 pp.

Dott, R. M., and R. L. Batten (1971). *Evolution of the Earth*, McGraw-Hill Book Company, New York, 649 pp.

Drewry, G. E., A. T. S. Ramsay, and A. G. Smith (1974). Climatically controlled sediments, the geomagnetic field and the trade wind belts in Phanerozoic time, *J. Geol. 82*:531–553.

Hallam, A., ed. (1973). *Atlas of Palaeobiography*, Elsevier, Amsterdam.

Hay, W. W., ed. (1974). *Studies in Paleo-oceanography, Soc. Econ. Paleontol. Mineral. Mem. 20*, 218 pp.

Hays, J. D., and W. C. Pitman III (1973). Lithospheric plate motion, sea level changes and climatic and ecological consequences, *Nature 246*:18.

Hughes, N. F., ed. (1973). Organisms and continents through time, *Spec. Paper in Palaeontology No. 12*, The Palaeontological Association, London, 334 pp.

Ingle, J. C., Jr. (in press). Cenozoic paleobathymetry and depositional history of selected sequences within the Southern California Continental Borderland, in O. L. Bandy Memorial Volume, R. L. Kolpack, ed., Spec. Pub., Cushman Foundation for Foraminiferal Research.

JOIDES Advisory Panel on Ocean Paleoenvironment (1977). A white paper in The Future of Scientific Ocean Drilling, a report by an *ad hoc* Subcommittee of the JOIDES Executive Committee, available from the JOIDES Office, University of Washington, Seattle, Wash. pp. 67–77.

Kennett, J. P., R. E. Houtz, P. B. Andrews, A. R. Edwards, V. A. Gostin, M. Hajos, M. A. Hampton, D. G. Jenkins, S. V. Margolis, A. T. Ovenshine, and K. Perch-Nielsen (1975). Cenozoic paleo-oceanography in the southwest Pacific Ocean, Antarctic glaciation and the development of the Circum-Antarctic Current, in *Initial Reports of the Deep Sea Drilling Project, Vol. XXIX*, J. P. Kennett, R. E. Houtz, *et al.*, eds., U.S. Government Printing Office, Washington, D.C., pp. 1155–1170.

LaBrecque, J. L. (1976). A study of the marine magnetic anomaly pattern employing techniques based on the fast Fourier transform algorithm, PhD thesis, Columbia U., New York, 273 pp.

Meyerhoff, A. A. (1970a). Continental drift: implications of paleomagnetic studies, meteorology, physical oceanography and climatology, *J. Geol. 78*:1–51.

Meyerhoff, A. A. (1970b). Continental drift II: high latitude evaporite deposits and geologic history of Arctic and North Atlantic Oceans, *J. Geol. 78*:406–499.

Meyerhoff, A. A., and C. Teichert (1971). Continental drift III: late Paleozoic glacial centers and Devonian–Eocene coal distribution, *J. Geol. 79*:285–321.

Minster, J. D., T. H. Jordan, P. Molnar, and E. Haines (1974). Numerical modelling of instantaneous plate tectonics, *Geophys. J. R. Astron. Soc. 36*:541–576.

Mitchum, R. M., Jr. (1976). Seismic stratigraphic investigation of West Florida Slope, Gulf of Mexico, *Am. Assoc. Petrol. Geol. Continuing Educ. Course Notes, Series 2*.

Payton, C. E., ed. (1977). *Seismic Stratigraphy—Applications to Hydrocarbon Exploration, Am. Assoc. Petrol. Geol. Mem. 26*, Tulsa, Okla., 516 pp.

Pitman, W. C., III, and M. Talwani (1972). Sea-floor spreading in the North Atlantic, *Geol. Soc. Am. Bull. 83*:619–646.

Rona, P. A. (1973). Worldwide unconformities in marine sediments related to eustatic changes of sea level, *Natural Phys. Sci. 244*:25–26.

Ruddiman, W. F., and A. McIntyre (1976). Northeast Atlantic paleoclimatic changes over the past 600,000 years, in *Investigation of Late Quaternary Paleooceanography and Paleoclimatology*, R. M. Cline and J. D. Hays, eds., *Geol. Soc. Am. Mem. 145*:111–146.

Ryan, W. B. F., and M. B. Cita (1971). Ignorance concerning episodes of ocean-wide stagnation, *Marine Geol. 23*:197–215.

Schlanger, S. O., and H. C. Jenkyns (1976). Cretaceous oceanic anoxic events: causes and consequences, *Geologie Mijnb. 55(3/4)*:179–184.

Shackleton, N. J., and J. P. Kennett (1975a). Paleotemperature history of the Cenozoic and the initiation of Antarctic glaciation; oxygen and carbon isotope analyses in DSDP sites 277, 279 and 281, in *Initial Reports of the Deep Sea Drilling Project, Vol. XXIX*, J. P. Kennett, R. E. Houtz, *et al.*, eds. U.S. Government Printing Office, Washington, D.C., pp. 743–755.

Shackleton, N. J., and J. P. Kennett (1975b). Late Cenozoic oxygen and carbon isotopic changes at DSDP site 284: implications for glacial history of the northern hemisphere and Antarctic, in *Initial Reports of the Deep Sea Drilling Project, Vol. XXIX*, J. P. Kennett, R. E. Houtz, *et al.*, eds., U.S. Government Printing Office, Washington, D.C., pp. 801–807.

Sieck, H. C., and G. W. Self (1977). Analysis of high resolution seismic data, in *Seismic Stratigraphy—Applications to Hydrocarbon Exploration*, C. E. Payton, ed., *Am. Assoc. Petrol. Geol. Mem. 26*, Tulsa, Okla.

Sleep, N. H. (1976). Platform subsidence mechanisms and eustatic sea-level change, *Tectonophysics 36*:45–56.

Smith, A. G., J. C. Briden, and G. E. Drewry (1973). Phanerozoic world maps, in *Organisms and Continents Through Time*, N. F. Hughes, ed., *Spec. Paper in Palaeontology No. 12*, The Palaeontological Association, London, pp. 1–42.

Thiede, J., and T. H. van Andel (1977). The paleoenvironment of anaerobic sediments in the late Mesozoic South Atlantic Ocean, *Earth Planet. Sci. Lett. 33*:45–56.

Vail, P. R. (1977a). Seismic recognition of depositional facies on slopes and rises, Am. Assoc. Petrol. Geol. Continuing Educ. Course Notes Series 5.

Vail, P. R. (1977b). Sea level changes and global unconformities from seismic sequence interpretation, a report of the JOIDES Subcommittee on the Future of Scientific Ocean Drilling, Woods Hole, Mass., Mar. 7–8, unpublished.

Vail, P. R., R. M. Mitchum, Jr., R. G. Todd, J. M. Widmier, S. Thompson III, J. B. Sangree, J. N. Bubb, and W. G. Halelid

(1977). Seismic stratigraphy and global changes of sea level, in *Seismic Stratigraphy—Applications to Hydrocarbon Exploration*, C. E. Payton, ed., *Am. Assoc. Petrol. Geol. Mem.* 26:49–212.

van Andel, T. H., and T. C. Moore, Jr. (1974). Cenozoic calcium carbonate distribution and calcite compensation depth in the central Equatorial Pacific, *Geology* 2:87–92.

van Andel, T. H., G. R. Heath, and T. C. Moore, Jr. (1975). Cenozoic history and paleooceanography of the central Equatorial Pacific Ocean: A regional synthesis of Deep Sea Drilling Project data, *Geol. Soc. Am. Mem.* 145:134.

Wegener, A. (1929). *The Origin of Continents and Oceans*, J. Biram, translator, Dover, New York, 1966.

Ziegler, A. J., K. S. Hansen, M. E. Johnson, M. A. Kelly, C. R. Scotese, and R. Van der Voo (1977a). Silurian continental distribution, paleogeography, climatology, and biogeography, *Tectonophysics* 40(1/2):13–51.

Ziegler, A. M., C. R. Scotese, W. S. McKerrow, M. E. Johnson, and R. K. Bambach (1977b). Paleozoic biogeography of continents bordering the Iapetus (Pre-Caledonian) and Rheic (Pre-Hercynian) Ocean, in *Paleontology and Plate Tectonics*, R. M. West, ed., Milwaukee Public Museum, Special Papers in Biology and Geology #2.

5 | Passive Margins

A. INTRODUCTION

About 200 million years ago, all the present continents were joined together into one supercontinent, Pangea (Figure 5.1). In a reconstruction of Pangea, there was only one small segment of passive margin. It extended from northwest Africa to northern Australia. The tectonic style of the entire remainder of the margin of the supercontinent was characterized by subduction of oceanic crust. Since that time, Pangea has been breaking up and reorganizing to form the global pattern of plate tectonics. Rifted margins have been forming and developing. A primary problem that confronts earth scientists is understanding the geological consequences of converting from a world with only one ocean and one continent into a world with several of each (Figure 5.1). No doubt, there were dramatic changes in the climates, affecting both the hydrosphere and the atmosphere, with concomitant major geological and biological effects.

The breakup of Pangea coincides with the formation of most passive margins of the world (Figure 5.2). (As a reminder, passive margins are located within a lithospheric plate and, within that plate, straddle the boundary between oceanic and continental crust.) Our task is to unravel the structure and trace the evolution of these margins from the early rifting stage to a mature stage.

Most of what we know about the physical history of the earth and the development of life comes from studies of the composition and structure of the sediments and sedimentary rocks that veneer the crust of the earth. We estimate that more than half of the sedimentary sequences younger than 200 m.y. have been deposited on passive margins since the breakup of the early Mesozoic supercontinent. These great reservoirs of sediments are truly monuments to 200 million years of slow subsidence of and deposition on the continental margins that originated in the interior of Pangea.

The ocean basins created during the breakup of Pangea have left behind partial records of their growth, their environment, and their disappearance. But records of their birth are well hidden. Only along passive margins can we find information on their genesis, and only passive margins document the history of the transitional boundary between continent and ocean basins.

Why are we so interested in the birth of oceans? As oceans are born, old continental rocks interact in unique ways with new, hot oceanic rocks and with waters invading from the oceans. This interaction may be one of the most salient processes of all in the accumulation of metallic mineral deposits. Similarly, sedimentation in the juvenile ocean basins often produces a combination of evaporites, source beds, and reservoir beds that constitute favorable circumstances for hydrocarbon accumulation.

Interesting and important as it is, the record of the birth of an ocean is only a start. Preserved accumulations of passive-margin sediments constitute a unique, long-term

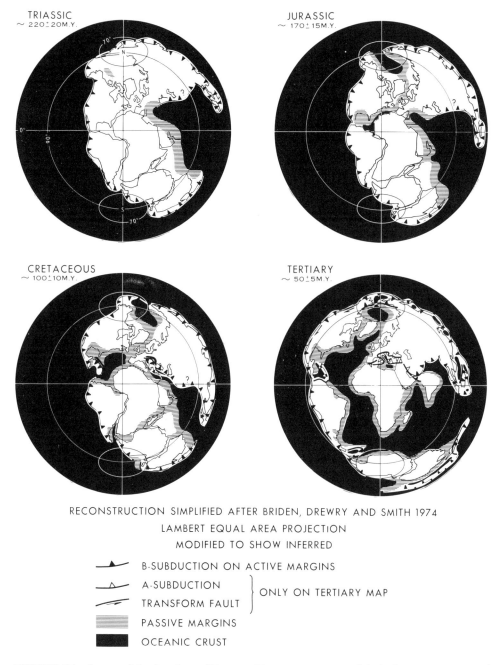

TRIASSIC
~ 220±20 M.Y.

JURASSIC
~ 170±15 M.Y.

CRETACEOUS
~ 100±10 M.Y.

TERTIARY
~ 50±5 M.Y.

RECONSTRUCTION SIMPLIFIED AFTER BRIDEN, DREWRY AND SMITH 1974
LAMBERT EQUAL AREA PROJECTION
MODIFIED TO SHOW INFERRED

B-SUBDUCTION ON ACTIVE MARGINS

A-SUBDUCTION

TRANSFORM FAULT ⎬ ONLY ON TERTIARY MAP

PASSIVE MARGINS

OCEANIC CRUST

FIGURE 5.1 Stages of the breakup of Pangea. (Reconstruction simplified after Briden *et al.*, 1974.)

(150–200 m.y.), homogeneous log of the oceanic and continental paleoenvironment.

Most investigations into the massive sediment wedges of passive continental margins have been either related to the search for hydrocarbon prospects or reconnaissance in nature. Our information is primarily surficial or indirect, obtained either from geophysical methods or by correlation with presumed analogous ancient deposits now uplifted into mountain ranges and exposed on land. The

JOIDES Deep Sea Drilling Project (DSDP) did not, in the main, attack continental-margin problems (IPOD Passive Margin Panel, 1977). DSDP entered the broad passive continental-margin zone in just a few places. To date, the international phase of DSDP drilling has done only some shallow reconnaissance in the outer parts of a few margins. The purpose of that drilling was to assist the planning for later phases of the International Program for Ocean Drilling (IPOD) that will more directly plunge into

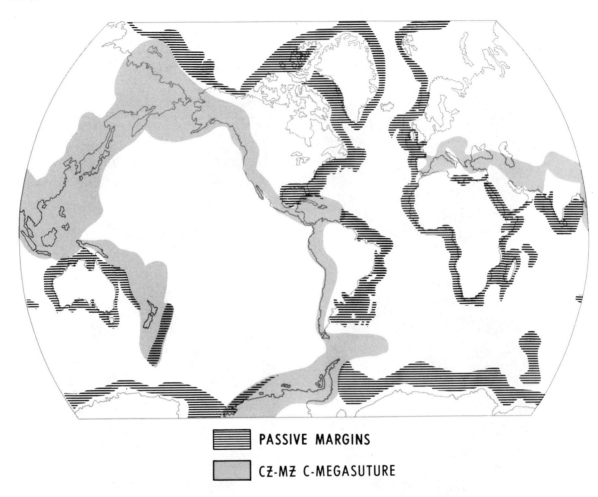

PASSIVE MARGINS

CZ-MZ C-MEGASUTURE

FIGURE 5.2   Passive margins of the world.

the important geological and geophysical problems of passive continental margins.

Although it is an essential tool, the scientific problems cannot be solved by drilling alone, any more than they could have been solved by surface studies and geophysics alone. Until just a few years ago, only the crudest estimates of the structure of continental margins could be made with seismic-refraction techniques. Multichannel seismic-reflection data have produced a quantum jump in our understanding of the structure of margins. The way is now open to build a widespread interpretation of the chronology and processes involved in the deposition of our greatest sediment reservoirs—*if drilling results are correlated with structures revealed by multichannel reflection records.*

B. HOW DO PASSIVE MARGINS FORM AND EVOLVE?*

Passive margins apparently were formed by rifting of continental crust, and their present configurations are the results of geological processes that, in turn, were affected by climate and ocean circulation. We are in dire need of deep structural and stratigraphic data to unravel the story and tell it in detail. Some pertinent data—particularly from Australia, Canada (Yorath *et al.*, 1975, and Given, 1977), and the North Sea (Woodland, 1975)—are being released by petroleum companies. IPOD garnered interesting data off Africa and Europe in preparing the plans

*Much of this discussion is based on the white paper of the IPOD Passive Margin Panel (1977).

for further drilling in these areas, and the IPOD holes were drilled on the basis of some reasonably comprehensive geological and geophysical surveys.

Several models for the evolution of passive margins have been proposed to explain particular margins. (For a review, see Bott, 1976a and in press.) The differences between these models suggest some real differences in the evolutionary history of each margin. However, a consensus exists that favors the following sequence of evolutionary phases (Figure 5.3):

1. Rifting;
2. Onset of drifting, i.e., separation of continental crust as oceanic crust accretes (displaying an array of magnetic stripes) in the gap between continental blocks; and
3. Postrift evolution, dominated by massive subsidence of the rifted margins and shaping of those margins by sedimentary and secondary (mostly gravity) tectonics.

C.   RIFTING

The great East African rifts (Pilger and Rösler, 1975, 1976; Baker *et al.*, 1972), the Rhine Valley graben (Illies and Fuchs, 1974), and the Rio Grande rift are believed to exemplify the basic rifting process. Rifts like these may—or may not—lead to the formation of an ocean. Sometimes, the rifting process continues; other times, it stops.

As shown by the classical sketches (Figures 5.4 and 5.5) of Cloos (1939), rifts are initiated on high domal uplifts. These uplifts are presumed to be thermal in origin. Extensive volcanism occurs. Erosion of the uplifted area and

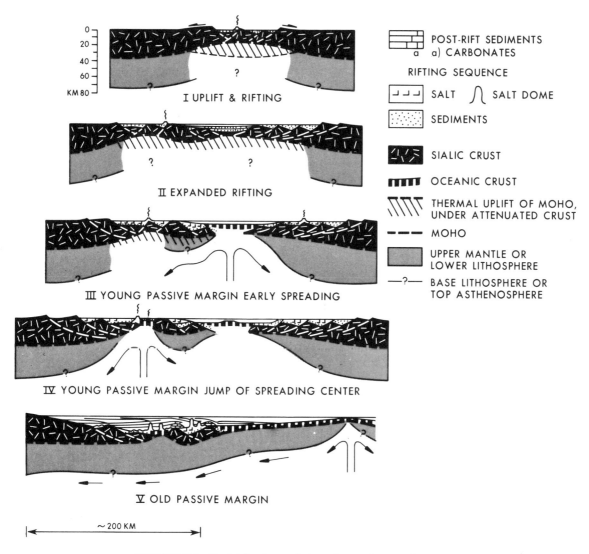

FIGURE 5.3   Model for the evolution of a passive margin.

FIGURE 5.4 The southwest corner of Germany. Rhine Valley trough (graben) between Vosges (foreground) and Black Forest, Kaiserstuhl volcano in its center. A low and broad shield-shaped dome fractured in the center. Swiss Jura folds to the right. Gneiss exposed on both sides of graben is overlain by flat-lying Triassic strata forming steps. [From Cloos (1953), reprinted by permission of Alfred A. Knopf, Inc.]

thermal metamorphism of the lower crust may cause crustal thinning. As the area of the uplift expands, the crestal portion cracks and splits into subparallel and trilete fault patterns that allow keystonelike segments to subside along normal faults.

Adjoining cratons may be further separated by crustal stretching along listric (i.e., curved, like sledrunners) normal faults (Lowell and Genik, 1972). Others (Mutter, in press) maintain that there is no significant crustal stretching during the rifting process and that broadening of the rift is caused by crustal movements that are dominantly vertical.

A different model by Falvey (1974) suggests that as the thermal center rises, the zone of uplift expands in an outward direction. With it, the crestal zone of normal faulting expands to form a wide zone of rifting. Thus, there will be a gradation from drastically altered continental crust in the domal crest region to less-altered crust away from the domal center. Falvey's model needs documentation. Refraction work on the South Australian margin (Mutter, in press) and on the Canadian Atlantic shelf (Keen and Keen, 1974) suggests an abrupt transition from typical continental crust to what might be interpreted as radically altered continental crust. These data contradict Falvey's model.

So far, our rifting models involve the breaking of normal continental crust that is substantially older (in excess of 150 m.y.) than the rifting process itself. Such models are based on the East African–Red Sea perspective.

There is mounting suspicion that some of the great rifts initiating the passive margins of the world are emplaced on relatively young, "orogenic" crust of folded belts. Figure 8.6 illustrates numerous basins that are called episutural basins because they lie within the megasuture discussed earlier. Some of these basins are rifted before

they spread to form marginal basins. Many of these basins are connected with strike-slip fault systems. For example and in a speculative vein, this concept predicts that western North America could drift away from the mainland, as shown on Figure 5.6.

Turning now to the material that fills rifts, it has been observed that volcanics alternate with and are covered by contemporaneous clastic sediments. Depending on the elevation of the rift-valley floor and the prevailing climate, repeated transgressions may result in the deposition of evaporites on the extending continental crust. The restricted environment within the graben systems can lead to the deposition and preservation of organic-rich source beds, either marine or lacustrine in origin. It should be kept in mind that the domal uplift drains streams away from the rift valley. Thus, the clastic sediment supply is somewhat scarce in the graben valley.

DSDP-IPOD drilling on the Rockall Plateau indicates that there was substantial subaerial relief at the end of rifting and that shallow-water marine volcanogenic conglomerates were deposited during rifting. However, the results from DSDP drilling in the Bay of Biscay and Galicia do not support the East African analog. In these areas, 2000 m of submarine relief were created by the end of rifting. The rifting process was apparently submarine and was not accompanied by widespread volcanism. These unexpected differences may reflect rifting of a Precambrian craton in the case of the Rockall Plateau and rifting of an epicontinental basin in the case of Biscay (Montadert *et al.*, 1977).

Simple rifts such as the Gulf of Suez (Figure 5.7B) can evolve into spreading oceans (as illustrated by the Red Sea, Figure 5.7A), or else the rifting process can be aborted to form deep-seated graben systems underlying large cratonic basins (e.g., the North Sea, Figure 5.7C). In the following section, we pursue the evolution of a rift system into a passive continental margin.

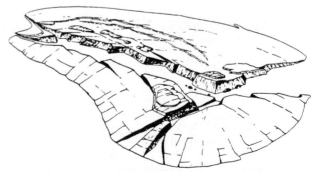

FIGURE 5.5 Red Sea and Gulf of Aden drawn as faulted trenches in the center of the uplifted and broken Nubian–Arabian shield and its basalt cover. View from southwest. Left foreground: Egypt and Abyssinia; right foreground: Somaliland; background: Arabian peninsula. [From Cloos (1953), reprinted by permission of Alfred A. Knopf, Inc.]

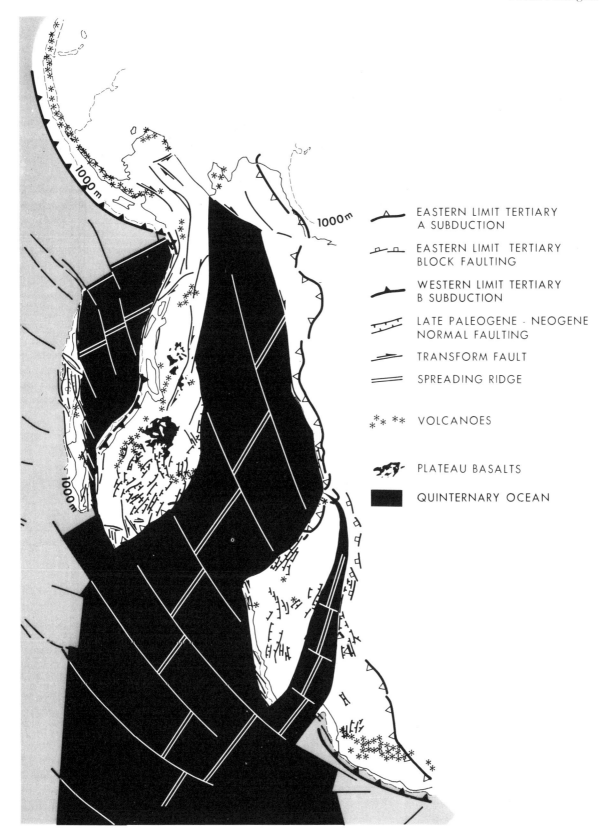

FIGURE 5.6   Quinternary tectonics, Western Cordillera. A speculative drifting prediction for western North America.

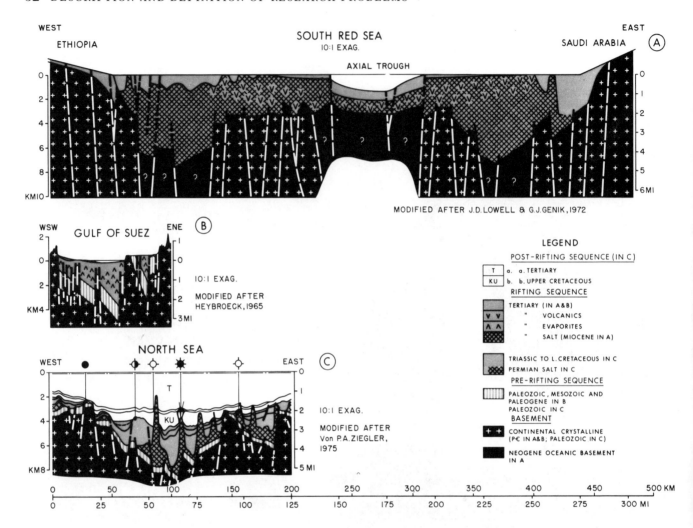

FIGURE 5.7   Examples of rifting. A, Rifting in the South Red Sea; B, rifting in the Gulf of Suez; C, aborted rift in the North Sea.

D. ONSET OF DRIFTING

During rifting, separation is accommodated by attenuation of continental crust (Rhinegraben Research Group for Explosion Seismology, 1974). Drift begins with the emplacement of oceanic crust between two continental fragments. Geologists determine the time of breakup by the age of the oldest seafloor immediately adjacent to the continent. Breakup times change along passive margins and may well progress from one end of the margin to the other, as has happened in the South Atlantic (Rabinowitz and LaBrecque, in press). The following is a list of passive margins for which the ages of the rift–drift transition (breakup time) are reasonably well known.

- The East African rift system: This may still be in the rifting stage.
- The Red Sea and Gulf of Aden (~10 m.y. B.P.; Laughton et al., 1970).

- The Gulf of California (~5 m.y. B.P.; Larson et al., 1968; Moore, 1973).
- The South Australian margin (~50 m.y. B.P.; Weissel and Hayes, 1972).
- The Labrador Sea and Europe; North America (~80 m.y. B.P.; Kristoffersen, 1977).
- The Norwegian Sea (~65 m.y. B.P.; Talwani and Eldholm, 1977).
- South America/Africa (~130 m.y. B.P.; Larson and Ladd, 1973).
- North America/Africa (~180 m.y. B.P.; Pitman and Talwani, 1972).

How, and for what reasons, drifting starts at a particular site is obscure. It is commonly assumed that it reflects a change from conductive heat flow in the mantle to convective heat flow driving the plates apart.

The line along which drift is initiated is not necessarily centrally located within the wide (up to 200 km) zone of rifting and blockfaulting. That line may be located toward one side, as it is between Africa and North America (Rabinowitz, 1974). The study of conjugate margins, that is, margins that were joined before they drifted apart, will shed much light on this problem.

In recent years, many predrift reconstructions have been made. They are based on matching coeval magnetic anomaly trends, fracture-zone directions, the presumed boundary between oceanic and continental crust, and major geological trends in continents that predate drifting. It is difficult to see how these continental fits can be refined much further on a regional scale, and yet, they indicate excessive overlaps or gaps between continental blocks, which must reflect erroneous positions in the ocean–continent boundary, stretching of the continental basement during rifting, or the nonrigidity of blocks that are currently presumed to be rigid. Three examples off eastern Canada illustrate the point:

1. The fit of Iberia against the Grand Banks, in which the inconvenient location of Flemish Cap and Galicia Bank causes either an overlap in the north or a gap in the south;

2. The fit of Rockall Bank against the northeast Newfoundland shelf is not well defined, which suggests that the ocean–continent boundary needs better definition, particularly southwest of the bank; and

3. Plate reconstructions between Greenland and North America involve an overlap within the Davis Strait region, or gaps in Baffin Bay and the Labrador Sea.

These and other areas of continental misfit are not adequately known, because many of them may have been subjected to poorly understood processes (e.g., stretching of, or "oceanization" of, continental crust) during initial rifting. If we could understand such processes better, we might be able to explain some of the mechanisms involved in the early history of continental margins. In particular, one could look for the continuation of prerift continental structures in marine areas where crustal thicknesses are similar to those usually associated with oceanic crust. Reconstructions require a precise definition and outline of the continent–ocean boundary. In many areas, magnetic anomalies and a characteristic isostatic gravity anomaly suggest that the continent–ocean boundary may be a narrow linear zone. However, multichannel reflection profiles and good seismic-refraction data across such boundaries are sparse and do not offer unambiguous interpretations. The data suggest many possibilities. Some involve the injection of dikes into continental crust; others, the presence of structural marginal highs.

Let us now consider topographic relief of the continent and ocean at the onset of drifting. The results of DSDP Leg 48 (Montadert *et al.*, 1977) suggest that the continent–ocean boundary west of Rockall Plateau may have been close to sea level at the onset of spreading. In other areas, multichannel profiles across the continent–ocean boundary show reflectors below the oceanic basement. These reflectors may indicate layers of sediments interbedded with lavas (e.g., Walvis Ridge; Lofoten Basin off Norway). Although occasional marine invasions occur during the earlier phases of rifting, real oceanic conditions do not begin until after the rift–drift transition. Immediately following that transition, there may be a brief period of dry seafloor spreading; that is, a period of time during which, although oceanic crust is forming, the configuration of the bounding block is such that the ocean cannot flood the spreading zone (e.g., the Afar Depression of Ethiopia).

An unconformity generally confined to the elevated horsts formed during rifting often separates the "rifting sequence" from the "drifting sequence" (for example, see Figure 5.8). Some scientists relate this unconformity to the major change in thermal regime from conductive to convective. During rifting, the heat source remained fixed beneath the rift axis, but on initiation of spreading, it moved away relative to the continent–ocean boundary, allowing the passive margin to cool. It has been suggested that secondary heating and thermal expansion may produce a short period of uplift. The results from drilling in Bay of Biscay and off Rockall Plateau (Montadert *et al.*, 1977) show that no emergence occurred—that subsidence began at the onset of spreading. The subsidence curves derived from the drilling results in these two areas are closely similar to each other. They fit empirical cooling curves constructed for the spreading oceanic crust. This indicates that the subsidence is independent of both the structure and the initial elevation of the continent. One effect of the subsidence is to allow wider transgression of the margin—often revealed as an unconformity with onlap separating the faulted sequence from the mostly unfaulted postrift sequence. The DSDP data from the Bay of Biscay suggest that, as expected, such an unconformity may not be present across the whole margin if it was already deep at the onset of spreading. Another important effect of subsidence is that the margin is gently tilted oceanward, reversing the drainage pattern. During the early spreading stage, fracture zone and aseismic ridges may strongly influence sedimentation.

E. POSTRIFT EVOLUTION: DRIFTING

Passive margins continue to drift until they are engulfed in subduction processes. Massive subsidence occurs during the rifting, and several models have been proposed for the subsidence of passive margins. Broadly speaking, the "drifting sequence" on passive continental margins is a function of age and a poorly understood interplay between subsidence, sedimentation, ocean circulation, and climate. Two types of margins are often contrasted: starved margins and mature margins.

Starved margins have a thin, prograding cover. They may be young (e.g., Neogene) or old (e.g., Mesozoic). Typical examples are the west margin of Rockall Plateau and the Bay of Biscay. So far, only starved margins have

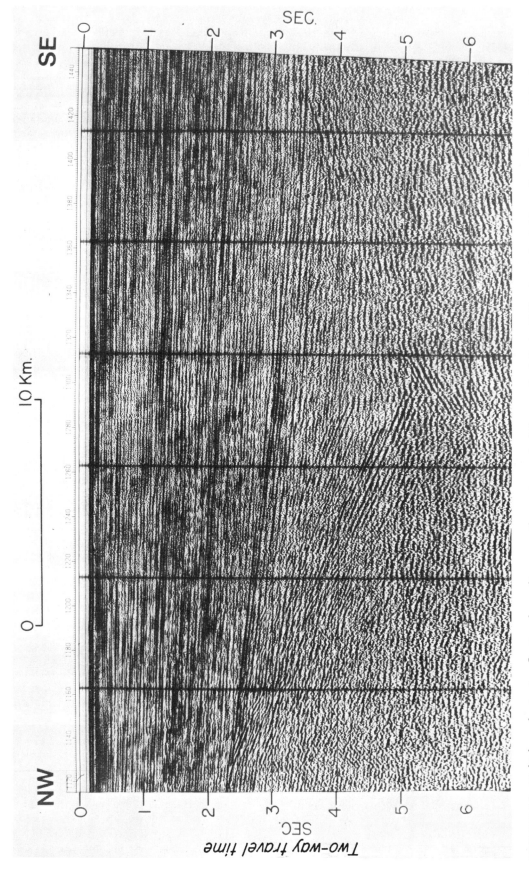

FIGURE 5.8 Multichannel seismic reflection along a line across the western end of Georges Bank Basin. Half graben is buried beneath gently seaward-dipping sequence of sedimentary rocks (upper and middle acoustic units). (From Schlee *et al.*, 1976, and courtesy of the U.S. Geological Survey.)

84

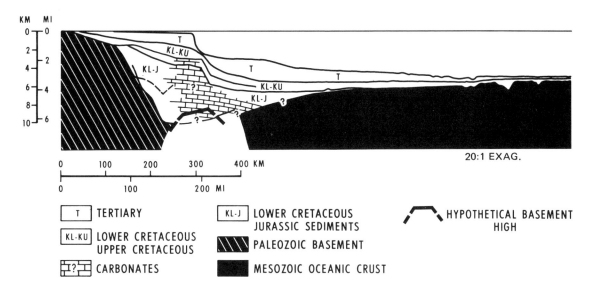

FIGURE 5.9   New York–Delaware Baltimore Canyon Trough–Atlantic Margin.

been drilled during IPOD-DSDP, because of the *Glomar Challenger*'s limited drilling capabilities.

An effect of subsidence is to bury deeply the first post-rift sediments, as the margin is prograding in a seaward direction. On the starved margin of west Rockall, this progression can be seen in the form of a transition from a clastic shelf sequence to a pelagic–bathyal sequence.

Mature margins are characterized by a thick (10 km) prograding wedge of shelf sediments, such as those on the U.S. Atlantic margin (Figures 5.9–5.13) and its conjugate margin off northwest Africa. These margins contain some of the thickest and probably most continuous sedimentary sections in the oceans. Together, the sedimentary strata of the shelf, slope, and rise provide a record of the subsidence of the rifted basement and the adjoining oceanic crust. These strata may also provide clues to other phenomena, such as eustasy, worldwide climatic changes, and variations in ocean circulation, chemistry, and local tectonic and sedimentary events. It is likely that near the shelf edge of many passive margins, the strata form a nearly continuous record of sedimentation from the earliest graben formation during rifting to the Pleistocene. The truly marine part of the section begins at the rift–drift transition. Evidence from borehole data (Sleep, 1971; Watts and Ryan, 1976), rock dredges (Fox *et al.*, 1970), and deep submersibles (W. B. F. Ryan, Lamont-Doherty Geological Observatory, personal communication) shows that, even at the shelf edge, there are deposits of shallow-water origin for nearly all strata—from the deepest to the shallowest layer. Tectonically, the shelf is usually relatively undisturbed.

The lithology and volume of the sediments that comprise the postrift sequence on starved and mature margins clearly depend on paleoenvironment, climate, sea level, and the size and geology of the continental hinterland.

Drilling results have provided some indication of the relation between these factors. Present margins distort the largely wind-driven latitudinal surface-water circulation to produce eastern and western boundary currents and major divergences. These boundary currents separate the stable central water masses from the highly variable coastal water masses over the margin. Changes in basin-water circulation have profoundly influenced the sediment dynamics along continental margins and have produced large hiatuses in the record. For instance, in the northeast Atlantic, gaps in the Lower Miocene (found in Rockall, Biscay, and in the record at other localities) may be associated with overflow across the Iceland–Faeroes Ridge. Fluctuations in the carbonate content of the sediments may be due to worldwide changes in the Calcite Compensation Depth (CCD).

F. ON MODELING THE SUBSIDENCE OF PASSIVE MARGINS

During their evolution, Atlantic-type continental margins are dominated by uplift and subsidence. The subsidence history of a margin is recorded in the sediments that accumulate soon after rifting and plate separation. Biostratigraphic data from commercial boreholes show that continental shelves are composed of substantial thicknesses of shallow-water sediments, which indicates that significant subsidence has occurred at these margins.

Walcott (1972) quantitatively modeled the contribution to margin subsidence made by loading sediments on the slope and rise regions. He used a simple model in which the lithosphere was represented by a thin elastic plate overlying a weak fluid layer. Assuming homogeneous elastic properties for the oceanic and continental plates,

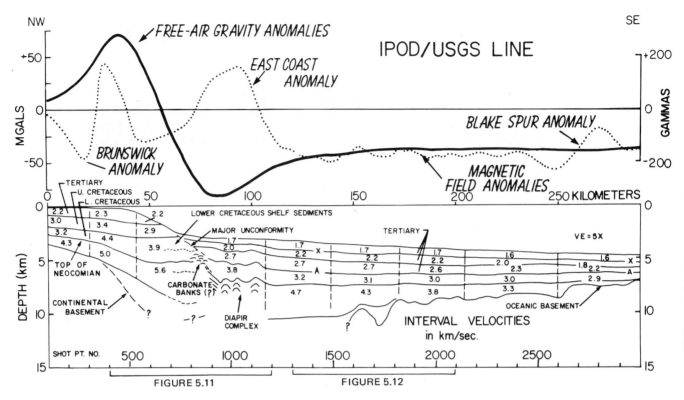

FIGURE 5.10 Composite geophysical profile along nearshore end of IPOD/USGS line off Cape Hatteras. (After Grow and Markl, 1977, and courtesy of the U.S. Geological Survey.)

Walcott showed that substantial thicknesses of sediments (up to 18 km) could accumulate. Although flexural models explain the shape and thicknesses of sediments on wide, Atlantic-type margins such as deltas, they do not explain the characteristics of narrow margins, such as the U.S. East Coast. There are three main problems with flexural models.

1. Atlantic-type margins are often characterized by narrower shelf widths than those produced in the flexural models.

2. The proportion of shallow-water to deepwater sediments in the flexural models is small (about 1:5) and does not explain the predominance of shallow-water sediments found in deep boreholes of the coastal plain and shelf regions of the Atlantic.

3. The rate of basement subsidence is dependent on the sedimentation rate in the flexural models and does not explain subsidence that occurs independently of sediment supply (for example, the Blake Plateau during the Tertiary). Turcotte *et al.* (in press) recently extended the sediment-loading model to account for different elastic properties between the oceanic and continental plates and to allow for the possibility of relative motion between them. Turcotte and his colleagues considered the possibilities of sediment loading in slope and rise regions as contributive to shelf subsidence, as Walcott did. They

also considered sediment loading of decoupled oceanic and continental crust, allowing vertical relative motion by faulting. They concluded that a decoupled model most satisfactorily explains the observed free-air gravity anomaly profiles.

Several scientists have considered other processes besides sediment loading as being contributors to subsidence. Most models include processes that thin the crust at a margin. When the crust is thinned, the isostatic equilibrium necessarily changes, and, as a result, the thinned crustal layers drop down, or subside.

Sleep (1971) quantitatively examined how thermal uplift and erosion could cause crustal thinning and subsidence of the shelf. The subsidence rates inferred from deep commercial boreholes off the U.S. East Coast and Gulf Coast fit a 50 m.y. decaying exponential, implying a thermally controlled mechanism for shelf subsidence. Recently, Sleep and Snell (1976) developed a thermally controlled model in which the viscoelastic behavior of the continental crust was also considered. They did not consider changes in the mechanical properties across the ocean–continent boundary. Their model showed that a decaying exponential rate, together with a viscoelastic relaxation of the lithosphere, could explain the stratigraphy of Atlantic-type coastal plain regions.

Bott (1971) proposed an alternative to thermal uplift

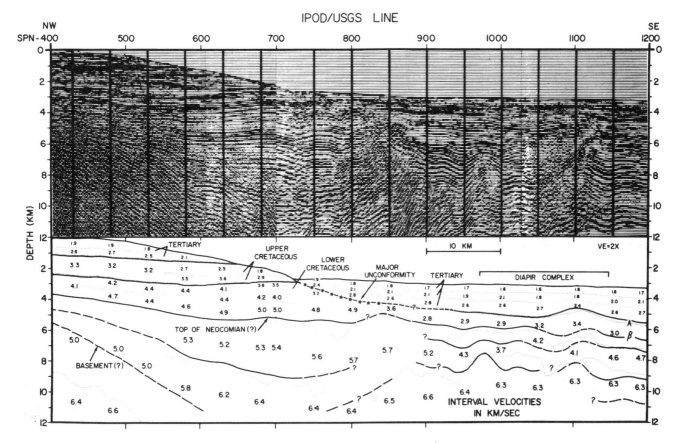

FIGURE 5.11 Depth section and line-drawing interpretation of Figure 5.10. Section between SPN-700 and SPN-1200 has also been migrated. Internal velocities used for depth conversion were averaged every 5 km along horizontal bands separated by dotted lines. Velocity values down to 4–5 km between SPN-400 and SPN-700 and 6–7 km between SPN-900 and SPN-1200 are considered reliable. Zone between SPN-700 and SPN-900 has complex topography and a major unconformity resulting in less reliable velocities and some artificial undulations of horizons. (After Grow and Markl, 1977, and courtesy of the U.S. Geological Survey.)

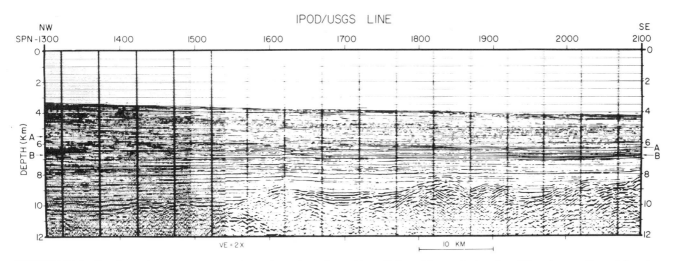

FIGURE 5.12 Depth section over inner continental rise (IPOD/USGS line off Cape Hatteras), showing 10-km-deep sediments lapping onto oceanic crust near SPN-1600. Reflectors A and B represent horizons A and Beta. (After Grow and Markl, 1977, and courtesy of the U.S. Geological Survey.)

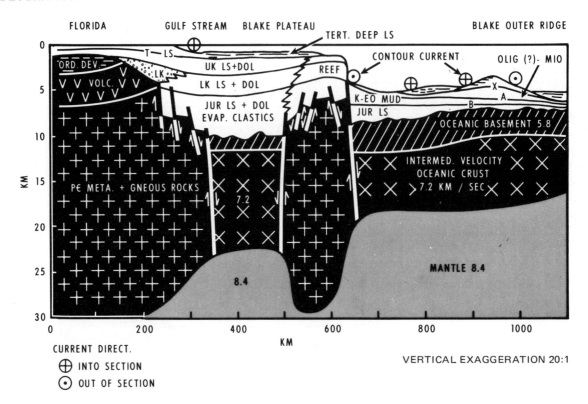

FIGURE 5.13   Blake Plateau. (Modified and redrawn from Sheridan and Osburn, 1975.)

and erosion as the principal causes of crustal thinning. Bott and Dean (1972) used the finite-element modeling technique to study the stress system at an Atlantic-type margin. They proposed that hot creep in the lower crust takes place as a result of stress differences that arise from the geometry of the transition zone between continental and oceanic crust. They suggested that shelf subsidence is caused by seaward flow of the lower crustal material.

Such studies have effectively outlined the problem of the cause of subsidence of Atlantic-type margins, but, without further research, a complete understanding cannot be achieved. Some potentially useful approaches to the problem follow:

1. Carefully analyze biostratigraphic data from deep commercial boreholes on the continental shelf for the information they contain of the subsidence history of Atlantic-type margins. We caution, however, that data from commercial boreholes have to be placed in a regional context. Most wells are drilled on some local structural anomaly. At best, they help to date some reflectors. Subsidence is best studied using continuous, dated horizons that were deposited close to sea level before they subsided. Such horizons should be documented by reflection seismic sections.

2. Study modern-day analogs of the earliest evolutionary stages of passive margins. There is a need to know more about the relative roles of phase changes in the upper mantle, necking of the lithosphere, and graben formation during the initial rifting stages. Studies of rift valleys, the crustal structure within young sedimentary basins, and studies at the margins of young ocean basins (such as the Red Sea) are necessary to document these initial conditions.

3. Construct models to explain the subsidence history in which the combined effects of sediment loading and crustal thinning are quantitatively considered.

G.   ATLANTIC-TYPE TRANSFORM MARGINS

The structural style and evolution of transform margins differ from those of rifted margins in the following ways:

1. The ocean–continent boundary is quite linear and is on the strike extension of an oceanic transform-fault system.

2. Several basins on the continent side of the margin strike at high angles to the transform zone, suggesting stress distribution characteristics for transcurrent faults.

3. On occasion, the transform zone forms a marginal high (Mascle, 1976).

Otherwise, Atlantic-type transform margins also display a clear separation of a rift stage from a drift stage. The Grand Banks offer a fine example of this (Figure 5.14).

In general, little detail is known about these transform margins. The most important examples are the southeast margin of South Africa (Scrutton and Dingle, 1976) and its conjugate margin off the Malvinas Plateau; the Sierra Leone–Liberia–Ivory Coast margin (Delteil *et al.*, 1976) and its possible conjugate offshore northeast Brazil (Kumar *et al.*, 1976); the northeast slope of the Bahama Bank; the southwest side of the Grand Banks; and, possibly, the margin facing the Canadian Abyssal Plain of the Arctic Ocean (Heezen *et al.*, 1975).

As stated earlier, there is a possibility that passive margins may evolve from pre-existing marginal basins and transform-fault systems that intersect active margins.

Recent studies on the rift–transform segmented patterns in young, spreading ocean basins and on active margins have revealed interesting early changes in the shapes of the plate edges. The rift–transform systems in the Gulf of California (which has been opening for approximately 4 m.y.) and in the Andaman Sea (which has been opening for slightly more than 10 m.y.) show a tendency to simplify into simpler patterns with longer rift-and-transform segments.

Initial separation and rifting may occur along irregular, older zones of weakness, with many short rift-and-transform segments. As time passes, some segments become inactive, while others extend in length to produce the broad, relatively simple patterns we see today in the Mid-Atlantic Ridge. In detail, therefore, histories of the transform segments of passive margins may be rather complex.

H. IGNEOUS ACTIVITY ASSOCIATED WITH PASSIVE-MARGIN EVOLUTION

Although magma generation occurs dominantly at plate margins, sufficient intraplate igneous activity that remains unexplained by plate tectonics exists to warrant more detailed study. Intraplate plutons and volcanics can have important implications in the evolution of passive margins. They may significantly affect the thermal history of the intruded sediments. They may form barriers for the dispersal of sediments (particularly turbidites) and change ocean-circulation patterns. Another problem in passive-margin evolution is the marginal "highs," inferred by some to be of igneous derivation in the early phase of drifting.

In Atlantic-type margins, rifting is dominated by igneous activity that is alkaline, tholeiitic, or bimodal. The best modern example is the rift system of Africa. The eastern rift shows dominantly alkaline volcanism since mid-Miocene (Baker *et al.*, 1972). The volcanics are stratigraphically complex, derived from numerous fissures and multicenter eruptions intercalated with thick accumulations of lavas and pyroclastics from central volcanoes. Alkali basalts, phonolites, trachytes, and rhyolites are common. This enormous alkaline volcanism may represent slow continental-crust distension with smaller-scale melting at greater depths (Baker *et al.*, 1972). This example of rifting may not be typical of other passive margins.

In the vicinity of the Red Sea, alkali basaltic eruptions

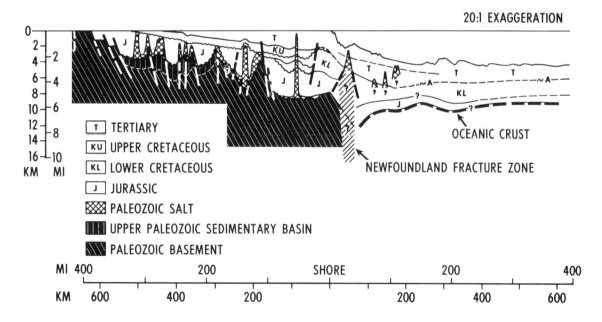

FIGURE 5.14 Newfoundland–Grand Banks: cross section. 20:1 exaggeration. [Modified from Bentley and Worzel (1956) and Amoco and Imperial (1973).]

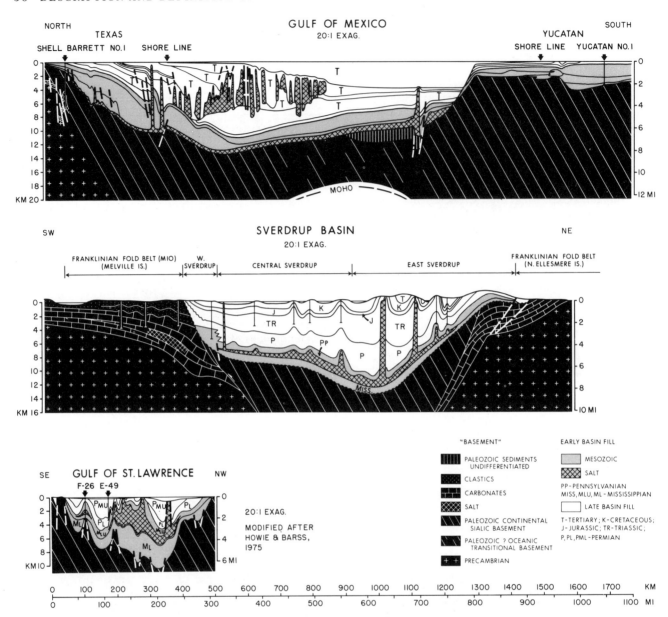

FIGURE 5.15 Geological cross sections for the Gulf of Mexico, the Sverdrup Basin, and the Gulf of St. Lawrence.

were widespread in the late Oligocene (southwest Saudi Arabia, Yemen, and Ethiopia). In the early Miocene, volcanic rift zones developed as a result of crustal thinning and warping, to be intruded by diabase dikes and other differentiates of subalkaline tholeiitic magmas. This was followed by Pliocene to Holocene eruptions of alkali basalts on a tilted rift zone (Coleman *et al.,* 1975). Basalts (subalkaline tholeiites) drilled (DSDP Leg 23) in the Red Sea axial trough are probably contemporaneous with alkali basalts of the coastal plain (Coleman *et al.,* 1975).

In the Brazilian margin, magmatic activity is bimodal

with extensive continental tholeiitic basalts and highly alkaline shallow plutons. The late Jurassic to early Cretaceous pulse, probably synchronous with continental rifting, is dominantly basaltic, whereas the late Cretaceous to Tertiary episodes are dominantly alkaline (Campos *et al.,* 1974).

In the rift basins peripheral to the Atlantic margin of North America, late Triassic to early Jurassic tholeiitic diabases have been described (Armstrong and Besancon, 1970). The bulk of the volcanics, as well as the continental–oceanic basement transition, is obscured by several kilometers of sediments and are beyond the reso-lution of seismic data. The early igneous history of most Atlantic-type margins remains enigmatic.

As seafloor spreading continues, the locus of igneous activity remains mainly along the midoceanic ridges, with extrusion of low-K tholeiitic basalt progressively farther away from the continental margin. The DSDP and recent IPOD Legs (46, 51–53) have increased our knowledge of the petrology, geochemistry, and geophysics of oceanic crust. These represent some of the first significant direct studies of oceanic layer-2.

Magmatic activity is not exclusively confined to spread-ing centers of divergent margins but persists on the conti-nent and in ocean basins. Although separation of North America from Africa began around 180 m.y. B.P. (Early Jurassic), igneous activity has persisted in the White Mountains of New England into the late Cretaceous (Zartman, 1977; Foland *et al.*, 1971; Armstrong and Stump, 1971). The New England seamount chain, farther offshore, was also active into the late Cretaceous (R. L. Houghton, Woods Hole Oceanographic Institution, per-sonal communication), and episodes of igneous activity are recognized onshore and offshore (Oligo-Miocene Bermuda volcanics) through the Tertiary. From the meager evidence available, much of the intraplate igne-ous activity is alkalic, with alkalic granites being the dom-inant rock type of the White Mountain magma series, al-kali basalts being typical of the seamount chain.

The origin of linear seamount chains is the subject of considerable debate. Several views are proposed for the New England seamount chain. One suggests motion of the North American plate over fixed mantle hot spots (Morgan, 1973; Coney, 1971; Vogt, 1973). Another pro-poses that the seamounts erupted along a seaward exten-sion of a Paleozoic or older structural lineament in the Appalachians (Drake *et al.*, 1959; Uchupi *et al.*, 1970). LePichon and Fox (1971) believe the seamounts erupted along an early transform fault. Sbar and Sykes (1973) show a zone of seismicity along an "extension" of the seamount chain into New England and Canada along the White Mountain magma trends, leading some to believe that the magmatism represents a continental extension of the seamount chain.

DSDP Leg 43 drilled the flanks of the Vogel and Nashville seamounts. Oldest sediments recovered above the volcanoclastic apron showed Coniacian age, and a K–Ar date of basaltic clasts in the breccia agrees with these dates for the eastern seamounts mentioned above. But data are insufficient to warrant a meaningful interpre-tation of the origin of the seamount chain or the oceanic basement associated with it. Deep drilling through the seamount into oceanic basement would be helpful to un-ravel the genesis, stratigraphy, and subsidence history of the seamounts.

More radiometric dating of the intraplate magmatic ac-tivity is necessary to correlate these stresses with interac-tion at plate margins. Change in the direction of plate motion, initiation of rifting or drifting, and/or migration over "hot spots" may be the cause of some of these intra-plate stresses and magmatism.

I. THE GULF OF MEXICO ENIGMA

A few basins in the passive margin family do not face a spreading ridge. They do not fit easily into the descrip-tion of passive margins, yet, they have a similar basin evolution. They are (Figure 5.15) the Gulf of Mexico; the Gulf of St. Lawrence; and, in a sense, the Sverdrup Basin in the Canadian Arctic Islands. The latter basin is mostly on land and was deformed during the Tertiary. These basins share a common ancestry: they were initiated by early rifting in Upper Paleozoic times, and they are all located within the worldwide Paleozoic fold belts.

Bally (1975, 1976) suggested that these deviant basins originated as backarc basins associated with continental collisions (e.g., Western Mediterranean and Pannonian Basin of Hungary).

Characteristically, these basins contain very thick evaporitic sequences, suggesting that they were isolated from the main oceans and thus had a chance to dry up. Thick clastic and carbonate sequences indicate massive subsidence. Structural deformation styles are dominated by gravity tectonics (i.e., salt or shale diapirs and exten-sive systems of listric normal growth faults).

J. SOME SPECIAL PROBLEMS CONCERNING PASSIVE MARGINS

THE OCEAN–CONTINENT TRANSITION

Definitive descriptions of the boundary zone between rifted continental crust and oceanic crust still elude us. Numerous alternative interpretations of the available data can be made. Some of these are shown in Figure 5.16. Keen *et al.* (1975), using data from seismic-refraction lines both parallel and perpendicular to the rise off Nova Scotia, have shown that the transition structure occurred near the magnetic slope anomaly. That anomaly could be explained by a magnetization contrast between oceanic and continental crust (edge effect). The width of the tran-sition zone is 50 km. The same transition occurs at the transform boundary south of Grand Banks, but off Nova Scotia, it is 30 km wide.

At the South Australian margin, there is a magnetically quiet zone 100–300 km wide that extends from Anomaly 22 north to beneath the rise. On the basis of seismic-refraction data obtained from long-range sonobuoys, Mut-ter (in press) interpreted this quiet zone to be transitional crust. He believes that the real ocean–continent transi-tion lies at the seaward edge of that zone. On the other hand, Ryan (Lamont-Doherty Geological Observatory, personal communication) believes that the quiet zone may be symptomatic of "dry" seafloor spreading. Rabinowitz (Lamont-Doherty Geological Observatory, personal communication) believes that the magnetic and gravity anomalies that lie on the landward side of the quiet zone near the slope and shelf edge provide a better definition of the continental margin. Much more work is needed to properly define the ocean–continent crustal transition zone. Its character may differ from margin to margin; moreover, it may vary along any particular margin.

One puzzling observation of the U.S. East Coast is that the shelf-edge gravity anomaly is not always coincident with the magnetic slope anomaly (Rabinowitz, 1974; Rabinowitz *et al.*, 1975; Schlee *et al.*, 1976). The shelf-edge gravity anomaly here, as well as along the margins of the South Atlantic and of South Australia, cannot be explained as a simple edge-effect phenomenon. Its explanation requires the presence of uncompensated high-density material either as intrabasement slivers or as a basement ridge. Seismic-refraction data (Drake *et al.*, 1959) and common-depth-point (CDP) reflection data (Schlee *et al.*, 1976) suggest the presence of a high-density ridge near the shelf edge of the U.S. East Coast. Whether this ridge is truly basement or volcanic material (Drake *et al.*, 1959; Rabinowitz, 1974) or whether it is basement material capped with a thick reefal complex (Sheridan, 1974) is not known. Knowing the origin of this ridge is important, since it probably will provide important clues to the rift–drift process.

A few available CDP lines traverse the shelf-edge, slope-rise area off the eastern United States. A few similar lines may be available off Australia, What is needed at a number of passive margins is a suite of CDP lines perpendicular to the bathymetric contours. These should be supplemented with seismic-refraction lines employing two ships, sonobuoys, ocean-bottom seismometers, and multichannel sensing arrays. The lines should cover the entire region from the shelf just landward of the shelf-edge gravity anomaly to a point sufficiently down the rise that oceanic crust may be identified unambiguously. These data should be tied to CDP and refraction lines that are parallel to the bathymetry. Supplementary gravity and magnetics data, also along lines perpendicular to the bathymetry, are necessary in order to interpret basement structure and composition.

Many investigators have limited their vision of differences between oceanic and continental structures to the crust. Differences presumably also occur in the upper mantle. Any evolutionary model of a continental margin should include at least that part of the upper mantle that lies within a plate. Differences in seismic shear-wave velocities between the oceanic and continental portions of the mantle extending to at least 400 km have been reported. Other differences, such as plate thickness, density, and viscosity, may also occur. We need to know whether the changes in physical properties are gradual or abrupt and, if gradual, how the differences relate to the age of the plate.

Studies should be undertaken of the deep structure of a major transform margin, such as that south of the Grand Banks, where the lithosphere presumably was once decoupled. Here, one might expect to find the simplest situation without the possible complications of attenuated continental material or a wide ocean–continent transition zone.

DRILLING THE OCEAN–CONTINENT TRANSITION

It is of critical importance that deep boreholes be drilled at carefully selected sites on the slope and rise. These boreholes should be drilled as deeply into the section as possible. Maximum sampling and a complete well-logging program should be made an indispensable part of any drilling project. These borehole data are necessary to tie geology to the geophysical data and to untangle the history of subsidence. Geophysical data, taken at a suite of margins of various ages and at conjugate margins, are necessary for (1) determining the oceanic–continent transition at various margins, and hence, learning about the rift–drift transition process; (2) determining stratigraphic

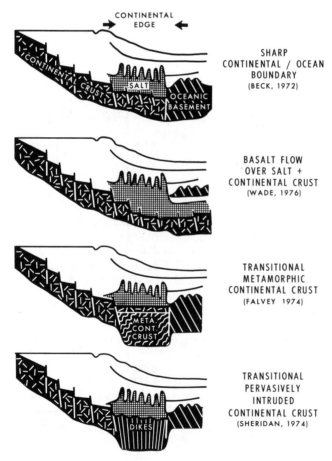

FIGURE 5.16 Alternative models for passive-margin ocean–continent boundary. (Courtesy of R. Sheridan, University of Delaware.)

and lithofacies relationships between the shelf, slope, and rise areas; (3) further defining the problem of the uncompensated mass excess necessary to explain the shelf-edge gravity anomaly; and (4) determining the detailed history of subsidence of the slope and rise area.

TECTONIC HISTORY AND STYLES OF DEFORMATION

(a) *Block Faulting:* Modern multichannel seismic-reflection sections commonly reveal substantial block faulting underlying a relatively undisturbed sedimentary package. The formation of fault blocks is presumed to be associated with the early rifting. In detail, the shape of the faults is not clear, and the subsidence history of the intervening grabens differs from place to place. Although of great structural and economic interest, these problems are at this time considered to be subordinate to the problems previously mentioned under "the ocean–continent transition" and "drilling the ocean–continent transition," but information obtained in the course of pursuing the top-priority objectives could significantly contribute to solving problems related to block faulting.

(b) *Transform-Fault Margins:* Structural style and evolution of transform margins have been described from the Grand Banks area, the east–west-trending portion of the Gulf of Guinea, and the southeast shelf of South Africa, but more detailed surveys are needed to properly describe the structural style of transform margins.

(c) *Diapirism:* Diapiric ridges, anticlines, and domes dominate the structure of some margins. A diapir is a fold or structure, the plastic, mobile core of which ruptures its overlying rocks. Some diapirs result from salt flowage; others from flowage of overpressured shales and mudstones. Diapirs are almost always exploration targets, because hydrocarbons are likely to accumulate near them. The nature and stratigraphic age of the mobile cores of diapirs are unknown.

(d) *Gravity Sliding:* Mass movements of sediments and sedimentary rocks are common on continental margins. Stability of sedimentary accumulations on a continental slope is a function of slope gradient, sediment type, rate of accumulation, and tectonic or oceanographic disturbance. Deepwater drilling, marine geological survey work, geophysical data, and surface mapping of ancient equivalents have shown that displaced sediments are common in outer-continental-margin deposits.

Major deltaic shelves, such as the Gulf of Mexico (Figure 5.15) and the Niger Delta (Delteil *et al.*, 1976), are underlain and deformed by complex systems of seaward-dipping faults. Whether or not these systems coalesce into a single, deep sole-fault is unknown, although seismic data suggest that they do. To resolve the problem will require deep drilling. It would be remarkable if the bulk of the shelf and the slope of one or more of these margins has slid a significant distance out to sea! They would represent the most gigantic landslides on earth.

QUESTIONS AND PROBLEMS CONCERNING CONTINENTAL FRAGMENTS

Orphan Knoll, in the Labrador Sea, lies at a depth of 1800 m and is separated from the continental shelf by a 200-km-wide, approximately 2500-m-deep basin. By contrast, Flemish Cap is 200 m deep, and Flemish Pass (50 km wide and 1000 m deep) separates Flemish Cap from the shelf. There is evidence that both the Knoll and the Cap were once part of the North American continent but were separated from it during initial rifting. Many questions can be posed concerning these fragments, the gaps between them, and the continent proper. Note also that these are the same areas that provide problems in fitting conjugate margins.

(a) Why do the fragments occur in some places and not in others? Many fragments (Orphan Knoll, Flemish Cap, Rockall Bank, South Tasman Rise) seem to be located near areas where seafloor spreading changed direction and where initial rifting was propagating with time along the line between continental masses. For example, the margins are younger as one progresses from the Grand Banks north to Baffin Bay. A systematic study of the distribution of fragments with respect to nearby plate motions might enable us to figure out their origin.

(b) What is the crustal structure beneath continental fragments? Is it, indeed, continental? Two continental fragments—Rockall Bank and the Seychelles—where crustal refraction data are available (Scrutton and Dingle, 1976) indicate the presence of a high-velocity crustal layer (about 7.0 km/sec) immediately above the Moho. Is this typical of continental fragments? If so, why?

(c) Were the spaces between the fragments and the shelves once sites of incipient spreading centers? Are the widths of the gaps related to the amount of attenuation or stretching of continental crust? Could stretching also have been active at continental margins, as van der Linden (1975) suggested for the Labrador Sea? Alternatively, are some of these fragments surrounded by oceanic crust, as appears to be the case for the Lomonosov Ridge? Processes that separated the fragments from the continents and caused a particular crustal structure to develop may be similar to those initially active along most passive continental margins.

THERMAL HISTORY OF PASSIVE MARGINS

In order to understand the subsidence of passive margins, it is necessary to understand their thermal history. To deduce the thermal history of oceanic areas, heat flow through the ocean bottom is usually computed by measuring the temperature gradient or the variation in temperature with depth in the upper several meters of the sediment column. These techniques may not be the best method with which to measure heat flow on the shelf and slope. The problem in these shallow-water areas is that the bottom-water temperature is very unstable. As a consequence, standard thermal gradient measurements are often not repeatable, and, hence, often not reliable (see the comments about heat-flow measurements in Chapter 11). Also, even if reliable data were available, the extreme variations in sediment types and lithologies with depth lead to ambiguous interpretations. One approach to this problem is to log continuous downhole temperatures in deep boreholes as well as logging appropriate bulk properties of the sedimentary strata. In addition, geochemical techniques being developed may be applied to drill samples to learn about the thermal history of buried strata.

PROBLEMS RELATED TO GENESIS AND EVOLUTION OF SEDIMENTARY SEQUENCES

(a) *Continental Hinterland and Climate:* Continental margins act as pathways to the ocean floor for terrigenous material that is delivered as particles or dissolved from the continental hinterland. The quantitative and qualitative composition of this terrigenous input is controlled by the geology of the continental hinterland, by the climatic processes above it, and by the size and configuration of the drainage basins that supply the terrigenous material to the continental edge. In terms of climatic systems, the following end-members need to be studied: (1) polar–subpolar, (2) transitional–temperate, (3) dry–subtropical, and (4) humid–tropical. This climatic zonation also controls the amount and composition of organic material swept from the land into the sea.

(b) *Oldest Clastic and Volcanic Deposits:* The clastic and/or volcanic rocks, thought to be the oldest materials deposited in the rifts, never have been drilled or sampled in a major ocean. They are too deeply buried and their regional distribution is restricted to narrow strips along the feet of passive margins. However, these deposits are present in recent rifts that have not yet developed into marine basins. Such early deposits can also be found in fossil grabens that failed before they could expand enough to contain an ocean basin. Volcanics underlying the halites of the Red Sea probably represent this stage (Lowell and Genik, 1972).

(c) *Early and Later Restricted Marine Paleoenvironments:* Though the clastic material that must fill the earliest rift has never been sampled, young passive continental margins often preserve records of the early restricted marine environments of the small ocean basins they surround. The narrow Mesozoic South Atlantic Ocean and Cenozoic Red Sea represent typical oceanic depositional settings of halite formation (as opposed to Mediterranean or continental basinlike situations). These salt deposits are frequently accompanied by sediments that reflect anoxic conditions. The origin of the organic matter in these sediments is not well understood. Such sediments are observed in the mid-Cretaceous South Atlantic and Pacific oceans. The distribution pattern of sediments that preserve an unusually high concentration of organic carbon along passive continental margins is virtually unknown. In several regions, the deposition of such sediments in the ocean basin was accompanied by the ac-

cumulation of large volumes of platform carbonates in shallow portions of the continental margin. The relationship of these diverse facies in space and time remains to be unraveled.

(d) *Ocean Surface-Water Circulation:* Once the ocean basins have matured, continental margins often distort the largely latitudinal surface-water circulation. Distortion of the wind-driven latitudinal surface-water circulation results in the western and eastern boundary currents along the continental edge. Distortion also results in major divergences that represent the most unstable and sensitive components of the oceanic surface-water circulation. Boundary currents separate stable, stratified, central water masses of the ocean from highly variable, turbulent coastal water masses over the inner part of the continental margin. In addition, continental margins provide possibilities for understanding the development of the pelagic paleoenvironment—an understanding that cannot be obtained in the central parts of the oceans themselves. (Microfauna cannot be found on a seafloor that did not previously exist! Also, calcareous tests dissolve at the deeper depths before they can reach the seafloor.)

(e) *Carbonate Platforms:* Specific paleooceanographic settings controlled the formation of the important carbonate platforms along passive continental margins. Carbonates were restricted to the relatively warm surface (and hence, shallow) water masses of the tropical–subtropical Late Mesozoic and Cenozoic climatic belts. These thick carbonate platforms are particularly interesting, because their upper surface must have stayed close to the sea surface as long as they continued to be built up. Therefore, their bulk accumulation rates monitor the past and present subsidence of continental margins in great detail.

(f) *Oceanic Water Column:* Transects of drill sites across passive margins provide insight into the vertical structure of the oceanic water column through time, because intermediate and deep oceanic water masses impinge now and in the past upon the rise and slope of the continents. We have no solid evidence for the mode of deepwater movement in the preglacial Mesozoic and Cenozoic oceans. We do know that, since the middle of the Tertiary, these water masses have been downwelled in the "glacial" polar regions near continents (the only accessible example to drilling is in the Norwegian–Greenland Seas).

(g) *Hiatuses and Eustatic Sea-Level Fluctuations:* The good correlation of the Late Mesozoic and Cenozoic Calcite Compensation Depth (CCD) with the sequence of eustatic sea-level fluctuations over the inner part of the continental margins proves that the oceanic deep was closely linked to the shallow margin paleoenvironment. Hiatuses in the stratigraphic record from passive margins mark oceanographic events that prevented sediment accumulation for a certain time or that removed portions of the sediment column. Indicators for transgressions and regressions have been traced on seismic records across many continental margins (see Chapter 4). The isochroneity and global occurrence of such indicators have led to the hypothesis that they were caused by eustatic sea-level fluctuations, which, in turn, may reflect major changes of seafloor-spreading rates and/or global tectonic events other than the pulsations of the Late Cenozoic glacial ice sheets.

(h) *Continental Margin Sediment Cover Due to Late Cenozoic Glacial Events versus "Typical" Sections:* In the youngest geological period, abnormal glacial events occurred that greatly impacted the paleoclimate and paleooceanography of this world. There were frequent eustatic and isostatic sea-level fluctuations caused by the waxing and waning of large glacial ice shields. Those variations of sea level distorted the distribution of shallow- and deepwater sediment facies along the passive margins. Now—20,000 years after the last glacial maximum—large portions of the continental shelves are still covered by glacial relict sediments. It is not clear when or how the continental margins will return to their nonglacial (basically pre-Quaternary) facies distribution.

Rates of pre-Quaternary transgressions and regressions were probably slower than those during the Quaternary. Since the transgression and regression rates were so different in the Quaternary, it is necessary to compare and contrast the more typical, earlier sequences deposited during a glacial–eustatic regime. Systematic coring of shelf margins will give us a chance to make these comparisons and to evolve a new "global stratigraphy."

EXPLORING CARBONATE PASSIVE MARGINS WITH SUBMERSIBLES

So far in the discussion, a basic assumption is made that the most fruitful way to calculate subsidence rates is to measure sedimentation rates. Borehole and multichannel seismic-reflection data are used to compute strata thicknesses. Corrections must be made for compaction (see Hamilton, 1976). Further, it must be assumed that the water depth of the original depositional setting can be quantitatively determined (changes in sediment thickness plus change in water depth). (See Van Hinte, 1978.) However, techniques using assemblages of benthic organisms for determining water depth in slope or basinal settings may have errors of 0.5 km or more.

A most promising approach is to investigate stratigraphic successions in carbonate banks, where (1) water depths were consistently in the photic zone, and (2) the carbonate has lithified more as a result of intergranular cementation than by compaction, so that volume changes are relatively minor and can be corrected for by using standard petrographic techniques. However, drilling in carbonate banks, such as that of the Bahamas (Andros well), has not provided adequately well-preserved fossil material. The paucity of such fossil material is because the present subaerial parts of the Bahamas correspond, in the subsurface, to deep environments that were heavily dolomitized or otherwise disturbed by diagenetic processes.

Recent deep (down to 3700 m) sampling of the seaward escarpments of Cat Island east of Andros by Heezen and Ryan (Lamont-Doherty Geological Observatory, personal communication), using the deep-submersible, *Alvin*, showed a practically continuous outcropping below 1900 m of horizontally bedded limestones. When examined, these rocks were found to contain highly fossiliferous material for a shallow fore-reef environment. No appreciable diagenetic disturbance was evident, except for manganese encrustations about 1–20 mm thick, and some organic borings. Heezen and Ryan discovered that photo reconnaissance of the escarpment front can provide a highly detailed profile of the rock column. Such a profile will record the bedding thicknesses, degree of relative induration, and the presence of such primary features as teepee structures and wavy stromatolitic lamination. Photo reconnaissance can also detect fossil shales and modular marlstones. Heezen and Ryan had no appreciable difficulty using the *Alvin*'s manipulator arm to get samples of up to 5 kg *in situ* from within the cliff face.

K. RECOMMENDATIONS

1. *We recommend an integrated study of the structural and stratigraphic evolution of selected passive-margin transects.* These should include the following:

- Young margins, such as the Red Sea and the Gulf of California;
- Young margins that were drowned in sediments, e.g., the Northern Gulf of California;
- Old starved margins and related microcontinents, such as the Galicia Plateau, Bay of Biscay, Rockall Plateau, Orphan Knoll, and the Seychelles;
- Old mature margins, such as the U.S. East Coast, Canada, Northwest Africa, and the Gulf of Mexico, including at least one carbonate margin (e.g., Florida–Bahama) and one clastic margin (e.g., Delaware Basin); and
- One or more transform margins, e.g., southwest Grand Banks.

We envisage the following general strategy:

- Compile previous work; continue reconnaissance on land and sea, with emphasis on multichannel seismic data and long refraction lines;
- Narrow down to regional studies on land and sea;
- Concentrate on one or more typical transects, using all geophysical tools available, and *only after* high-quality geophysical surveys are done; and
- Complete the study with a deep-drilling program, using current drilling capabilities on shelves and eventually deep riser-drilling (*Glomar Explorer*) for slopes and rises.

In planning, an attempt should be made, if the science

justifies it, to tie the traverse into related programs such as the old Transcontinental Geophysical Traverse Program, the Transatlantic Geotraverse (TAG) program, and the long-range Consortium for Continental Reflection Profiling (COCORP) plan for a U.S. traverse.

2. *We recommend that a systematic geophysical–geological study of the transition between ocean and continent be made.* Such a study would search for an adequate description of the boundary zone, based on CDP seismic-refraction work and other geophysical measurements. In addition to this, bedrock mapping using submersibles would help to calibrate some of the seismic data.

This program is separated from the previous program because, at this time, this transition has to be studied on a worldwide basis in order to gain perspective on the variations of the theme. Surveys of the different margin types described in the first recommendation would have to be made in order to help to select drilling sites.

3. *We recommend that every effort be made to obtain reliable temperature data from deep boreholes.* These should be complemented by measurements of the rock properties (e.g., conductivity, velocity) of the strata penetrated.

4. Earthquakes do occur on passive margins. *We recommend that a comprehensive study of the seismicity on passive margins be undertaken.* Such a study may lead to a better understanding of the dynamics of passive margins.

5. *We recommend a systematic study of the geochemistry of igneous rocks occurring on passive margins.* These rocks provide valuable information on deep processes that occur during the formation of passive margins.

REFERENCES AND BIBLIOGRAPHY

American Association of Petroleum Geologists (1977). Geology of continental margins, *Continuing Educ. Course Notes Series 5.*

Amoco & Imperial (1973). Regional geology of the Grand Banks, *Bull. Can. Petrol. Geol. 21*:479–503.

Armstrong, R. L., and J. Besancon (1970). A Triassic time scale dilemma: K-Ar dating of Upper Triassic mafic igneous rocks—Eastern U.S.A. and Canada and Post-Upper Triassic pluton, western Idaho, U.S.A., *Eclog. Geol. Helv. 63*(1):15–28.

Armstrong, R. L., and E. Stump (1971). Additional K-Ar dates, White Mountain Magma Series, New England, *Am. J. Sci., Spec. Paper 132,* 67 pp.

Baker, B. H., P. A. Mohr, and L. A. J. Williams (1972). Geology of the Eastern Rift System of Africa, *Geol. Soc. Am. Spec. Paper 136,* 67 pp.

Bally, A. W. (1975). A geodynamic scenario for hydrocarbon occurrences, *Proc. of the Ninth World Petroleum Congress, Tokyo, Vol. 2 (Geology),* Paper PD-1, Applied Science Pub., Ltd., Ripple Road, Barking, Essex, England, pp. 33–44.

Bally, A. W. (1976). Canada's passive continental margins—A review, *Mar. Geophys. Res. 2*(4):327–340.

Behrendt, J. C. (1977). U.S. Geological Survey programs of resource assessment, geologic environmental studies, and marine geology investigations of the continental margin adjacent to deep sea areas in the Atlantic and Gulf of Mexico, *Open File #77-320,* U.S. Geological Survey, Woods Hole, Mass., March.

Bentley, C. R., and J. L. Worzel (1956). Geophysical investigations in the emerged and submerged Atlantic Coastal Plains, *Geol. Soc. Am. Bull. 67*(1):1–18.

Boeuf, M. G., and H. Doust (1975). Structure and development of the southern margin of Australia, *Aust. Petrol. Expl. Assoc. J.,* pp. 33–43.

Bott, M. H. P. (1971). Evolution of young continental margins and formation of shelf basins, *Tectonophysics 11*:319–327.

Bott, M. H. P. (1976a). Mechanisms of basin subsidence—an introductory review, *Tectonophysics 36*:1–4.

Bott, M. H. P., ed. (1976b). Sedimentary basins of continental margins and cratons, Spec. issue, *Tectonophysics 36*:1–3.

Bott, M. H. P. (in press). Subsidence mechanisms at passive continental margins, *Bull. Am. Assoc. Petrol. Geol.*

Bott, M. H. P., and W. S. Dean (1972). Stress systems at young continental margins, *Nature Phys. Sci. 235*:23.

Briden, J. C., D. J. Drewry, and A. G. Smith (1974). Phanerozoic equal-area world maps, *J. Geol. 82*:555–574.

Campos, C. W. M., F. C. Ponte, and K. Miura (1974). Geology of the Brazilian continental margin, in *The Geology of Continental Margins,* C. A. Burk and C. L. Drake, eds., Springer-Verlag, New York, pp. 447–461.

Cloos, H. (1939). Hebung-Spaltung-Vulkanismus: Elemente einer geometrischen Analyse irdischer Grossformen, *Geol. Rundsch. 30*(4a):405–527.

Cloos, H. (1953). *Conversation with the Earth,* E. Cloos and C. Dietz, eds. (E. B. Garside, translator), Alfred A. Knopf, New York, 413 pp.

Coleman, R. G., R. J. Flek, C. E. Hedge, and E. D. Chent (1975). The volcanic rocks of Southeast Saudi Arabia and the opening of the Red Sea, *U.S. Geol. Surv. Saudi Arabian Project Report 194.*

Coney, P. J. (1971). Cordilleran tectonic transitions and motion of the North American Plate, *Nature 233*:462–465.

Craddock, C., compiler (1975). Earth science investigations, U.S. Antarctic Research Program (USARP), A summary report prepared for the Working Group on Geology of the Scientific Committee on Antarctic Research (SCAR), International Council of Scientific Unions (ICSU), National Academy of Sciences, Washington, D.C.

Curray, J. R. (1977). Modes of emplacement of prospective hydrocarbon reservoir rocks of outer continental marine environments, in *Am. Assoc. Petrol. Geol. Continuing Educ. Course Notes Series 5.*

de Almeida, F. F. M., ed. (1976a). Continental margins of Atlantic type, *Ann. Acad. Brasil. Cien. 48,* Proceedings of the International Symposium on Continental Margins of Atlantic Type, held in São Paulo, Brazil, October 1975.

de Almeida, F. F. M. (1976b). The system of continental rifts bordering the Santos Basin, Brazil, in Continental Margins of Atlantic Type, *Ann. Acad. Brasil. Cien. 48*:15–26.

Delteil, J. R., F. Rivier, L. Montadert, V. Apostoles, J. Didier, M. Goslin, and P. H. Patriat (1976). Structure and sedimentation of the continental margin of the Gulf of Benin, in Continental Margins of Atlantic Type, *Ann. Acad. Brasil. Cien. 48*:51–66.

Drake, C. L., M. Ewing, and G. H. Sutton (1959). Continental margins and geosynclines; The East Coast of North America north of Cape Hatteras, in *Physics and Chemistry of the Earth, Vol. 3,* Pergamon Press, Elmsford, N.Y., pp. 110–198.

Emery, K. O. (1977). Structure and stratigraphy of divergent continental margins, in *Am. Assoc. Petrol. Geol. Bull. Cont. Educ. Course Note Series 5*.

Emery, K. O., E. Uchupi, U. D. Phillips, C. O. Bowin, E. T. Bunce, and S. T. Knott (1970). Continental rise off eastern North America, *Am. Assoc. Petrol. Geol. Bull. 54*(1):44–108.

Falvey, D. A. (1974). The development of continental margins in plate tectonic theory, *Aust. Petrol. Expl. Assoc. J. 14*:95–106.

Foland, K. A., A. W. Quinn, and B. J. Giletti (1971). K–Ar and Rb–Sr Jurassic and Cretaceous ages for intrusives of the White Mountain Magma Series, northern New England, *Am. J. Sci. 270*(5):321–330.

Fox, P. J., B. C. Heezen, and G. L. Johnson (1970). Jurassic sandstone from the tropical Atlantic, *Science 170*:1402–1404.

Given, M. M. (1977). Mesozoic and early Cenozoic geology of offshore Nova Scotia, *Bull. Can. Petrol. Geol. 25*(1):63–91.

Grow, J. A., and R. G. Markl (1977). IPOD-USGS multi-channel seismic reflection profile from Cape Hatteras to the Mid-Atlantic Ridge, *Geology 5*:625–630.

Grow, J. A., R. E. Mattick, and J. S. Schlee (in press). Multichannel seismic depth sections and interval velocities over outer continental shelf and upper continental slope between Cape Hatteras and Cape Cod, in *Geological Investigations of Continental Margins*, J. S. Watkins, L. Montadert, and P. W. Dickerson, eds., Am. Assoc. Petrol. Geol. Mem.

Hamilton, E. L. (1976). Variations of density and porosity with depth in deep-sea sediments, *J. Sed. Petrol. 46*:280–300.

Hathaway, J. C., *et al.* (1976). Preliminary summary of the 1976 Atlantic margin coring project of the U.S. Geological Survey, November, U.S. Geol. Surv., Washington, D.C.

Heezen, B. C., M. Tharp, and M. Pinther (1975). Map of the Arctic Regions, prepared by the Cartographic Division of the American Geographic Society, New York.

Heybroeck, R. (1965). The Red Sea Evaporite Basin, in *Salt Basins Around Africa*, Inst. Petrol. London, pp. 17–40.

Howie, R. D., and M. S. Barss (1975). Paleogeography and sedimentation in the Upper Paleozoic, Eastern Canada, Canada's Continental Margins, in C. J. Yorath, E. R. Parker, and D. J. Glass, eds., *Can. Soc. Petrol. Geol. Mem. 4*:45–57.

Illies, J. H., and K. Fuchs, eds. (1974). Approaches to taphrogenesis, *Inter-Union Commission on Geodynamics, Scientific Rep. No. 8*, E. Schweizerbart'sche Verlagsbuchhandlung, Stuttgart.

IPOD Passive Margin Panel (1977). Proposal for IPOD Drilling 1980–81, in *The Future of Scientific Ocean Drilling*, a report by an *ad hoc* subcommittee of the JOIDES Executive Committee, available from the JOIDES Office, U. of Washington, Seattle, Wash., pp. 32–49.

Keen, C. E., and M. J. Keen (1974). The continental margins of eastern Canada and Baffin Bay, in *The Geology of Continental Margins*, C. A. Burk and C. L. Drake, eds., Springer-Verlag, New York, pp. 381–389.

Keen, C. E., M. J. Keen, D. L. Barrett, and D. E. Heffler (1975). Some aspects of the ocean–continent transition at the continental margin of eastern North America, in *Offshore Geology of Eastern Canada, Can. Geol. Surv. Paper 74–30 2*:189–197.

Kinsman, D. J. J. (1975). Rift valley basins and sedimentary history of trailing continental margin, in *Petroleum and Global Tectonics*, A. C. Fisher and S. Judson, eds., Princeton U. Press, pp. 83–128.

Kristoffersen, Y. (1977). Labrador Sea: A geophysical study, PhD Thesis, Columbia U., 150 pp., unpublished.

Kumar, N., G. Bryan, M. Gorini, and J. Carvalho (1976). Evolution of the continental margin off northern Brazil: sediment distribution and hydrocarbon potential, in Continental Margins of Atlantic Type, F. F. M. de Almeida, ed., *Ann. Acad. Brasil. Cien. 48*:131–143.

Larson, R. L., and J. W. Ladd (1973). Evidence for the opening of the South Atlantic in the Early Cretaceous, *Nature 246*:227–266.

Larson, R. L., H. W. Menard, and S. M. Smith (1968). Gulf of California: A result of ocean-floor spreading and transform faulting, *Science 161*:781–784.

Laughton, A. S., R. B. Whitmarsh, and M. T. Jones (1970). The evolution of the Gulf of Aden, *Phil. Trans. R. Soc. London A 267*:227–266.

LePichon, X., and P. J. Fox (1971). Marginal offsets, fracture zones and early opening of the North Atlantic, *J. Geophys. Res. 76*:6294–6308.

LePichon, X., J.-C. Sibuet, and J. Francheteau (1977). The fit of the continents around the North Atlantic Ocean, *Tectonophysics 38*(3/4):169–209.

Lowell, J. D., and G. J. Genik (1972). Sea-floor spreading and structural evolution of southern Red Sea, *Am. Assoc. Petrol. Geol. Bull. 56*:247–259.

Lowell, J. D., G. J. Genik, T. H. Nelson, and P. M. Tucker (1975). Petroleum and plate tectonics of the southern Red Sea, in *Petroleum and Global Tectonics*, A. G. Fisher and S. Judson, eds., Princeton U. Press, pp. 129–158.

Mascle, J. (1976). Atlantic type continental margins: distinction of two basic structural types, in Continental margins of Atlantic type, F. F. M. de Almeida, ed., *Ann. Acad. Brasil. Cien. 48*:191–198.

Milanovsky, Ye. Ye. (1974). Rift zones of the geologic past and their associated formations—Report 1, *Mosk. Obshch. Isptateley Prir. Byull., otd., geol. 49*(5):51–71.

Montadert, L., D. G. Roberts, G. Auffret, W. Bock, P. A. DuPeuble, E. A. Hailwood, W. Harrison, H. Kagami, D. N. Lumsden, C. Muller, D. Schnitker, R. W. Thompson, T. L. Thompson, and P. P. Timofeev (1977). Rifting and subsidence on passive continental margins in the North East Atlantic, *Nature 268*(5618):305–309.

Moore, D. G. (1973). Plate edge deformation and crustal growth, Gulf of California Structural Province, *Bull. Geol. Soc. Am. 84*:1883–1906.

Morgan, W. J. (1973). Plate motions and deep mantle convection, in *Studies in Earth and Space Sciences, a Memoir in Honor of Harry Hammond Hess*, R. Shagam, R. B. Hargraves, W. J. Morgan, F. B. Van Houten, C. A. Burk, H. D. Holland, and L. C. Hollister, eds., *Geol. Soc. Am. Mem. 132*, pp. 7–22.

Mutter, J. (in press). Structure and evolution of the magnetic quiet zone in the continental margin south of Australia, in *Geological Investigations of Continental Margins*, J. S. Watkins, L. Montadert, and P. W. Dickerson, eds., Am. Assoc. Petrol. Geol. Mem. (in press).

Pelletier, B. R., ed. (1974). Offshore Geology of Eastern Canada: Vol. 1: Concepts and Applications of Environmental Marine Geology, 160 pp.; Vol. 2: Regional Geology, W. J. M. van der Linden and J. A. Wade, eds. (1975), 258 pp., Geol. Surv. Can. Paper 74–30, Ottawa, Canada.

Pilger, A., and A. Rösler, eds. (1975). Afar Depression of Ethiopia, Inter-Union Commission on Geodynamics Sci. Rep. No. 14, E. Schweizerbart'sche Verlagsbuchhandlung, Stuttgart, 416 pp.

Pilger, A., and A. Rösler, eds. (1976). Afar between Continental and Oceanic Rifting, Inter-Union Commission on Geodynamics Sci. Rep. No. 16, E. Schweizerbart'sche Verlagsbuchhandlung, Stuttgart, 216 pp.

Pitman, W. C., III, and M. Talwani (1972). Sea-floor spreading in the North Atlantic, *Bull. Geol. Soc. Am.* 83:619–646.

Ponte, F. C., and H. E. Asmus (1976). The Brazilian marginal basins: Current state of knowledge, in Continental margins of Atlantic type, F. F. M. de Almeida, ed., *Ann. Acad. Brasil. Cien.* 48:215–216.

Powell, D. E. (1976). The geological evolution of the continental margin off northwest Australia, *Aust. Petrol. Expl. Assoc. J.*, pp. 13–23.

Puzyrev, N. N., M. M. Mandelbaum, S. V. Krylov, B. P. Mishenkin, G. V. Krupskaya, and G. V. Petrick (1973). Deep seismic investigations in the Baikal rift zone, *Tectonophysics* 20:85–95.

Rabinowitz, P. D. (1974). The boundary between oceanic and continental crust in the western North Atlantic, in *The Geology of Continental Margins*, C. A. Burk and C. L. Drake, eds., Springer-Verlag, New York, pp. 67–84.

Rabinowitz, P. D., and J. L. LaBrecque (in press). The Mesozoic South Atlantic Ocean and evolution of its continental margin, in *Geological Investigations of Continental Margins*, J. S. Watkins, L. Montadert, and P. W. Dickerson, eds., Am. Assoc. Petrol. Geol. Mem. (in press).

Rabinowitz, P. D., S. C. Lande, and J. L. LaBrecque (1975). The Falkland Escarpment and Agulhas Fracture Zone: The boundary between oceanic and continental basement at conjugate continental margins, in Continental margins of Atlantic type, F. F. M. de Almeida, ed., *Ann. Acad. Brasil. Cien.* 48:240–252.

Rhinegraben Research Group for Explosion Seismology (1974). The 1972 Seismic Refraction Experiment in the Rhinegraben—First Results, in *Approaches to Taphrogenesis*, J. H. Illies and K. Fuchs, eds., E. Schweizerbart'sche Verlagsbuchhandlung, Stuttgart, pp. 122–137.

Sbar, M. L., and L. R. Sykes (1973). Contemporary compressive stress and seismicity in eastern North America: An example of intra-plate tectonics, *Geol. Soc. Am. Bull.* 84:1861–1882.

Schlee, J., J. C. Behrendt, J. A. Grow, J. M. Robb, R. E. Mattick, P. T. Taylor, and B. J. Lawson (1976). Regional geologic framework of northeastern United States, *Bull. Am. Assoc. Petrol. Geol.* 60(6):926–951.

Schlee, J. S., R. G. Martin, R. E. Mattick, W. P. Dillon, and M. M. Ball (1977). Petroleum geology on the United States Atlantic-Gulf of Mexico margins, in Proc. of Southwestern Legal Foundation, *Exploration and Economics of the Petroleum Industry*, Vol. 15, Matthew Bender Co., New York, pp. 47–93.

Scholle, P. A., ed. (n.d.). Geological studies on the COST No. B-2 well, U.S. Mid-Atlantic Outer Continental Shelf Area, *U.S. Geol. Surv. Cir. 750*, available from the U.S. Geological Survey, Washington, D.C.

Sclater, J. G., R. N. Anderson, and M. L. Bell (1971). Elevation of ridges and evolution of the central Eastern Pacific, *J. Geophys. Res.* 76(32):7888–7915.

Scrutton, R. A., and R. V. Dingle (1976). Observations on the processes of sedimentary basin formation at the margins of southern Africa, *Tectonophysics* 36:143–156.

Sheridan, R. E. (1974). Atlantic continental margin of North America, in *The Geology of Continental Margins*, C. A. Burk and C. L. Drake, eds., Springer-Verlag, New York, pp. 391–406.

Sheridan, R., and W. L. Osburn (1975). Marine geological and geophysical studies of the Florida-Blake Plateau–Bahamas Area, in *Canada's Continental Margins and Offshore Petroleum Exploration*, C. J. Yorath, E. R. Parker, and D. J. Glass, eds., Can. Soc. Petrol. Geol. Mem. 4.

Sial, A. N. (1976). The post-Paleozoic volcanism of northeast Brazil and its tectonic significance, in Continental margins of Atlantic type, F. F. M. de Almeida, ed., *Ann. Acad. Brasil. Cien.* 48:299–311.

Sleep, N. H. (1971). Thermal effects of the formation of Atlantic continental margins by continental break-up, *Geophys. J. R. Astron. Soc.* 24:325–350.

Sleep, N. H. (1976). Platform subsidence mechanisms and "eustatic" sea-level changes, *Tectonophysics* 36:45–56.

Sleep, N. H., and N. S. Snell (1976). Thermal contraction and flexure of midcontinent and Atlantic marginal basins, *Geophys. J. R. Astron. Soc.* 45:125–154.

Talwani, M., and O. Eldholm (1977). Evolution of the Norwegian-Greenland Sea, *Bull. Geol. Soc. Am.* 88:969–999.

Trehu, A., J. G. Sclater, and J. Nabelek (1976). The depth and thickness of the ocean crust and its dependence upon age, *Bull. Soc. Geol. France* 7(4):917–930.

Turcotte, D. L., J. L. Ahern, and J. M. Bird (in press). The state of stress at continental margins, *Tectonophysics* 42:1–28.

Uchupi, E., J. D. Phillips, and K. E. Prada (1970). Origin and structure of the New England Seamount chain, *Deep-Sea Res.* 17:483–494.

van der Linden, W. J. M. (1975). Mesozoic and Cenozoic opening of the Labrador Sea, the North Atlantic, and the Bay of Biscay, *Nature* 253:320–324.

Van Hinte, J. E. (1978). Geohistory analysis—Applications of micropaleontology in exploration geology, *Am. Assoc. Petrol. Geol. Bull.* 62(2):201–222.

Veevers, J. J., ed. (1975). Deep sea drilling in Australasian waters, *Challenger* Symposium, Sydney, Macquarie U., 37 pp.

Vogt, P. R. (1973). Early events in the opening of the North Atlantic, in *Implications of Continental Drift to the Earth Sciences*, D. H. Tarling and S. K. Runcorn, eds., Academic Press, New York, pp. 693–712.

Walcott, R. I. (1972). Gravity, flexure and the growth of sedimentary basins at a continental edge, *Bull. Geol. Soc. Am.* 83:1845.

Watts, A. B., and W. B. F. Ryan (1976). Flexure of the lithosphere and continental margin basins, *Tectonophysics* 36:25–44.

Weissel, J. K., and D. E. Hayes (1972). Magnetic anomalies in the southeast Indian Ocean, in *Antarctic Oceanology II, the Australian-New Zealand Sector*, Antarctic Research Series 19, Am. Geophys. Union, Washington, D.C., pp. 165–196.

Willcox, J. B., and N. F. Exon (1976). The regional geology of the Exmouth Plateau, *Aust. Petrol. Expl. Assoc. J.*, pp. 1–11.

Woodland, A. N., ed. (1975). *Petroleum and the Continental Shelf of Northwest Europe*, Vol. I, Halsted Press Book, John Wiley and Sons, New York, 501 pp.

Yorath, C. J., E. R. Parker, and D. J. Glass, eds. (1975). Canada's continental margins and offshore petroleum exploration, *Can. Soc. Petrol. Geol. Mem. 4* (45 papers), Calgary, Alberta, Canada, 898 pp.

Zartman, R. E. (1977). Geochronology of some alkalic rock provinces in eastern and central United States, *Ann. Rev. Earth Planet. Sci.* 5:257–286.

Ziegler, P. A. (1977). Geology and hydrocarbon provinces of the North Sea, *Geojournal* 1(1):7–32.

6 | Cratonic Margins

A. INTRODUCTION

Cratonic basins are located on cratons, on older continental crust. Cratons (sometimes referred to as shields or platforms) are the stable, usually central, regions of continents. Marine geologists are strangely aloof when approached on the subject of cratonic basins, even though so many such basins are today under the sea. (Figure 6.1 shows submerged cratonic basins together with their land cousins.) Hudson Bay (Figure 2.6), the South Arctic Basin, the Grand Banks Basin, the Jones–Lancaster, and the Foxe Basin—all in northern Canada—the Barents Sea, the Baltic Sea, and the North Sea (with its oil and gas riches) are basins still under water.

Cratonic basins on land are mostly well exposed on the continental portions of the lithosphere and, because of their hydrocarbon potential, they have been avidly studied. Consequently, for these basins, a wealth of information is available.

Basins on cratons may be essentially undeformed. Metamorphism may be restricted to changes that accompany burial of up to 5 km, to the subgreenschist facies. These basins record vertical movements of the continental interiors of lithospheric plates and are potential sources of information about the rheological properties of the lithosphere and asthenosphere. In principle, they might yield this information for extended periods of time, since they developed throughout the Phanerozoic, unlike the undeformed continental margins, which are wholly Mesozoic and Tertiary in age. Cratonic basins also act as "tidemarks"; they record the elevation of continental masses in relation to sea level, again, as a function of extended periods of time.

B. CRATON CENTERS: THEIR CHARACTERISTICS

Cratonic basins formed in regions that were relatively "positive," and their sedimentary thicknesses are thin by comparison with other sedimentary basins, for example: Hudson subbasin, 2500 m; Moose River, 600 m; Williston, 3500 m. Stable margins may have as much as 15,000 m of sediments. Basins or subbasins are separated from each other by epeirogenic arches, over which there may once have been thin sediments, now largely eroded away. Many craton-centered basins formed in the Paleozoic. For example, sediments in Hudson Bay range in age from Ordovician to Pennsylvanian. Some, such as the Williston and others lying east of the Cordillera, may have Mesozoic sediments as a "second-cycle" cover (often called "exogeosynclines" or "foredeeps"), such sediments having been derived from the adjacent rising mountain ranges. Others (e.g., the Moose River Basin Cretaceous) may have later continental sediments. The basins are subcircular in plan, have a symmetrical cross

99

section, and are structurally undeformed. The sediments characteristically contain basal clastics, succeeded by shallow-water carbonates, evaporites, and only minor clastics.

It is necessary to explain (1) the initiation of a depression; (2) the continual subsidence, with the basin being filled by sediment almost the entire time; (3) the apparent decrease in size of each basin over time, with the centers being occupied by younger rocks; and (4) the time scale for a basin's development (about 100 million to 200 million years).

• Haxby *et al.* (1976) modeled the initiating mechanism as a disk-shaped load applied at an instant of time in an elastic—not viscoelastic—lithosphere. Sediments were used in his model only to provide a restoring buoyancy force.

• Beaumont (1978), taking the problem a step further, investigated the response of a *viscoelastic* lithosphere overlying an inviscid asthenosphere, in which the load (caused by sediments following initiation) changes both areally and temporally as the basin changes. Beaumont showed that the kind of basin that develops depends on the time and spatial constants of the initiating mechanism; and he derived those constants for isostatic adjustment. Hudson Bay Basin, in particular, is a relatively shallow basin that exhibits the expected characteristics of thermal uplift and contraction as an initiating mechanism and that exhibits loading (by sediments) on a viscoelastic half-space. A basin that develops in this manner will deepen with time because of the initiating mechanism and loading. A peripheral bulge develops as a result of the load, and the bulge, itself, migrates inward.

C. THEORETICAL DEVELOPMENT

The shallow-water facies of the sediments indicate that we do not merely have a deep hole filled up with sediments (unlike the ocean basins, which are, on a grand scale, holes not yet filled up with sediments). We need (1) an initiating mechanism, creating a depression, which will have its own time scale for development and decay

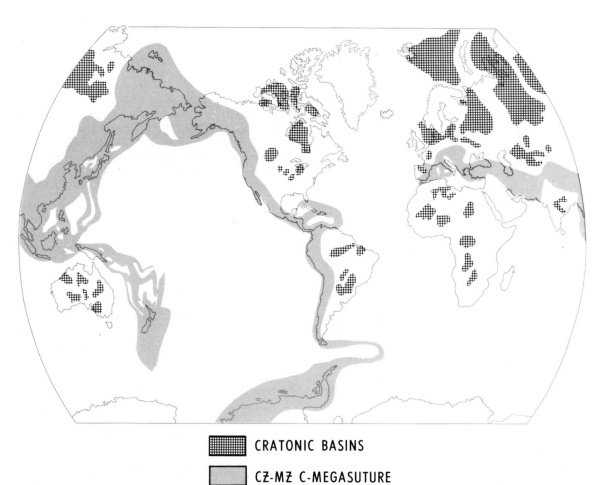

▦ CRATONIC BASINS

▢ CZ-MZ C-MEGASUTURE

FIGURE 6.1   Cratonic basins of the world.

and (2) the response of the earth to the subsequent loading by sediments that fill the basin as it develops.

1. Several initiating mechanisms can be proposed:

- Thermal uplift (followed by erosion and thermal contraction)
- Graben formation (with subsequent isostatic adjustment)
- Intrusion of dense material into the lithosphere (mantle diapirs) or, the equivalent in effect, "necking" of the lithosphere
- Phase changes to more dense materials at depth
- Downwelling "plumes"

Rifting and spreading leads, as we know, to ocean basins and sedimentary loads on continental margins. The history of basin formation after initiation can be modeled by continuous sediment filling, which, of itself, changes the size of the basin and makes room for more sediment.

2. With regard to the subsidence mechanism for cratonic basins, Sleep and Snell (1976) assumed that the decay of a thermal "event" dominated—that the effect of sedimentary fill merely changed the isostatic response. But thermal events decay exponentially, with time constants in the vicinity of 50 million years, and subsidence in the Sverdrup Basin (for example) is not exponential; it lasted 200 million years (Pennsylvanian to Tertiary). However, whether or not the Sverdrup Basin is, in fact, a cratonic basin is still a matter of debate. (See "The Gulf of Mexico Enigma" in Chapter 5.)

D. FUTURE WORK

In the near future, work on cratonic basins will be concerned primarily with proper analyses of available data, theoretical work, and the properties of the underlying crust.

Data analysis will involve investigating basins of different sizes on different cratons to see how their dimensions and sedimentary loads changed with time. Gravity anomalies, corrected for sedimentary fill, will reveal what anomalous mass, if any, exists beneath a basin. Theoretical studies may produce self-consistent sets of values for the rheological properties of the lithosphere, information as to whether these differ from craton to craton or change with time, and if the development of a craton in one location affects craton development in others. Topographic effects of a developing basin are widespread and by no means restricted to its immediate area.

Most of all, evidence on the initiating mechanisms needs to be gathered. Even if we understood the isostatic adjustment process, dynamic modeling is nearly useless if we do not know how things get started. Clues about the initiating mechanism will be found in the basement and crust beneath the basin. For example, do the anomalously high-velocity zones found at the base of the crust beneath the Sverdrup and other basins constitute evidence of phase change—or are they intrusives of the kind postulated by Haxby *et al.* (1976)? Seismic refraction will show whether such zones are ubiquitous. If there is no evidence of an erosional interval before the onset of subsidence, the thermal expansion–erosion–cooling process that Sleep and Snell adopted would be wrong. Basins floored by crustal sections (such as that beneath the Rhine Graben) would reinforce arguments that basins develop when grabens fail to rift. Again, only carefully chosen and carefully processed multichannel seismic records will reveal the presence or absence of a "rifting" phase in the bottoms of cratonic basins.

REFERENCES AND BIBLIOGRAPHY

Beaumont, C. (1978). Lithospheric flexure and sedimentary basins, Abstract, *EOS, Trans. Am. Geophys. Union* 59(4):372.

Bott, M. H. P., ed. (1976). Sedimentary basins of continental margins and cratons, a report based on the Symposium on Sedimentary Basins of Continental Margins and Cratons, held at Durham, England, April 5–9, 1976, *Tectonophysics* 36, (spec. issue), 314 pp.

Haxby, W. F., D. L. Turcotte, and J. M. Bird (1976). Thermal and mechanical evolution of the Michigan Basin, *Tectonophysics* 36:57–75.

Klemme, H. D. (1975). Giant oil fields related to their geologic setting: A possible guide to exploration, *Bull. Can. Petrol. Geol.* 23:30–66.

McCrossan, R. G., and J. W. Porter (1973). The geology and petroleum potential of the Canadian sedimentary basins—A synthesis, in *Future Petroleum Provinces of Canada*, R. G. McCrossan, ed., *Can. Soc. Petrol. Geol. Mem.* 1:589–720.

Sanford, B. V., and A. W. Norris (1973). Hudson platform, in *Future Petroleum Provinces of Canada*, R. G. McCrossan, ed., *Can. Soc. Petrol. Geol. Mem.* 1:387–410.

Sleep, N. H., and N. S. Snell (1976). Thermal contraction and flexure of mid-continent and Atlantic marginal basins, *Geophys. J. R. Astron. Soc.* 45:125–154.

Sloss, L. L., and W. Scherer (1975). Geometry of sedimentary basins: Applications to Devonian of North America and Europe, in *Quantitative Studies in the Geological Sciences, A Memoir in Honor of William C. Krumbein*, E. H. T. Whitten, ed., *Geol. Soc. Am. Mem.* 142:71–88.

Tillement, B. A., G. Peniguel, and J. P. Guillemin (1976). Marine Pennsylvanian rocks in Hudson Bay, *Bull. Can. Soc. Petrol. Geol.* 24:418–439.

7 | Continental Margins of the Arctic Ocean

A. INTRODUCTION

The importance of Arctic margins is put into perspective as Geer (1975) reviews the distribution of U.S. continental margins from the shoreline out to a water depth of 2500 m: 421,000,000 acres (47 percent) lie adjacent to the conterminous United States and the south coast of Alaska. 469,000,000 acres (53 percent) are located north of the Alaskan Peninsula. The economic potential of the submarine Arctic margins is virtually unknown. Plans for resource exploration and exploitation have to be preceded by studies of and based on an understanding of the offshore geology and the environment of these areas. A large part of the U.S. Arctic margin falls into the active margins, but the margin facing the Arctic Ocean is believed, in essence, to be a passive margin. In the past two decades, Canadians have made a considerable effort to explore this margin in the Beaufort Sea and near the Arctic Islands. Their experience is valuable.

Scientific problems that relate to the continental margins bordering the Arctic Ocean must be assessed differently from those of other continental margins. So little is known of the area, and the working conditions are difficult. We know at least one order of magnitude less about Arctic marine geology and geophysics than about any other areas. Not even the broad tectonic framework of the area has been adequately described. We urgently need to collect basic geological and geophysical data on the mar-

gins and in the deep basins. Collecting data in the Arctic is complicated by the extreme difficulty of working in a hostile environment, much of which is ice-covered. More than anywhere else, there is need for cooperative programs in the Arctic, not just between individuals and organizations but also between nations.

B. STRUCTURE AND SCIENTIFIC PROBLEMS

Figure 7.1 shows the main physiographic provinces of the Arctic area. The more accessible areas, including Baffin Bay, the Labrador Sea, the Norwegian Sea, and the Iceland Basin, are the best understood. These areas are not discussed here, as their margins are detailed elsewhere in this report. The Arctic proper can be considered as two basins—the Eurasian Basin and the Amerasian Basin—separated by the Lomonosov Ridge. Vogt and Avery (1974) and Yorath et al. (1975) provide a good review of Arctic knowledge and problems.

LOMONOSOV RIDGE

The Lomonosov Ridge is generally considered to be a sliver of continental crust, split off from the Eurasian shelf when spreading began in the Eurasian Basin. It differs from most other continental fragments in its linearity, length, and narrowness. However, the continental nature

102

of the Lomonosov Ridge has never been unambiguously demonstrated. We need to determine its crustal properties and the nature of its margins. They probably are different on either side. We need to know how it terminates against the Greenland margin at one end and the Siberian margin at the other end.

EURASIAN BASIN AND MARGINS

The Eurasian Basin appears to be relatively simple tectonically, having been created during the last 60 m.y. by spreading about the Nansen Ridge—an extension of the Mid-Atlantic Ridge system. A unique feature of the area is that the active plate boundary approaches its own locus of rotational opening. Spreading rates have always been small, only 5–8 mm/year, half-rate.

Some of the more outstanding problems of the Eurasian Basin margins are the following:

1. What is the structure where an actively spreading ridge intersects a continental block?

2. How does the Eurasian margin, formed at slow separation rates, compare with other rifted margins that formed faster? The comparison may reveal the effects of different thermal regimes.

3. How does the Eurasian margin, now significantly sediment-loaded, compare with the formerly contiguous Lomonosov margin, which has relatively little sediment? Comparison could provide information on postrifting tectonics and loading effects.

4. What are the structures of the Yermak Plateau and Morris-Jessup Rise, apparently volcanic and formed after rifting? How do these structures affect the continental margin in their areas? Why did they form?

AMERASIAN BASIN AND MARGINS

The Amerasian Basin is more surveyed than the Eurasian Basin, but there is no consensus in the literature on its tectonic history. For example, the Alpha-Mendeleev Ridge has been considered to be an active spreading ridge during Late Cretaceous–Early Tertiary; thickened crust along a subduction zone at the same time; and foundered continental crust. Estimates of the age of the Amerasian Basin range from upper Paleozoic to Tertiary, but there is no direct evidence. Understanding the margins is obviously difficult when the whole area is so little understood.

The margins of the Amerasian Basin include the north slope of Alaska, the MacKenzie Delta area, the Queen Elizabeth Islands (Canadian Arctic Archipelago), the Lomonosov Ridge, the East Siberian Fan, and the Chukchi Sea. Much of what we do know about these margins comes from the more accessible contiguous land geology and from the information gained as a result of industry's

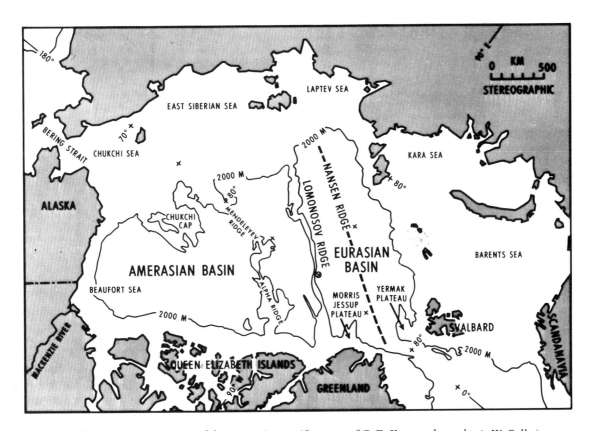

FIGURE 7.1 Province map of the Arctic Ocean. (Courtesy of C. E. Keen, redrawn by A. W. Bally.)

increasing search for petroleum products on the North Slope, the MacKenzie Delta, and the Canadian Archipelago. The relationship of the known continental geology to possible evolutionary models for the Amerasian Basin is reviewed by Bally (1976). It is obvious from his and others' discussions (Vogt and Avery, 1974) that evidence for the timing and geometry of continental breakup is inadequate and sometimes conflicting.

Some serious questions that emerge from these studies are as follows:

1. Has there been subduction, transform motion, and/or rifting along the Alaskan and North Slope–Beaufort Sea margin? What might be the time frame of these events? How do they relate to the configuration of the Amerasian Basin?

2. Vertical tectonics in the Canadian Archipelago suggests a complex history of subsidence and uplift over a period of 300 m.y. How does this reflect the plate-tectonics history of the deep basin and its margins?

3. Paleozoic fold belts surround much of the Amerasian Basin. Does this suggest, for example, that at least part of the basin may be an old interarc basin or marginal sea?

4. What is the structure of the Chukchi Cap? It is thought to be continental, but its original position is not known. Its large size inhibits realistic geometrical reconstructions of the area until its genesis and history are known.

C. CURRENT AND FUTURE PROJECTS

Some waters on the Arctic continental shelves are navigable by ships. Drilling platforms can operate for at least a small part of the year. Mostly, however, continental margins and the deep Arctic Basin have been studied from ice islands that move along uncontrollable paths at very slow speeds, by using the ice as a "land" base in conjunction with helicopters and other aircraft as means of transportation, and by aeromagnetic studies. The Russians, Americans, and Canadians are putting major efforts into Arctic studies of this sort. Recently, Canadians have used unmanned submersibles (on a trial basis) in ice-infested waters in order to study the morphology and surficial geology of the shelf area.

The Nansen Drift Project is a major U.S. proposal, with some international participation, to "freeze" an icebreaker into the ice, thus allowing it to drift slowly down the Nansen Ridge—collecting, on the way, all possible oceanographic measurements, including geological and geophysical data.

The United States and Canada are engaged in a systematic aeromagnetic survey of the Arctic area (under the Naval Research Laboratory, the Naval Oceanographic Research and Development Activity, and the Geological Survey of Canada), at a 10-mile (19-km) line spacing flown at 600 feet (200 m). This survey involves the use of long-range P3 aircraft equipped with inertial navigation.

This work has already provided valuable clues to the age and tectonic patterns of the ocean floor beneath the ice (Vogt et al., in press).

Earth scientists at the Pacific–Arctic Branch of the U.S. Geological Survey (USGS) plan to continue their investigations into geo-environmental processes and the formation of the structural and stratigraphic framework of the Chukchi and Beaufort seas and adjacent margins. USGS vessels equipped to support high-resolution geophysical surveys, detailed bottom sampling, and offshore multichannel and related geophysical studies and bottom sampling will routinely visit these areas during the ice-free months. Efforts will also be made to continue joint USGS–U.S. Coast Guard studies aboard Coast Guard ice breakers.

Canadian investigators (organized by the Earth Physics Branch and Polar Continental Shelf groups of Canada) are planning a study of the Lomonosov Ridge in 1979. Complementary to that study, an ice island may be established by the United States in order to obtain long crustal lines from the Canadian station to Svalbard and possibly Nord and Alert. This study will include crustal seismic-refraction, gravity, and other measurements that pertain to the underlying crustal structure (H. Weber, Earth Physics Branch, Ottawa, personal communication). Analysis of gravity data collected by the Canadian Earth Physics Branch in the Beaufort Sea area has given scientists a preliminary picture of the crustal structure and the position of the ocean–continent boundary in that vicinity. Marine data obtained by the Geological Survey of Canada (GSC) in the same area include magnetics, shallow seismic reflection, and surficial sediment sampling. These data soon will be published. The west coast arm of the GSC is continuing this work in conjunction with hydrographic surveys of the area. Norwegian investigators have embarked on a multichannel seismic program around Svalbard, and some data have been collected along the northern continental margin of the archipelago.

D. OPERATIONAL ASPECTS

The major difficulties with doing field research in the Arctic are usually considered to be, first, the physical difficulty of collecting the data and, second, the cost. However, as shown below, Arctic costs are not significantly different from costs elsewhere.

Costs of ship-based Arctic operations are comparable with marine costs elsewhere, although lost time because of ice conditions can be expensive. Aeromagnetic costs are comparable with costs elsewhere, although long transit times to some areas due to lack of suitable airfields is a problem. Costs of on-ice work are frequently considered excessive, but an examination of past and future Arctic project costs suggests that this is incorrect. The AIDJEX project (which involved 20 months of on-ice fieldwork over 5 years) cost about $20 million. There were about 30 separate field projects, each of which would entail at least one year's research, so that the costs came to $0.6 million per project, or $1 million per field-month. Nansen Drift estimates of $30 million for 27 months of fieldwork by many investigators give similar costs. The Canadian Lomonosov Ridge project, which is much smaller (only two months on the ice, 10–14 scientists), will also cost about $1 million per field-month, but costs per project (e.g., the seismic-refraction experiments) will be around $0.5 million, as the projects are small.

In these Arctic projects, the operations costs (e.g., transportation) are about half of the total, or $0.5 million per field-month. By comparison, a conventional marine operation in the Labrador Sea for three months with three projects costs about $1.2 million, or $0.4 million per field-month. Each completed project would require at least $0.6 million. The cost of a one-month IPOD site survey has been estimated at $0.5 million. In conclusion, the costs of Arctic on-ice operations and projects are not substantially different from research costs elsewhere. Cost should not be considered a stumbling block.

The physical problem of collecting data in the Arctic is a major difficulty. Only restricted areas can be studied by conventional ship-based programs, and even then, project completion is uncertain. For example, the period of good working conditions in the Beaufort Sea over the last six years has varied from three weeks to almost four months. Ship programs have to be very flexible, since it is never certain that work can be done in the preferred area.

On-ice projects require different techniques and instruments. The detailed spatial coverage needed in order to collect geophysical data is not easily achieved from even a large number of points on ice. New through-the-ice and under-the-ice instrumentation would help. Many margins of the Arctic basins are particularly hard to study because they are located in zones of dynamic ice–shear. These areas are not readily accessible by ship, nor are they easily worked from ice stations.

Costs of a geological–geophysical project that would satisfy *minimum* requirements for studies on Arctic continental margins over the next five to seven years can be projected on the basis of the above estimates. First, let us consider the cost, time, and personnel involved in completing a single traverse across an Arctic continental margin. This study would span from the shelf to the deep-ocean basin, typically a distance of 150 km, and extend along the margin for 150 km.

The minimum survey should consist of bathymetry and gravity on a 5 km × 5 km grid, heat flow and coring on a 20 km × 20 km grid, magnetotellurics at two or three stations, four seismic-refraction lines 150 km in length parallel to the margin, ten short refraction lines perpendicular to the margin, and aeromagnetic coverage at 5 km line spacing. Continuous seismic-reflection lines are also required, but those require subice vehicles, which are not yet available.

We suggest that at least four traverses, including the above measurements, be undertaken over a five-year period as a joint Canadian–American project. Each traverse should be situated on a different portion of the Arctic margin. The positions of the traverses should be selected to provide maximum information concerning the history of tectonic events in each area. In order of priority, these should be (Figure 7.1):

1. The Alaskan North Slope
2. The Queen Elizabeth Islands Margin
3. The East Siberian Sea Margin (excluding Chukchi Cap area)

one of
4. The Chukchi Cap Area
5. The North Greenland Margin–Lomonosov Ridge Area
6. The Beaufort Sea Margin

The first three are fairly simple, while the latter three are in more complicated areas. It is also important to bear in mind that some measurements are now available in areas 1, 2, and 6, thanks to the impetus provided by petroleum exploration. The best use possible of this previous work should be made.

We stress the importance of developing the technology to carry out continuous, under-ice measurements in addition to airborne and from-the-ice-station work, because the total costs in the operations would not significantly increase by increasing the number of projects. However, the costs of under-ice instrument and vehicle development are not considered here. In the following list of costs, people, and time, we have assumed the following:

(a) That regional aeromagnetic coverage will already be available;
(b) That about 2 months/year could be spent in fieldwork; and
(c) That it is sensible to obtain basic reconnaissance information—gravity, bathymetry, aeromagnetics—during the first year and, based on the results of these measurements, carry out the seismic-refraction, heat-flow, and coring programs during the second year. The latter measurements all require either many visits to one site or several camps to be set up. Hence, this division into two groups of measurements makes logistical sense.

Given these assumptions, the following are estimates for each transect of the margin:

YEAR 1

A. *From-the-ice measurements*

Bathymetry: 5 km × 5 km grid = 900 stations
Gravity: 5 km × 5 km grid = 900 stations
Magnetotellurics: Two or three stations
Requires helicopter landings only. Several visits over 2-month period to monitor magnetotelluric equipment.

B. Aeromagnetics

150-km lines at 5-km spacing = 4500 km

YEAR 2

C. Through-the-ice measurements

Heat Flow: 20 km × 20 km grid = 50 stations
Coring: 20 km × 20 km grid = 50 stations

Requires winch and through-the-ice work at each site. May require a camp at each site.

D. On-ice seismic profiles

Four 150-km refraction lines
Ten 25-km refraction lines

Requires two camps for each long line and many visits to receiving points. Long lines would involve two shot-points and many (about 30, minimum) receiving points.

PERSONNEL ESTIMATES

A. Six
B. Eight
C. Fifteen
D. Three

This includes cook, helicopter pilot, and other field-support personnel, as well as scientific staff.

COST ESTIMATES

A. $0.5 million
B. $1.5 million
C. $3.0 million
D. $50,000 per transect, provided more than one area is done in one year.

TOTAL COST/TRANSECT

$5.0 million

NOTES

- This does not include data analysis.
- In order to complete four transects in five years, more than several areas should be simultaneously under investigation.

In conclusion, we again stress that the total cost of this five-year project, $25 million, is not excessive when compared with the expense of a conventional marine program of similar scope.

E. RECOMMENDATIONS AND PRIORITIES

1. Aeromagnetic studies are the most cost-effective means of making useful geophysical measurements over the more inaccessible areas of the Arctic Ocean. The trends, amplitudes, and signatures of the magnetic anomalies should indicate directions of plate motions and possibly their timing, leading in turn to basic information about the genesis of continental margins. Magnetic anomalies over the margins, proper, may provide clues to the ocean–continent transition. Aeromagnetic surveys should be a prerequisite to almost all geophysical surveys. *We recommend that aeromagnetic surveys of the type currently being pursued by Canadian and American scientists be extended to cover the entire Amerasian Basin and surrounding continental margins. When the technology for airborne gravity surveys is realized, we recommend that these measurements be included in the surveys.*

2. *We recommend carefully planned crustal seismic-refraction experiments, together with multichannel reflection lines and gravity measurements, to determine the nature of the crust in Arctic basins (particularly the Amerasian Basin).* These studies should focus on the location and nature of the ocean–continent transition.

3. *We recommend that borehole information be related to stratigraphy as it is shown on seismic-reflection lines.* This effort will permit scientists to determine rates of subsidence and periods of uplift. Such information then can be related to initiation of continental separation and subduction processes. Shallow drilling (say, 20–100 ft) from a surface ship should be attempted in areas that suggest outcrop of diagnostic seismic horizons. This should be done on both the shelf and the slope.

4. A determined assault should be made to establish the validity of present interpretations that claim certain features to be continental fragments or ancient spreading centers. Refraction methods would assist such studies.

5. *We recommend that efforts be made to develop instruments that will make geophysical measurements both under the ice with tethered or remote vehicles and through the ice.*

6. *We recommend more paleomagnetic studies of rocks from the Amerasian Basin, the surrounding land masses, margins, and continental fragments.* These would be extremely valuable in reconstructing the past geometry of the area.

7. *We strongly recommend that the effort in the Arctic be a joint project of Canadian and American scientists. Furthermore, Russian and Scandinavian participation is highly desirable and should be diligently sought.*

REFERENCES AND BIBLIOGRAPHY

Bally, A. W. (1976). Canada's passive continental margins—a review, *Marine Geophys. Res.* 2(4): 327–340.

Committee for the Nansen Drift Station (1976). Scientific plan for the proposed Nansen Drift Stations, Polar Research Board, National Research Council, Washington, D.C., 247 pp.

Congressional Research Service, Library of Congress (1976). *Polar Energy Resources Potential*, A report prepared for the Subcommittee on Energy Research, Development, and Demonstration (Fossil Fuels) of the Committee on Science and Technology, U.S. House of Representatives, Ninety-Fourth Congress, Second Session.

Geer, R. L. (1975). National needs, current capabilities and engineering requirements, offshore petroleum industry, National Technical Information Service (NTIS), No. PB 249110, prepared for Committee on Seafloor Engineering of the Marine Board, National Research Council, Washington, D.C.

Pitcher, M. G. (1973). *Arctic Geology*, Proceedings of the Second International Symposium on Arctic Geology, San Francisco, Feb. 1971, *Am. Assoc. Petrol. Geol. Mem. 19*, 747 pp.

Vogt, P. R., and O. E. Avery (1974). Tectonic history of the Arctic Basin: Partial solution and unsolved mysteries, in *Marine Geology and Oceanography of the Arctic Seas*, Y. Hermen, ed., Springer-Verlag, Berlin, pp. 83–117.

Vogt, P. R., P. T. Taylor, L. C. Kovacs, and G. L. Johnson (in press). Detailed aeromagnetic investigation of the Arctic Basin, *J. Geophys. Res.*

Yorath, C. J., E. R. Parker, and D. J. Glass, eds. (1975). Canada's continental margins and offshore petroleum exploration, *Can. Soc. Petrol. Geol. Mem. 4*, Calgary, Alberta, Canada, 898 pp.

8 | Active Margins

A. INTRODUCTION

Active margins are not easily defined. Elsewhere in this report, we use the term "Megasuture," to embrace all products of subduction-related processes during Mesozoic–Cenozoic times. Active margins comprise that *portion* of the Megasuture that is active today, as is clearly shown on earthquake-distribution maps. On such maps, a narrow band with great earthquake density delineates the position of subduction zones (Figure 1.4). Within the continents, that band becomes more diffuse. Active margins coincide with convergent plate boundaries and, on occasion, they are bounded by transform-fault systems. At convergent plate boundaries, the lithosphere of one plate is subducted at depth under the lithosphere of another plate. Earthquakes that emanate from within the sinking slab allow us to trace it deep into the mantle. Volcanism, crustal deformation, and the opening of backarc basins are the more superficial expressions of the subduction process. Igneous activity and metamorphism accompany the process at depth.

The seismologic definition of an active margin excludes large areas occupied by marginal seas. Therefore, most marine geologists prefer to define active margins as the transition zone between the active subduction zone and the continental mainland. This view, however, completely excludes the active subduction zones within continents proper.

Another limited conception of active margins led to the bizarre situation that exists within the IPOD Active Margins Panel. This group has excluded the Caribbean, Scotia Sea, Mediterranean, and Gulf of Arabia from their consideration. Fortunately for science, special subpanels for some of these areas found a safe and friendly hearing in the IPOD Passive Margins Panel. Geologists realize that active margins are that part of the Megasuture that is mostly under the sea and associated with high earthquake frequency, but they generally prefer to study the Megasuture as a whole. This allows them to transcend the ephemeral seashore and study the earth under both land and water over longer time intervals.

Figure 8.1 shows the Cenozoic–Mesozoic Megasuture, its subduction boundaries, and the distribution of continental and oceanic crust. Note that around the Philippine Sea the subduction zone bifurcates. Note further that large segments of the active margins are underlain by old continental crust. Table 8.1 classifies the various kinds of subduction margins.

Geophysical observations on active margins are often difficult to interpret because not many wells for calibrating the geophysical measurements have been drilled. (Currently, the *Glomar Challenger* is drilling some rele-

---

Much of the discussion in this chapter is based on the white paper of the IPOD Active Margins Panel (1977).

vant sites in the Western Pacific.) Observations of rock outcrops on islands or on the mainland are not often integrated with marine geological and seismic observations. Such an integration is a tough task, because the outcropping rocks are largely the product of processes that happened at depth during earlier subduction phases. Comparison of ancient and recent sediments is often difficult because of diagenetic and metamorphic overprints and alteration due to uplift and subaerial exposure.

The scientific interest in active margins coincides with some considerably practical concerns. Major among these is the need to make predictions of earthquakes and volcanic eruptions and to promote understanding of the genesis of ore deposits and hydrocarbon accumulations.

Numerous ore deposits are located in mountain ranges that were formed during the subduction process. Some scientists directly link massive sulfide and porphyry cop-

per deposits to subduction, remelting of oceanic rocks, and subsequent emplacement of the ores in the overriding plate. Modern economic geologists are trying to trace the distribution of metallogenetic zones in folded belts. Across the Andes, that distribution suggests derivations of components of magmatic–hydrothermal systems through partial melting of subducted oceanic lithosphere at progressively greater depths across a stable Benioff Zone. The extensive discussion of plate tectonics and metallogenesis in Strong (1976) indicates that postulated relations are still tenuous and need further research and documentation.

The search for hydrocarbons in basins related to active margins has been frustrating. Some areas like Indonesia, Borneo, and California offer prolific hydrocarbon deposits, while others like Japan, Taiwan, and Central America are almost barren of them. Current explanations

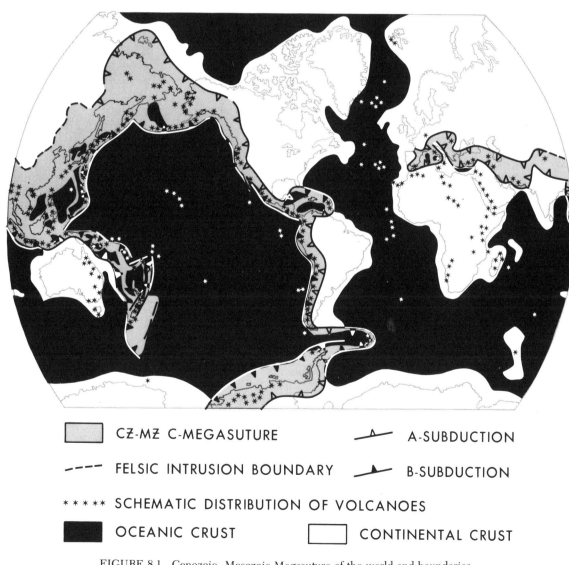

FIGURE 8.1 Cenozoic–Mesozoic Megasuture of the world and boundaries.

Table 8.1 Classifications of Active Margins

Types	Examples
1. Convergent active margins	
11. B-Subduction margins, subduction of oceanic lithosphere	
111. Oceanic crust subducted below continent	Peru–Chili, Guatemala, Honduras, Eastern Aleutian, Java–Sumatra
112. Ocean crust subducted below island arc	New Hebrides, Ryukyu, Japan, Kurile, Andaman, Bonin Mariana, Tonga, Carribean, Scotia Arc
113. Oceanic plateau subducted below island arc	Ontong Java Plateau under Solomon Islands
12. A-Subduction margins, subduction of continental lithosphere	
121. Subduction of continental crust below island arc	Timor-Ceram
122. Subduction of continent below continent (continent collision)	
1221. Without oceanic marginal basins	Gulf of Arabia
1222. With formation of marginal basin	Eastern and Western Mediterranean
2. Transform active margins	
11. Continent slipping past ocean	Queen Charlotte margin
12. Transform–rift system intersecting continent	Gulf of California
13. Ocean slipping past ocean	Hunter Fracture Zone
14. Continent slipping past continent	San Andreas Fault

for these differences in hydrocarbon productivity are superficial and unconvincing.

B. THE EVOLVING ACTIVE-MARGIN MODEL

In the 1920's and early 1930's, F. A. Vening Meinesz, the great Dutch geophysicist, discovered belts of large negative gravity anomalies on the outside of island arcs. He interpreted these anomalies to reflect a symmetrical down-buckling of the sialic crust. Arthur Holmes, a Scottish geologist, in 1928, had proposed a convection-current mechanism for continental drift. Convection currents were subsequently adopted as the mechanism that formed the orogenic root that Vening Meinesz postulated. Originally symmetrical, the model was soon made asymmetric to better fit the details of geological cross sections across ocean trenches and island arcs. Dynamic models were set up in laboratories to underscore the plausibility of the model, which was named "tectogene."

The high seismicity of the trench areas and the associated Benioff Zones seemed to fit that tectogene concept. In addition to this, the calc-alkalic volcanism of island arcs was explained in the context of relative thrusting of ocean crust under continents, as occurs along the circum-Pacific "ring of fire."

The tectogene concept further evolved into the converging-margin model as soon as it was recognized that the earth was fragmented in several lithospheric plates that often carry both continents and oceans within a single unit. Along subduction zones, the oceanic lithosphere is flexed downward, underthrusts the adjacent plate, and finally descends as a slab into the deep mantle. At a depth of about 100 km, the descending slab partially melts and the melt rises to form a volcanic arc.

On the seafloor, a trench forms over the depressed oceanic crust and is flanked by the continental slope (also called the inner trench wall) if the subduction occurs between oceanic and continental crust. Depending on the sediment supply, some trenches are filled with sediments, while others are virtually empty. The inner wall of the trench often consists of sediments that were scraped off the oceanic crust; these appear as ill-defined, structurally deformed layers on seismic-reflection sections. It is inferred from seismic sections that structural imbrications of sediments accrete on the inner trench wall by thrust-faulting and tectonic stacking. In some cases, even the top of the oceanic lithosphere is imbricated. In other cases, it appears that sediments are carried with the oceanic crust to subcrustal depths.

Unless the earth is expanding, seafloor spreading requires the disposal of great amounts of oceanic lithosphere continuously being produced at midocean ridges. That disposal occurs in subduction zones, where oceanic lithospheric slabs sink deep into the mantle. Soon after this was recognized, the subduction model was applied to explain the origin and development of mountain belts. Earlier in this century, subduction of sialic crust had been invoked by some alpine geologists. The modern subduction model led geologists to restudy the contact relations of outcropping ultramafic rocks. These are now known to be frequently underlain by thrust faults, suggesting that the rocks were moved great distances from their original sites. Strangely, most of these occurrences imply that the ultramafics were thrust over sialic crust (a process called obduction), an altogether different situation from simple underthrusting of oceanic lithosphere (subduction).

Summaries of active margins are given by Sugimura and Uyeda (1973), Miyamura and Uyeda (1972), Gorai and Igi (1973), and Talwani and Pitman (1977).

C. THE FOREARC REGION

The forearc region extends from the deep-sea trench to the volcanic area and has a minimum width of 100 km. According to Seely and Dickinson (1977), this minimum width is due to the fact that volcanoes usually stand between 90 and 150 km above an inclined seismic Benioff Zone. As shown in Figure 8.2, such a geometry implies that no arc will be located immediately next to the trench and that, in most cases, some oceanic crust can be expected to underlie the forearc basin. Figure 8.3 is a generalized model of forearc areas and terminology applied to the model.

Figure 8.4 is a drawing of a typical geophysical section near a trench. Occasionally, the top of the oceanic basement can be followed for considerable distances under the continental slope. On seismic sections like that shown in this figure, the whole area lying between the bedded sedimentary section and the top of the landward-dipping oceanic crust is often devoid of any reliable reflectors. On rare occasions, there are hints of an imbricate structure. Landward, there are virtually no reflection or refraction data below the forearc basins. On the arc itself, there is no geophysical information, but there are some surface geological data and wells. In essence, the onland informa-

tion sits as an island linked to the original subduction concept by a thin web of inference.

For example, the maximum age of undeformed sediment has been paleontologically dated only in the Aleutian Trench. Only three sites (East and Central Aleutian Trench, Nankai Trough) in the accretionary sedimentary wedge of the subduction complex have been sampled very deeply below the surface. To answer certain questions, we need to determine the ages of rocks and decipher the structures of the subduction complex by drilling. Among the questions are the following: What proportion of sediment is accreted in the subduction zone, and what portion is subducted? What is the volumetric balance between sediments involved in the subduction complex and those deposited on or near the active margin during the subduction process? How does the thickness and alteration of lithology of subducted sediments influence the structural style of deformation?

We do not know whether oceanic crust is sliced up in the subduction zone. Ancient fold belts suggest that outcropping ultramafic rocks and their associated pillow lavas and cherts (ophiolites) are remnants of old ocean floors that were obducted over the continental crust. In some regions, midocean ridges have been subducted. DeLong and Fox (1977) have speculated over the consequences of this process, but so far, data are lacking to test these hypotheses. Kelleher and McCann (1976) suggest that uplifted regions and aseismic ridges are areas of buoyant oceanic lithosphere that resist subduction when they collide with an active trench. Cady (1975) and his co-workers show that, in the Olympic Mountains of northern Washington, an Eocene seamount was involved in the subduction process.

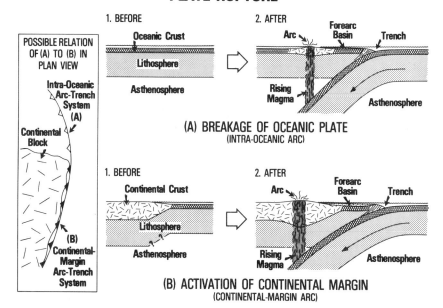

FIGURE 8.2 Models for inception of active margin. (After Seely and Dickinson, 1977.)

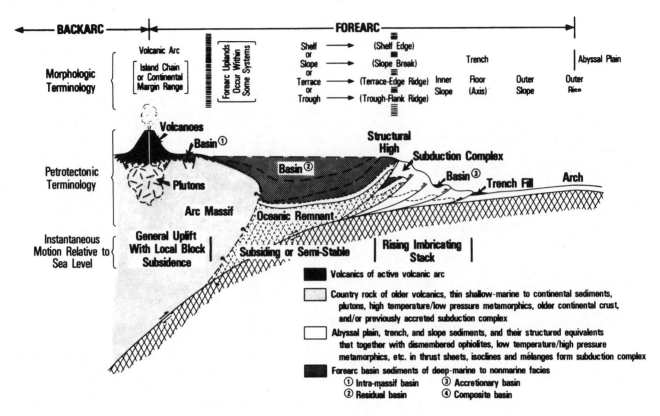

FIGURE 8.3 Generalized instantaneous forearc model and terminology. (After Seely and Dickinson, 1977.)

The formation and topographic relief of structural highs (i.e., trench–slope breaks) that border the common forearc basins remain an enigma. Some of these features are completely missing on some active margins (see Figure 8.5).

D. THE ARC REGION, ITS VOLCANOES AND GEOCHEMISTRY OF IGNEOUS ROCKS

Most features of active margins are quite diverse. This includes the arc region itself. The simplest arc types are intra-oceanic, arc–trench systems such as the Marianas. In these, a volcanic arc separates the main ocean from an oceanic backarc basin. Other arc systems, like those adjacent to the North and South American Cordilleras, are complex crustal accretions on older fold belts. Arcs such as the islands of Japan or New Zealand are continental fragments, detached from the mainland.

Volcanoes are characteristic of the arc region. Their evolution and chemistry should give clues to deep subduction processes. Magmas and their volcanic and plutonic rock products in active margins areas fall predominantly into two compositional categories, the *calc-alkalic* and *tholeiitic* series. Calc-alkalic rocks generally show greater iron enrichment relative to magnesium than to tholeiitic rocks during differentiation, as they crystal-

lize from magma. The division between these two rock groups in active margins is not always clear, however. Rock types that do not correspond directly to either group, and transitional types, are by no means uncommon.

Calc-alkalic rocks, in general, characterize the continental portions of active margins; tholeiites are more common in oceanic portions. Continental arcs and island arcs with continental-type roots are characterized by calc-alkalic granites and andesites, whereas oceanic island arcs contain primarily basalts and basaltic andesites of the tholeiitic or high-alumina types. Both lateral variation and temporal change of magma types in volcanic arcs are reported. Although the division between these two rock suites is not entirely well defined, studies of their temporal and spatial relationships to each other provide an important *chemical* view of continental margins complementary to the *physical* view provided by plate tectonics.

Ringwood (1977) emphasizes the role of water contained in the subducted mafic oceanic crust. He summarizes magmatic processes as follows:

The primary magmas produced near the Benioff zone at depths of 80–100 km consist of hydrous, tholeiitic basalts, close to silica saturation. They are *not* andesitic. The tholeiitic magmas fractionate as they rise, principally by olivine separation, thereby producing a spectrum of basaltic andesite to andesite magmas at shallow depths. Further shallow fractionation by separation of amphibole, pyroxene and plagioclase produces dacites and rhyolites. The differentiation trend of the volcanic suite is tholeiitic, possessing defined petrochemical characteristics, and is responsible for the tholeiitic stage of development of island arcs.

Thus, it might be considered that, during subduction, no material other than basalt or peridotite need be melted in order to produce the volcanic products of the arc. The role that contamination by molten sediments plays in these magnetic processes is puzzling.

Experimental petrologists consider all the various types of material that might be subducted. Studies of geochemical tracers in the end products of volcanism in the magmatic arc are useful in designing petrology experiments. Successful tracers are the alkalies, transition metals, rare-earth elements, and isotope ratios of hydrogen, oxygen, sulfur, and, in particular, strontium and lead. The chemical differences that exist in the subducting oceanic and subcontinental areas may have an influence on the composition of the erupted magmas. If they do, then it would be sensible to determine the degree of inhomogeneities in the descending oceanic plates.

For example, Kay (1977) concludes for the Aleutian Arc that processes controlling its magma composition include crystal settling, partial melting, and contamination of "oceanic" mantle and crust by admixture of a "continental component" rich in Ba, Rb and K, and radiogenic Pb and Sr.

There are variations in the volume of volcanic products in the deposits of arc massifs (e.g., Oregon–Washington Cascades) and in ash deposits found in deep-sea sediments. The chemical composition of ash deposited in the deep sea has changed over relatively short (2.5 m.y.) intervals of time. Such changes may be cyclical and represent trends different from those affecting the overall composition of magmas associated with evolving island arcs.

Some problems that are worth pursuing are the following:

1. Determine what the variations in magma composition were over time at pertinent portions of volcanic arcs. It has been found that the potassium content of andesitic rocks increases in a backarc direction. This empirical relation serves as a guide to reconstruct paleoseismic subduction zones. It is important to determine whether the relation of the K_2O content to the depth of the Benioff Zone is always valid or if it varies in space and with time. Reconstruction of fossil subduction zones depends on this concept.

2. In the evolution of active margins, it is not known

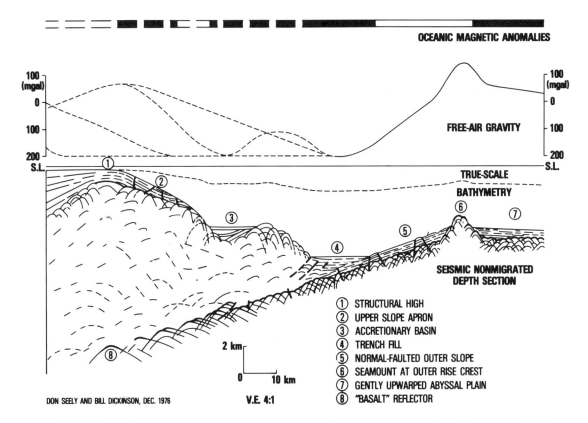

FIGURE 8.4 Composite diagram of active-margin trench features. (After Seely and Dickinson, 1977.)

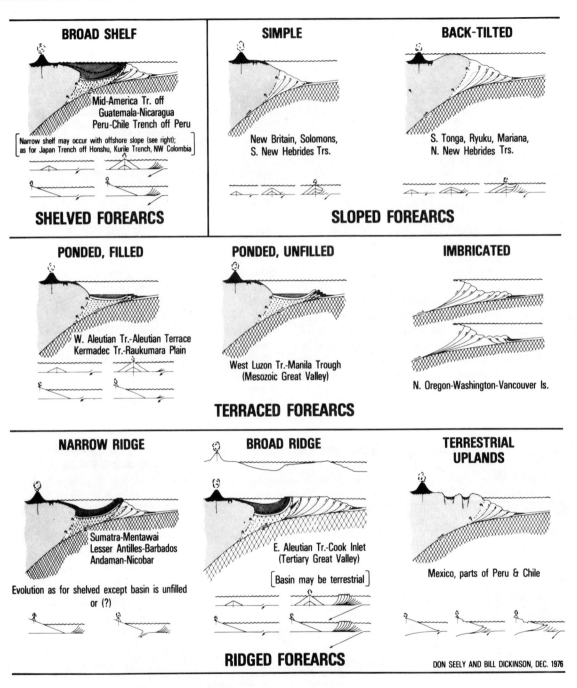

FIGURE 8.5 Models of modern forearcs. (After Seely and Dickinson, 1977.)

whether volcanism and plutonic activity have been relatively constant or episodic through time. Since magma beneath active arc systems is thought to originate along the slip surface of subducting lithospheric slabs, if it could be established that widespread episodicity of igneous activity did occur, it would provide insights into the subduction process itself. Interpreting the history of volcanic arcs from land geology alone is not always satisfac-

tory, because gaps in the record due to erosional unconformities or superposition of later volcanics cannot always be distinguished from periods of volcanic quiescence. Repeated intrusions, especially at deeper levels, often reset or alter the radiometric "clocks" used to date these rocks.

Maximum-penetration drill cores behind an active volcanic arc could provide the information necessary to link the ashes and volcanoclastic debris deposited in the submarine sediments to the extrusive and erosional events indicated on exposed portions of the arc system. Direct comparison between subareal- and submarine-deposited volcanic material could also help to clarify the influences of seawater–volcanic rock interaction, burial diagenesis, and hydrothermal activity on the chemistry and mineralogy of volcanic material.

For a given arc, we must relate eruption patterns in terms of time, volume, and composition to the marine-ash record both on land and in the ocean. Recent studies suggest that several episodes of explosive volcanic activity are synchronous on a worldwide basis. The cause of and reasons for this synchroneity are not known. Are discontinuous plate motions manifested in the magmatic evolution of the arc?

3. Test the models for the formation of island-arc volcanoes. It should be possible to evaluate, through isotope studies (e.g., strontium and lead), the relative contributions made by the subducted oceanic lithosphere and sediment to the magmas in island-arc settings.

4. Metamorphic rocks whose origin is affected more by high pressure than by temperature are associated with the forearc portions of arc systems (Ernst, 1975). They are thought to form at depth within the subduction complex above the descending lithospheric slab. Because these high-pressure rocks, often termed blueschists, are rare in other environments, they are useful for identifying inactive subduction zones, which lack features of active zones such as seismicity and low heat flow. These rocks are less common in Paleozoic and Precambrian rocks than in later sequences. Whether this is related to subsequent destruction of blueschist minerals or to changing tectonic or thermal conditions through time is still unknown.

High-temperature/low-pressure metamorphism occurs beneath volcanic arcs because of the heating associated with magma as it rises to the surface. Behind continental volcanic arcs, subduction-related metamorphism may also be of the high-temperature/high-pressure type. Both high-pressure/high-temperature and high-temperature/high-pressure metamorphism suggest conditions within the crust that depart from a "normal" geothermal gradient. These departures from "normal" geothermal conditions are expressed in active convergent margins as generally low heat flow in forearc areas and high flow in the volcanic-arc and backarc areas. If, as various authors have proposed, these conditions arise as a response to subduction of lithospheric plates beneath active margins, the extent of these anomalous metamorphic conditions may indicate the quantity of crust that is affected thermally by the subduction process.

E. THE BACKARC REGION

Earthquakes, volcanoes, and the deep trenches are the spectacular expressions of the subduction process. The slower basin-forming processes in the backarc regions attract less attention, but Figure 8.6 reveals that these basins occupy most of the active-margin area. Most of the basins shown on this map are Tertiary; they began to form in Late Paleogene time and subsided during Neogene. Table 8.2 offers a tentative classification.

Marginal basins *sensu stricto* or backarc basins floored by oceanic crust (121 of Table 8.2 and the Andaman Sea on Figure 8.7) have recently been explained by Toksöz and Bird (1977) as a necessary consequence of the subduction of oceanic lithosphere that results in an induced convective circulation in the wedge above the slab (Figure 8.8). They propose to classify marginal basins into underdeveloped, active-spreading, mature, and inactive. The assumed time interval of 20–40 m.y. between the beginning of subduction and secondary spreading is consistent with the time it takes for a gradual warming and weakening of the overriding lithosphere by convection. Toksöz and Bird believe that stress induced by convection is minor but that tension exerted on the lithosphere by the downgoing slab probably plays a role in starting secondary spreading.

Some unanswered questions concerning marginal basins are the following:

- Are marginal-basin basalts composed of the same type of material that forms seafloor at midocean ridges?
- What is the nature of the crust and/or mantle that underlies the marginal-basin basalts?
- What is the relation between the marginal-basin lithosphere and the ophiolite sequences found on land?
- Do marginal basins form by the same mechanical processes that form the midocean ridges?
- Are marginal basins favorable locations for development of ore bodies by hydrothermal emanations?

Backarc basins floored by continental crust are important to the petroleum industry, because a number of these basins bear oil. Some of them (1221 of Table 8.2) can be viewed simply as mini-passive-margins facing small ocean basins. Others (1222 of Table 8.2), however, are floored almost entirely by continental crust. Finally, there are some backarc basins associated specifically with A-subduction (123).

All backarc basins floored by continental crust are characterized by a block-faulted bottom that often forms paleostructures that preferentially trap oil. The bounding faults may be shaped like sled runners (listric faults) and suggest crustal attenuation by extension. The basins sometimes display high heat flow (e.g., the Pannonian and Tyrrhenian basins).

Despite the fact that backarc basins floored by continental crust have been explored by the industry in many areas of the world, available data are woefully incomplete. Industry has released some, but not many, data, published multichannel sections are rare, regional geological sections integrating surface and subsurface in-

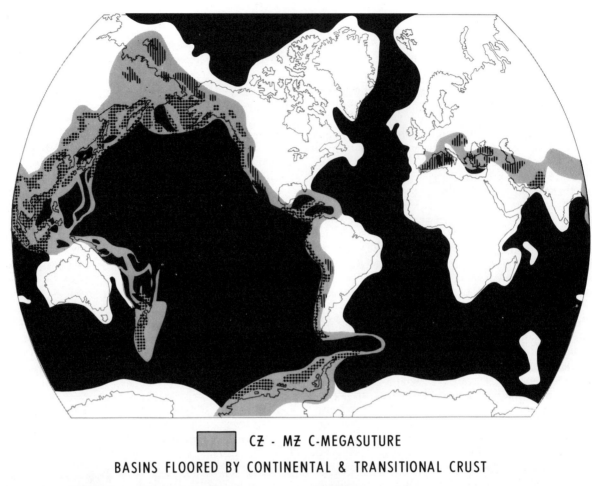

FIGURE 8.6   Epi-sutural basins. (For a detailed classification, see Table 8.2.)

formation virtually do not exist, and refraction surveys that give insight on the nature of the crust and the upper mantle are hard to come by.

More heat-flow measurements are needed to trace the thermal evolution of backarc spreading. It is encouraging to see the geothermal data of Southeast Asia compiled in the SEAPEX Project.

F.   PERI-SUTURAL BASINS

Table 8.2 and Figure 8.9 show that peri-sutural basins form moats on the rigid lithosphere adjacent to the Megasuture. On oceanic lithosphere, the classical deep-sea trenches are developed. Much has been written about

deep-sea trenches. Areas of interest include the presence or absence of sedimentary fill (Scholl and Marlow, 1974; Scholl, 1974), lithospheric flexure at outer (seaward) rises of trenches (Watts and Talwani, 1974; Dubois *et al.*, 1977), and the initiation of trenches (McKenzie, 1977).

On continents, foredeeps accompany the A-subduction boundary of fold belts. Montecchi (1976) illustrated some fine examples of incipient foredeep formation related to the collision of the Australian continent with the Indonesian island arc. In Figure 8.9, one group of foredeeps hugs the Cenozoic–Mesozoic Megasuture; the remaining foredeeps are associated with Paleozoic fold belts. Both foredeeps and deep-sea trenches reflect the differences between the loading history of continental and oceanic crust near the subduction zone. There is also some

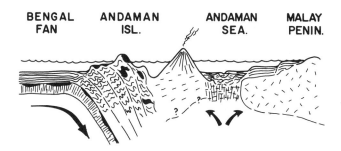

BENGAL FAN **ANDAMAN ISL.** **ANDAMAN SEA.** **MALAY PENIN.**

FIGURE 8.7 Diagrammatic present-day cross section from the Bengal Fan to the Malay Peninsula. (From Curray and Moore, 1974.)

geological evidence that sialic crust may be "softened" in the process, as illustrated by flow structures in reactivated sialic basement in folded belts.

Whether oceanic or continental crust is subducted along a subduction zone determines structural style, sediment distribution, local stratigraphy, and the presence or absence of hydrocarbons. It is necessary to compare and contrast the subduction of continental segments with the subduction of oceanic crust. Places where this can best be done are the offshore areas of Iran (Makran Coast versus Gulf of Arabia, the Bay of Bengal versus the Ganges Plains, and the Java Trench versus the Timor Trench. These studies should involve gravimetric surveys, reflection seismology, refraction seismology, wells to obtain accurate subsidence rates, and surface geological studies on stratigraphy and sedimentation.

G. TRANSFORM MARGINS

Transform margins are conspicuous plate boundaries. Some involve slipping of ocean crust against ocean crust, e.g., the Hunter fracture zone at the south margin of the Fiji Plateau. Others, like the Queen Charlotte fault zone, separate continental from oceanic crust. Most of these are not single faults but are complex fault systems dominated by strike-slip faulting. For example, Young and Chase (1966, 1977) have mapped another strike-slip fault—the Sandspit Fault—some 50 km to the east of the Queen Charlotte Fault.

A dramatic display of the complexities of transform faulting is seen onshore and offshore Southern California. (See Kovach and Nur, 1973; Vedder *et al.*, 1974; and Blake *et al.*, 1978.) The California Borderland is but a part of a complex that includes all Tertiary structures associated with the San Andreas Fault system, extending

Table 8.2   Classifications of Basins Related to the Cenozoic–Mesozoic Megasuture

Types	Examples
1. Basins inside the megasuture (episutural basins; Bally, 1975)	
11. Forearc basins: Current nomenclature and different models have been illustrated by Seely and Dickinson (1977)	
12. Backarc basins:	
121. Backarc basins floored by oceanic crust and associated with B-subduction (marginal sea *sensu stricto*) subdivided according to Toksöz and Bird (1977)	
1211. Actively spreading basins, with high heat flow	Mariana, Lau-Havre, Scotia Sea
1212. Inactive basins, with high heat flow	S. Fiji, West Philippines, Sea of Japan
1213. Inactive mature basins, with normal heat flow	Fiji Plateau, Sea of Okhotsk, Parece–Vela Basin
1214. Undeveloped basins, involving oceanic basement that has not yet reached the spreading stage, following its capture by an island arc	E. Bering Sea, maybe Caribbean
122. Backarc basins floored by continental or intermediate crust, associated with B-subduction	
1221. Continental margins of 121 basins	S.E. China Shelf, W. Thailand Shelf
1222. Basins entirely on continental crust	Bering Sea, Java–Sumatra Basin
123. Backarc basins, associated with A-subduction	
1231. On continental crust	Pannonian Basin
1232. On continental and oceanic crust	West Mediterranean
13. California-type basins, associated with intersecting rift transform system	
2. Basins outside the megasuture (perisutural basins; Bally, 1975)	
21. Basins on oceanic crust: Deep-sea trenches	
22. Basins on continental crust associated with A-subduction margin: Foredeeps	
23. Basins associated with felsic intrusion boundary: Chinese basins	

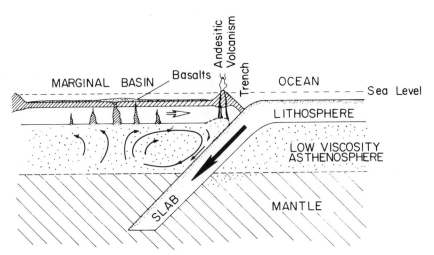

FIGURE 8.8 Genetic model of the formation of a marginal basin. (After Toksöz and Bird, 1977.)

from the Gulf of California to Cape Mendocino, north of San Francisco. The California Borderland is dominated by northwest-trending fault blocks, which are intercepted in the north by the east–west trending Transverse Ranges and their offshore continuation. Although much work has been done by U.S. Geological Survey (USGS) and industry in preparing for offshore sales, there is no synthesis of the geology of the California Borderland. This is a typical case in which pragmatic efforts satisfied the needs of government and industry but did not lead to a scientific understanding of the genesis of a transform margin.

H. INCIPIENT SUBDUCTION

There are no generally recognized areas of incipient subduction, only active and extinct subduction zones. We do not understand how a subduction zone becomes inactive or extinct.

In a recent paper, McKenzie (1977) addresses the problem of initiating trenches. The abstract of his paper says:

Kinematic plate evolution continuously reduces the number of plates, and therefore there must be mechanisms which create new plate boundaries. Creation of new ridges is simple, but producing new trenches is not. Simple plate models are used to demonstrate that a compressive stress of at least 800 bars and a rate of approach of at least 1.3 cm/yr are required to form a new trench. Neither condition is easily satisfied. To illustrate these ideas three regions where new trenches may be starting are compared with the simple theoretical models.

The three regions McKenzie cites are (1) southeast of Gorrenge Bank, west of Gibraltar; (2) the Aegean Arc in the Eastern Mediterranean; and (3) a diffuse line of epicenters that crosses the Indian Ocean from Ceylon (Sri Lanka) to Australia. [Sykes (1970) had suggested that a new island arc may be developing in this last area.]

The inception of subduction zones remains a mystery. Scientists will need to investigate the youngest subduction zones formed on active margins.

I. THE MEDITERRANEAN AND THE CARIBBEAN: TWO SPECIAL CASES

The Caribbean (see Nairn and Stehli, 1975) and the Mediterranean (see Biju-Duval and Montadert, 1977; Dewey et al., 1973; Closs et al., 1978) undoubtedly qualify as areas with active margins. A look at earthquake distribution (Figure 1.4) and Recent and Teritary volcanism (Figure 8.1) demonstrates this. These areas have been through a complex structural evolution, as is revealed by geological and topographic maps. Understanding these areas is made more difficult both because of the customary separation between land geologists and seafaring earth scientists and because of the different nationalities at work in both regions.

There are basic differences between the two areas.

1. The Mediterranean is the product of a continental collision, as was so brilliantly perceived by Argand in 1924. That collision led to the closing of the Tethys, an old ocean located between the northern Eurasian continents and the continents of the southern hemisphere. Only the margin stretching from southern Tunisia to Israel and Syria and, possibly, a spur extending from Libya to the Adriatic Sea are remnants of the passive margin of the southern continent. The remainder of the Mediterranean, i.e., the Western Mediterranean and the Aegean Sea, are backarc basins that developed during mid-Tertiary times behind the subduction zones of the Apennines-Atlas and the Hellenides and their eastern extension.

2. The Caribbean has a different history (see Malfait and Dinkelman, 1972; Mattson, 1978). North America separated from South America. As the continents moved along a path that included clockwise rotation of both North and South America, space for the Caribbean was created. At present, the continental blocks are slowly converging and, in the process, are squeezing the Caribbean.

Understandably, the Caribbean exhibits some unusual

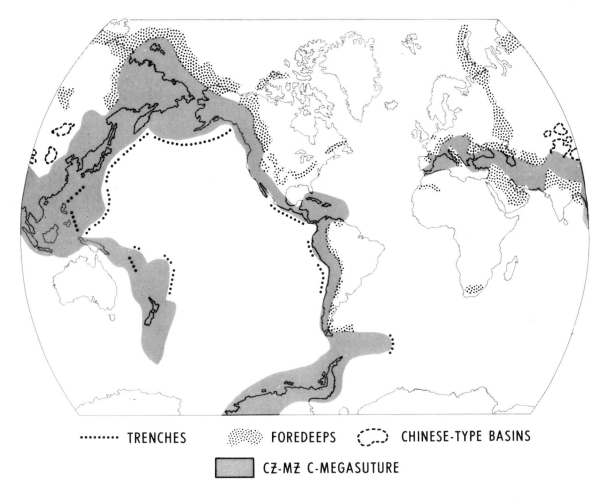

········· TRENCHES   FOREDEEPS   CHINESE-TYPE BASINS

CZ-MZ C-MEGASUTURE

FIGURE 8.9 Peri-sutural basins. Foredeeps not related to Cenozoic–Mesozoic Megasuture are associated with Paleozoic Megasuture.

structures. The Isthmus of Panama is flanked to the north and south by fold belts. This suggests conjugate, landward-dipping subduction zones. The Greater Antilles are bounded by fold belts and trenches. The well-known Puerto Rico Trench is to the north, and the lesser-known Muertos Trough is to the south. The Venezuelan Borderland exhibits structures that indicate both subduction and transform movement. The Central Caribbean crust appears to be drifting eastward with respect to both North and South America, while being squeezed in a constriction formed by the northward protuberance of the South American continental crust. The Lesser Antilles is perhaps the least anomalous area of the Caribbean. It represents normal island-arc subduction but with pronounced displacement of the vocanic axis in mid-Tertiary time.

Both the Mediterranean and the Caribbean abound in local geological problems, but there are two especially significant problems.

1. Each area is in serious need of a comprehensive synthesis. There are already many published papers that summarize the status of our knowledge. But what is needed is a summary that defines the problems, puts them in hierarchic order of importance, and explains the geological rationale for such an order. Committees being singularly unsuitable for the purpose, science will have to wait for a gifted individual to perform the task.

2. A simpler problem, but one calling for an expensive solution, requires that some key holes be drilled in the center of selected deep Caribbean and Mediterranean basins. A hole in the Colombian Basin is needed to determine the age of the oceanic basement. Another hole or two are required in the Venezuelan Basin to penetrate and determine the nature of seismic reflections that were observed below the top of the basaltic layer. Additional holes are needed on the Aves Ridge and on the leading edge of the Antilles subduction zone. In the Mediterranean, holes need to be drilled to penetrate the section below the Miocene salt that fills its deep basins.

J. THE ROLE OF DRILLING ON ACTIVE MARGINS

INTRODUCTION

Ten years of DSDP drilling verified ocean-floor spreading. Thus, a basic tenet of global tectonics was dimly established. But intensive study of another aspect of global tectonics—subduction—was avoided. IPOD drilling currently under way in the Western Pacific is addressing this problem, but this effort is severely limited by the drilling capabilities of the *Glomar Challenger*. Verifying subduction is not so simple as verifying spreading because the descending ocean crust is soon out of drilling range. Because subduction is thought to be a necessary consequence of spreading, its existence is not much in doubt. Our questions are more concerned with subduction mechanisms and their consequences.

For future studies of active margins, four requisites are emphasized here.

1. *Greater drilling capability,* including both increased drill-stem length and blowout prevention, that permits safe crustal drilling in the landward wall of trenches and backarc basins is essential to reach the scientific targets.

2. *Multidisciplinary study.* A combined geophysical and geological approach will be the most productive and definitive, because of the complexity of active margins. Recent advances in the multichannel seismic-reflection techniques, ocean-bottom seismographs, and deep-tow geophysics should be coordinated into a program of investigation. The use of submersibles should be considered. Drilling should be planned to provide three-dimensional pictures of active margins by placing a network of holes in areas selected by both geologists and geophysicists.

3. *Downhole geophysics.* In order to retrieve the maximum amount of information from expensive deep holes, downhole logging of sonic velocity, density, magnetic, electrical, thermal, and other properties of sediment and igneous rock must be measured. *In situ* stress measurements, made by such techniques as hydrofracturing and overcoring, should be attempted in subduction zones and backarc regions. Holes could also be instrumented for long-term monitoring. In subduction zones, direct detection of dynamic processes such as seismic and aseismic displacements, crustal shortening, and shear-stress heating could open an entirely new range of investigation. (See Geller *et al.,* 1977.) The basic technology for data acquisition and transmission in ocean-bottom interplate geodesy is already developed and is being used in other applications, making such a project entirely feasible in the near future.

4. *Ancient and modern analogs.* There are differences between ancient margins now exposed on land and modern margins that are largely covered by water. Drill cores would provide young rocks that are currently in the initial stages of metamorphism and deformation seen in the rocks that outcrop in coastal mountains. Drill cores would give scientists modern analogs for ancient environments and, conversely, allow them to compare drill samples with the products from ancient margins. Thus, an effort would be made to provide coherence between marine and terrestrial investigations.

CONCERNING PRESENT DRILLING CAPABILITY

Unfortunately, present technology is inadequate to sample rocks at those depths where the subduction model is based largely on inference. There are, however, many shallower objectives of considerable significance that can be and that currently are being reached with *Glomar Challenger* capability (Scientists aboard *Glomar Challenger* Leg 57, 1978).

1. Dating sediments in the accretionary wedge answers questions about rates of accretion and enables us to make estimates of relative volumes between accreted and subducted sediment. For instance, it was surprising to find deformed Miocene sediment in cores from the midslope terrace of the central Aleutian Trench wall on Leg 19 of the DSDP, when deformed sediments from a similar site along the eastern Aleutian Trench were much younger (Pleistocene or upper Pliocene). The former sediments suggest little accretion and mostly subduction; the latter suggest just the opposite.

2. Establishing the source from which the accreted sediment and volcanic rock came helps to interpret seismic-reflection records and to make geological cross sections. Is the deformed sediment scraped off the ocean floor, or has it come down the landward trench wall from terrigenous sources? Have parts of the continent been involved in subduction?

3. To study the mechanism of subduction quantitatively, we have to know what the physical properties of sediments are and how diagenetic processes have affected them. Some sediments sampled from accretionary wedges during DSDP showed that they were subjected to very high pressure without much diagenesis. Thus, cores from drill holes can provide information on tectonic stress, thermal environments, fluid migration, and chemical changes. Complex structures lead to complex velocity distributions in subduction zones. The true depth interpretation of seismic-reflection lines is dependent on correct velocity assumptions. Therefore, we need accurate velocity measurements on the rocks penetrated by wells. Gravity data cannot be analyzed quantitatively without knowledge of rock density, and magnetic data similarly suffer from lack of measured magnetic susceptibility.

4. Probing into shallow depth ranges where geophysical techniques are blind (the "acoustic basement") is important to resolve geological structure. Seismic-reflection data at DSDP Site 181 (off the Aleutians) showed "acoustic basement" at a very shallow depth. Subsequent drilling

sampled highly de-watered and tightly deformed Pleistocene muds below a regularly bedded layer of slope deposits. "Acoustic basement" is commonly seen at shallow depths in seismic records across landward trench walls, even in modern multichannel seismic-reflection records. Along island-arc margins, this basement is likely to be, though not always, igneous rock.

5. Penetrating through the toes of accretionary wedges and into subducting igneous oceanic crust can be achieved in such areas as the Antilles. Here, oceanic sediment appears to be accreting above a shallow and nearly horizontal thrust fault.

CONCERNING FUTURE DRILLING CAPABILITIES

Further development of the subduction model requires drilling capabilities that exceed current technology and new geophysical measurements.

1. The present model of subduction processes is highly inferential. The products of ancient accretionary prisms exposed on land appear to have been subjected to environments 10–40 km deep. Technology in the foreseeable future may never allow sampling from such depths. However, some 5–6 km sampling in a few critical areas, with long-term downhole instrumentation and postdrilling geophysical surveys, could help us to elucidate at least the initial stages of deformation, shear-heating, diagenesis, and stress accumulation. Time-dependent behavior of plate convergence and subduction, rates of seismic and aseismic displacements, and the relation between episodic subduction and episodic spreading on the scale of tens of years would be directly observed. With the increase of drilling-depth capability, the scope of knowledge regarding the deep tectonic processes will increase and the gap between the model at depth and observations in ancient active margins on land will be better bridged.

2. It has often been inferred from seismic-reflection records that in the forearc basins, gently deformed upper Cenozoic sediments are unconformably underlain by Mesozoic basement. What is the significance of such a structure in relation to the geological evolution of trench–arc systems and orogenic mountain belts? Deeper drilling through thick sedimentary sequences in landward slopes of trenches will enable us to unravel these basic questions. Effective blowout-preventing devices are needed for such a study.

3. Deeper drilling in backarc basins will provide the most basic information to determine age, crustal composition, and possibly the origin of backarc basins. Based on a wealth of geophysical data, such as gravity, magnetics, heat flow, and seismic structure, various hypotheses on the origin of backarc basins have been presented. But the key information to test these hypotheses is lacking; drilling will provide many scientific answers and also significant information related to the problem of hydrocarbon maturation.

Deep drilling should start with the simplest and most typical continental and island-arc-type margins, where subduction rates are estimated to be high and where the land geology is well known. Because of the high costs of deep drilling, only a few such areas should be selected to ensure a detailed multidisciplinary program that would consist of densely spaced series of holes in transect from the ocean basin to the entire backarc basin (when it exists), with several networks of holes in critical sites. In this regard, considering also the weather conditions and faunal abundances in sediments, Japan and Tonga–Kermadec trench–arc–backarc systems and Middle to South American trench–arc systems would be best suited for the first target. Domestic considerations may, however, underscore the need for studying the Alaskan active margin.

K. TRANSECTS ON ACTIVE MARGINS

INTRODUCTION

If there ever was a case for combined studies on land and in the oceans, it is on active margins. Institutions typically are concerned with either land research or ocean research. Inadequate cooperation between the two kinds of institutions presents conspicuous obstacles to scientific advances regarding active margins. To correct this situation, such institutions need to begin working closely together. Active margins do continue on land, that is, on the island arcs and onto the continental mainland. Geological observations on land simply must be reconciled with the many new geophysical and oceanographic measurements. The whole thing has to be viewed as one entity concerned with present and past subduction processes or, in more traditional terms, mountain building.

The preceding pages have revealed how we are just barely beginning to preceive the complex consequences of what is so simply—and maybe naively—summed up as subduction. We cannot tackle every subduction margin of the world. We have to be selective. We will fall back on the old geodynamics standby, the transect (or traverse, or corridor, as others call it).

A transect that is 100–200 km wide would allow a three-dimensional geological and geophysical analysis of the crustal structure and stratigraphy of this complex zone. Furthermore, the frequent occurrence of igneous activity (from both island-arc volcanism and backarc spreading) in a broad zone provides a means of studying the nature and origin of the igneous crust in a number of convergent-zone settings such as oceanic–continental and intraoceanic subduction boundaries.

All appropriate geophysical tools and techniques (deep-penetration seismic reflection and refraction, gravity, magnetics, for example) will be required to elucidate the crustal framework (structure) of each transect. Geological information from the oceans can be obtained by sampling rock outcrops by means of bottom-transponder-navigated gear, submersibles, and drilling and on land by conventional outcrop sampling and terrestrial drilling techniques. Careful integration of all information across a transect should begin to give us a better image of the nature and evolutionary framework of active margins.

Figure 8.10 sketches a synthesis of a transect. Note that all the geological information is based on surface mapping, as there are no wells in the area. Note that refraction information is limited to the offshore because of difficult access on land. Note also large blank areas that have neither geological nor geophysical information.

We assume that a general strategy (such as the one outlined in the recommendations for passive margins, pp. 95–96) will prevail in the future. Assuming this, we cite some specific targets for transects to make sure that some broad classes of problems will be covered. Our listing of these examples is not intended to prejudge a future decision on traverse sites. That decision should come from a grass-roots level—from people most familiar with specific areas. The decision should follow a thoughtful review of all scientific and logistic aspects. Ours are "such as" examples.

ALASKA

Alaska is the subject of many industry and government surveys. There are principles and concepts to be recognized by research in Alaska. There, a swirl of mountain ranges connects the North American Cordillera with the mountains of Eastern Siberia. They extend from northern Yukon to the Bering Sea, which is, in essence, a complex of marginal seas.

The active margin of southern Alaska is the youngest of many accreting subduction zones that extend from the Brooks Range in the north to the Alaska Trench in the south. Old ocean crust outcrops as the ophiolites of the Brooks Range. All of Alaska was active throughout the Tertiary, with folds deforming young Tertiary sediments near the northeast shore of Alaska, with young strike-slip faults (the Kaltag, Denali, and Castle mountain fault systems), and finally, with the subduction phenomena of southern Alaska.

A geodynamic traverse on land, with deep crustal refraction and reflection work, complemented by other geophysical measurements and surface geology, would be exciting.

The southern portion of this traverse could be designed to investigate the tectonic and structural–stratigraphic development of the continental margin and adjacent mountain ranges (see also Figure 8.10). The rocks underlying these ranges and the flanking margin record a complex, generally seaward, migration of an active Pacific margin during the past 200–250 m.y. The traverse crosses the present active margin, where Mesozoic subduction complexes and less-deformed beds of superimposed successor basins are exposed, and the adjacent active margin and trench and then terminates on early Cenozoic oceanic crust of the Gulf of Alaska. Global information suggests that since the late Paleozoic, as much as 10,000–15,000 km of Pacific oceanic lithosphere have underthrust southern Alaska.

CROSS SECTION, ALASKA PENINSULA-KODIAK ISLAND-ALEUTIAN TRENCH

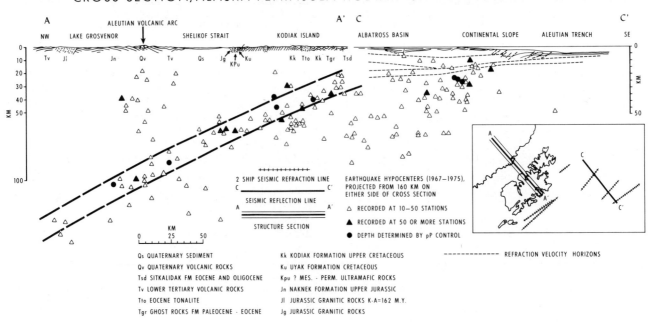

FIGURE 8.10 Structural section across Southern Alaska continental margin. (Courtesy of R. Von Huene, G. W. Moore, and J. C. Moore. Redrawn by A. W. Bally.)

BERING SEA AREA

A Bering Sea traverse investigates the seaward migration of the Alaskan–Asian margin and the formation of continental crust bordering the North Pacific during the past 300 m.y. In this area, the history of marginal growth appears to involve the accretion and superposition of rock terranes that form at active margins. In the Cenozoic, a mammoth seaward displacement of the accretion site and the formation of the Aleutian Ridge (Bering Sea) may be forming new continental crust.

The traverse should begin over the northern Bering shelf, where Paleozoic miogeosynclinal rocks (i.e., passive-margin sequences) are exposed, and trend southward across the shelf normal to the largely submerged terranes of magmatic arcs and subduction complexes of late Paleozoic and Mesozoic age that structurally connect North America and Asia. Continuing southward, the traverse should cross the Beringian margin and the very large Cenozoic successor basins superimposed upon an older, extensionally rifted, Mesozoic subduction complex. Farther south, the traverse crosses the abyssal depths of the Aleutian Basin (Bering Sea; presumably underlain by a fragment of Early Cretaceous Pacific lithosphere), the Aleutian Ridge, Trench, and Benioff Zone of Cenozoic age, and the bordering Pacific crust of latest Cretaceous and earliest Tertiary age.

U.S. WEST COAST TRANSECTS

The Plate Boundaries Group of the U.S. Geodynamics Committee, under the vigorous stewardship of John Maxwell, has assembled a set of cross sections that incorporate geological and geophysical data soon to be published. They are most useful to define remaining problems. The obvious common denominator of many of these sections is the lack of deep crustal reflection and refraction data. Many of them do not extend into the submarine continental margins, because the authors did not have access to modern multichannel sections. Several traverses are needed on the U.S. West Coast continental margin, but our comments are limited to two examples.

1. *Oregon–Washington Area:* This traverse coincides with a thick section of accreted late Cenozoic sediments that form the outer part of the continental margin. This section is on the landward side of a filled trench that includes two large, superimposed submarine fans. Several different types of accretionary mechanisms are suggested by the structural and stratigraphic framework of the outer margin. An older Cenozoic crust forms the core of the adjacent coastal mountains. This traverse is designed to investigate the history of complex marginal growth, subduction, and unusually thin (20 km) crust beneath western Oregon and Washington.

2. *Southern California and the California Borderland:* This is a classic area for a complex transform margin. If existing USGS multichannel lines and nonproprietary in-

dustry [e.g., holes drilled by the consortium of petroleum companies, the Continental Offshore Stratigraphic Test Group (COST)] data are integrated with newly acquired detailed crustal measurements, we will get a more comprehensive, new conception of this area. The details of California's land geology would be much better understood if this marine perspective were available.

MIDDLE AMERICA TRENCH TRANSECTS

This trench exhibits two strongly contrasting structural styles. Northwest of the Tehuantepec Rise in southern Mexico, very few accreted sediments are found on the inner trench wall, but southeast of the rise, seismic data indicate great thicknesses of accreted sediments. Throughout the southeastern region, the trench appears distinctly segmented, i.e., broken by transverse faults into blocks. Where the trench has been carefully investigated (off Costa Rica), scientists found that the blocks differ in levels of volcanic activity, slopes of Benioff Zones, and topography.

The existing transects are being investigated off southern Mexico, Guatemala, and Cost Rica to study the accretionary mechanism, subduction dynamics, forearc basin structure, and segmentation. Methods include onshore and offshore seismicity studies, multichannel seismic reflection, wide-angle seismic reflection, seismic refraction, collecting magnetic and bathymetric data, coring, and dredging. Narrow-beam echo-sounding and drilling techniques are included in the plans for the Guatemala and southern Mexico transects. Future plans are contingent on the results of the present efforts.

CARIBBEAN TRANSECTS

IDOE-IOC sponsored a meeting of Caribbean scientists in 1974. They recommended a series of intersecting transects across the Caribbean and its borderlands. A collation of data obtained along a transect from the Atlantic (north of Puerto Rico) to the Venezuela craton was presented at the Caribbean Congress in Curacao in the summer of 1977. This transect showed the structure of the Puerto Rico Trench, the Greater Antilles, the Muertos Trough and its associated fold belt, the Venezuelan Basin, the fold belts of the Venezuelan margin, and the Venezuelan craton. Figures 8.11–8.13 from Ladd and Watkins (in press) show the foot of the Venezuelan margin folded sequence.

Although no U.S. Government funds have been made available to support the transect work directly, a few U.S. and foreign investigators are pursuing and coordinating research in the transect areas.

SOUTH AMERICAN TRANSECTS

Peru–Chile traverses should be located in each of two regions: one of accretion over a long period of time (Chile) and one of possible tectonic erosion or consump-

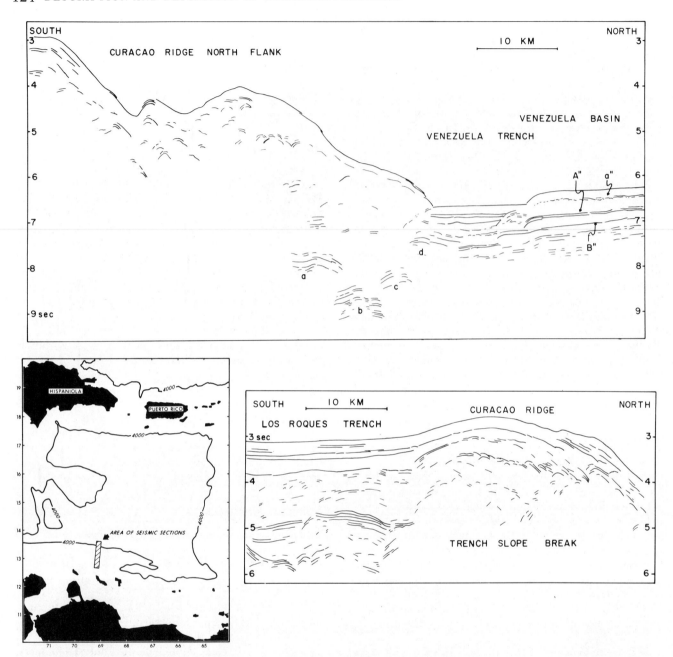

FIGURE 8.11 Line drawing of a seismic section showing structure of Venezuela Trench, Curacao Ridge, and Los Roques Trench. A″, a″, and B″ are prominent basinwide reflectors; a, b, c, and d are thought to be reflections from ocean crust being subducted beneath the margin. [From Ladd and Watkins (1978 in press), with permission of the American Association of Petroleum Geologists.]

tion (central Peru). The arc massif consists of a complex series of metamorphic rock ranging in age from Precambrian to Paleozoic. The dip of the Benioff Zone is relatively shallow in Peru and much steeper in Chile. Shallow and deep crustal studies could resolve the differences in structural styles of these two regions and also determine what factors control accretion or tectonic erosion in a given area. While the crystalline continental block exhibits both uplift and subsidence, the reasons for this large-scale tectonism are not fully understood.

NORTHWEST PACIFIC PLATE DYNAMICS TRAVERSES

A principal program of the U.S.–U.S.S.R. bilateral agreement in oceanography is the North Pacific Plate Dynamics Project. Five geological–geophysical traverses

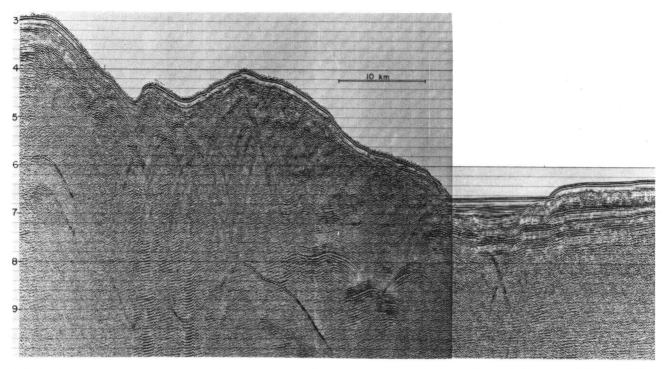

FIGURE 8.12   Seismic section corresponding to upper part of Figure 8.11, Venezuela Basin, Venezuela Trench, and Curacao Ridge North Flank. [From Ladd and Watkins (in press), with permission of the American Association of Petroleum Geologists.]

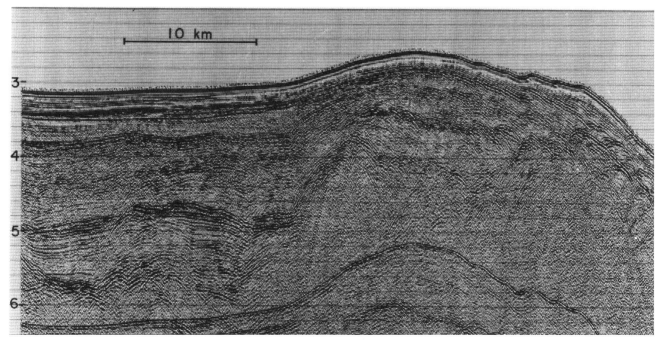

FIGURE 8.13   Seismic section corresponding to lower part (Curacao Ridge and Los Roques Trench) of Figure 8.11. [From Ladd and Watkins (in press), with permission of American Association of Petroleum Geologists.]

will be laid out across the Asian continental margin. This study aims to determine the geological evolution of the Asian continental margin from the Paleozoic to the Recent.

A geological–geophysical traverse has been designed

that starts inland of the Okhotsk Sea; moves along the Amur River, crossing the Paleozoic–Mesozoic sequence of the Asian continent; crosses the Cenozoic fold zone of the Sakhalin Island into the Kuril Deep; extends across the Kuril Island Arc sequence and the Kuril Trench across

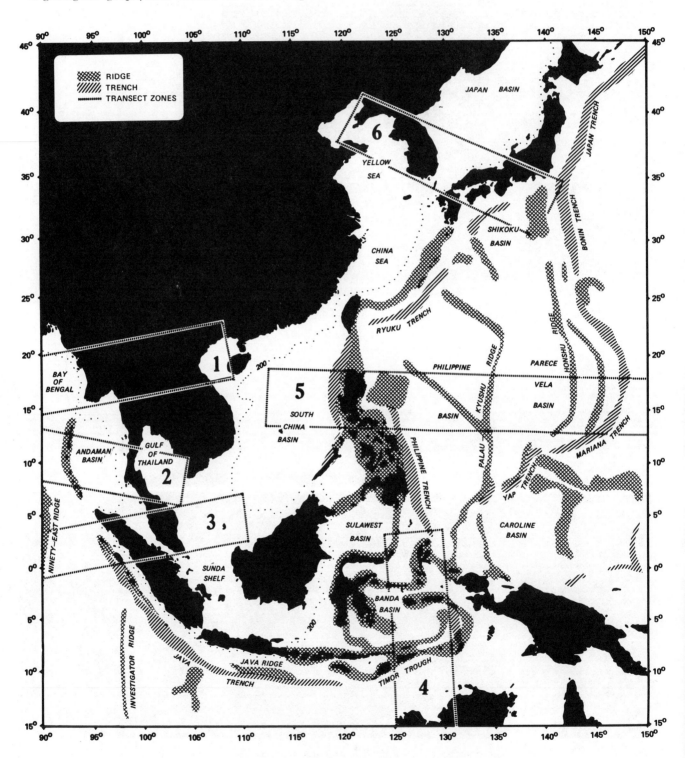

FIGURE 8.14 East and Southeast Asia transects. (From Hayes, 1975.)

the Jurassic–Cretaceous ocean floor of the Pacific; and terminates at the Shatsky Rise.

Involved in this project are the USGS, the National Oceanic and Atmospheric Administration, U.S. academic institutions, and the U.S.S.R. Academy of Sciences. Ships from both the United States and the Soviet Union will lay out long lines of ocean-bottom seismometers and electromagnetic and heat-flow arrays on the ocean floor. Magnetic and gravity traverses will also be made and extended to land-based geological and geophysical traverses. The project has been approved by both the United States and the Soviet Union and will start in 1978–1979, with two ships from the United States and two from the Soviet Union participating. Land-based seismic observations will be made by both nations on Sakhalin Island and in the Kuril Archipelago to supplement the marine seismic traverses and to use the natural seismicity of the Kuril Islands to study deep lithospheric structure.

EAST AND SOUTHEAST ASIA TRANSECTS

The transect concept was adopted in September 1973, at a combined CCOP–IOC–IDOE workshop in Bangkok. As part of a larger program, it was decided to concentrate on six transects (shown in Figure 8.14), with U.S. participation worked out by a steering committee. The major elements of the current program are the following:

1. A synthesis of existing marine geological and geophysical data;
2. A plan for a two-ship refraction program and other geophysical studies of the Banda Sea area;
3. A study of igneous and metamorphic processes in part of the Philippines (in cooperation with the Philippine Government);
4. Island (e.g., Nias and Sumatra) geological studies (largely in cooperation with Indonesia); and
5. Seismicity studies of plate boundaries and intraplate tectonics.

The total program extends over five years to the currently scheduled end of the IDOE. Scientists involved in the program are from Cornell University, Lamont-Doherty Geological Observatory, Scripps Institution of Oceanography, and Woods Hole Oceanographic Institution.

IDOE pioneered the transect concept in several of its Seabed Assessment Programs. The Southeast Asian program in progress is a fine example of the potential of a joint effort involving many nations. Further plans may be formulated when the results of this work are in and when the current active-margin drilling plans of the IPOD-DSDP Project are completed.

SEISMOLOGIC STUDIES ON ACTIVE MARGINS

Seismologists developed the concept of the sinking lithospheric slab, a major contribution to the plate-tec-

tonics revolution. Working to predict earthquakes, much of their effort is concentrated on active margins. Chapter 10, "Seismicity and the Deep Structure of Continental Margins," reviews active-margin problems. Referring to that chapter, we here flag three main areas of interest:

1. Stress and strain studies of passive margins;
2. The fate of the subducting slab; and
3. Long-term monitoring of subduction dynamics.

L. RECOMMENDATION

We recommend integrated transect studies of active margins. The strategy for such a study is the same as outlined in the first passive-margin recommendations, i.e.,

1. Compile previous work; continue reconnaissance on land and sea, with emphasis on multichannel seismic data and long refraction lines;
2. Narrow down to regional studies on land and sea;
3. Concentrate on one or more typical transect bands; and
4. Complete the study with deep-drilling programs, using current drilling capabilities on shelves and, eventually, deep riser-drilling (i.e., *Glomar Explorer*) for slopes and rises.

For active margins, we add the following:

5. Include emplacement of long-term monitoring devices in the subducting and overriding plates of convergence zones to assess and evaluate the dynamic processes associated with plate convergence. Temporal observations of seismic events, the stress–strain field, fluid distribution, and crustal displacements all will assist in developing more precise evolutionary models for active margins and for interpreting the geological and geophysical data in these areas.

We note that such traverse plans are well under way in Southeast and Eastern Asia and that they are planned in the Northwest Pacific as part of the U.S.–U.S.S.R. bilateral agreement. We look forward to the synthesis of the work undertaken in the Nazca Plate Project off South America. Also, we support and add our encouragement to the efforts of U.S. scientists to instigate a transect program in the Caribbean and Central America.

We are deeply troubled that there is so little effort to work on U.S. active margins. Alaska and the U.S. West Coast offer some uniquely interesting scientific opportunities. The work of the U.S. Geodynamics Committee's Plate Boundaries Group is a useful first step in the right direction, but new geophysical measurements are needed in order to understand our own active margins. We recommend that such measurements be made with dispatch.

REFERENCES AND BIBLIOGRAPHY

American Association of Petroleum Geologists (1977). Geology of continental margins, *Continuing Educ. Course Notes Series 5.*

Bally, A. W. (1975). A geodynamic scenario for hydrocarbon occurrences, *Proceedings of the Ninth World Petroleum Cong.*, Tokyo, Geology 2, Applied Sci. Pub., Ltd., Essex, England.

Beck, R. H., P. Lehner, with collaboration of P. Diebold, G. Bakker, and H. Doust (1975). New geophysical data on key problems of global tectonics, *Proceedings of the Ninth World Petroleum Congress*, Tokyo, Panel Discussion 1, Geology 2, Applied Sci. Pub., Ltd., Essex, England.

Biju-Duval, B., and L. Montadert (1977). Structural history of the Mediterranean basins. Symposium International Editions Technip., 448 pp.

Biju-Duval, B., J. Letouzey, and L. Montadert (in press). Structure and evolution of the Mediterranean basins, *Initial Reports of Deep Sea Drilling Project*, National Science Foundation, Washington, D.C.

Blake, M. C., Jr., R. H. Campbell, T. W. Dibblee, Jr., D. G. Howell, T. H. Nilsen, W. R. Normark, J. C. Vedder, and E. A. Silver (1978). Neogene basin formation in relation to plate tectonic evolution of San Andreas fault system, California, *Am. Assoc. Petrol. Geol. Bull. 62*(3):344–372.

Bowin, C. (1976). Caribbean gravity field and plate tectonics, *Geol. Soc. Am. Spec. Paper 169*, 79 pp.

Cady, W. M. (1975). Tectonic setting of the Tertiary volcanic rocks of the Olympic Peninsula, *U.S. Geol. Surv. J. Res. 3*, Washington, D.C., pp. 573–582.

Closs, H., D. Roeder, and K. Schmidt, eds. (1978). *Alps, Apennines, and Hellenides.* Inter-Union Commission on Geodynamics Sci. Rep. 38, E. Schwelzerbart'sche Verlagsbuchhandlung, Stuttgart, 620 pp.

Curray, J. R., and D. G. Moore (1974). Sedimentary processes in the Bengal deep-sea fan and geosyncline, in *Geology of Continental Margins*, C. A. Burk and C. L. Drake, eds., Springer-Verlag, Berlin, pp. 617–627.

Curray, J. R., G. C. Shor, Jr., R. W. Raitt, and M. Henry (1977). Seismic refraction and reflection studies of crustal structure of the eastern Sunda and western Banda arcs, *J. Geophys. Res. 82*:2479–2489.

DeLong, S. E., and P. J. Fox (1977). Geological consequences of ridge subduction, in *Island Arcs, Deep Sea Trenches, and Back-Arc Basins*, M. Talwani and W. C. Pitman III, eds., Maurice Ewing Series 1, Am. Geophys. Union, Washington, D.C., pp. 221–228.

Dewey, J. F., W. C. Pitman III, W. B. F. Ryan, and J. Bonnin (1973). Plate tectonics and the evolution of the Alpine system, *Geol. Soc. Am. Bull. 84*:3137–3180.

Dubois, J., J. Dupont, A. Lapouille, and J. Recy (1977). Lithospheric bulge and thickening of the lithosphere with age, examples in the South-West Pacific, Geodynamics in the South-West Pacific, Editions Technip., pp. 371–380.

Ernst, G. W. (1975). Systematics of large-scale tectonics and age progressions in Alpine and circum-Pacific blueschist belts, *Tectonophysics 26*:229–246.

Fukao, Y. (1973). Thrust faulting at a lithospheric plate boundary—the Portugal earthquake of 1969, *Earth Planet. Sci. Lett. 18*:205–216.

Geller, R. J., S. Stein, and S. Uyeda (1977). Proposal for downhole measurements of tilt, strain and stress near a subduction boundary, unpublished memo, Seismologic Laboratory, California Institute of Technology, Pasadena, 5 pp.

Gorai, M., and S. Igi, eds. (1973). The crust and upper mantle of the Japanese area, *Part II—Geology and Geochemistry*, Japanese National Committee for Upper Mantle Project, Geol. Survey of Japan, Kawasaki, p. 176.

Hawkins, J. W., Jr. (1977). Petrologic and geochemical characteristics of marginal basin basalts, in *Island Arcs, Deep Sea Trenches, and Back-Arc Basins*, M. Talwani and W. C. Pitman III, eds., Maurice Ewing Series 1, Am. Geophys. Union, Washington, D.C., pp. 355–365.

Hayes, D. E. (1975). U.S. proposal: East and Southeast Asia, *Geotimes*, December, pp. 22–24.

Hilde, T. W. C., S. Uyeda, and L. Kroenke (1977). Evolution of the western Pacific and its margin, *Tectonophysics 38*(1/2):145–165.

Intergovernmental Oceanographic Commission (1975). Report of the CCOP/SOPAC/IOC/IDOE international workshop on geology mineral resources and geophysics of the South Pacific, *Workshop Rep. No. 6*, Suva, Fiji, Sept. 1–6.

IPOD Active Margins Panel (1977). A white paper from the IPOD Active Margins Panel, in *The Future of Scientific Ocean Drilling*, a report by *ad hoc* Subcommittee of the JOIDES Executive Committee, available from the JOIDES Office, U. of Washington, Seattle, Wash., pp. 50–56.

Karig, D. E., J. G. Caldwell, and E. M. Parmentier (1976). Effects of accretion on the geometry of the descending lithosphere, *J. Geophys. Res. 81*(35):6281–6291.

Kay, R. W. (1977). Geochemical constraints on the origin of Aleutian magmas, in *Island Arcs, Deep Sea Trenches, and Back-Arc Basins*, M. Talwani and W. C. Pitman III, eds., Maurice Ewing Series 1, Am. Geophys. Union, Washington, D.C., pp. 229–242.

Kelleher, J., and W. McCann (1977). Bathymetric highs and the development of convergent plate boundaries, in *Island Arcs, Deep Sea Trenches, and Back-Arc Basins*, M. Talwani and W. C. Pitman III, eds., Maurice Ewing Series 1, Am. Geophys. Union, Washington, D.C., pp. 115–122.

Kovach, R. L., and A. Nur, eds. (1973). *Proceedings of a Conference on Tectonic Problems of the San Andreas Fault System*, Stanford U. Publ., *XIII*.

Ladd, J. W. (1976). Relative motion of South America with respect to North America and Caribbean tectonics, *Bull. Geol. Soc. Am. 87*(7):969–976.

Ladd, J. W., and J. S. Watkins (in press). Tectonic development of trench-arc complexes on the northern and southern margins of the Venezuela basin, *Am. Assoc. Petrol. Geol.*, Tulsa, Okla.

Laubscher, H., and D. Bernoulli (1977). Mediterranean and Tethys, in *The Ocean Basins and Margins IV: Mediterranean*, A. E. M. Nairn, W. H. Kanes, and F. G. Stehli, eds., Plenum, New York, pp. 1–28.

Malfait, B. T., and M. G. Dinkelman (1972). Circum-Caribbean tectonic and igneous activity and the evolution of the Caribbean Plate, *Geol. Soc. Am. Bull. 83*(5):251–271.

Mattson, P. H., ed. (1978). *West Indies Island Arcs*, Benchmark Papers in Geology, Vol. 33, Dowden Hutchinson and Ross, Inc., Stroudsburg, Pa.

McKenzie, D. P. (1977). The initiation of trenches: A finite amplitude instability, in *Island Arcs, Deep Sea Trenches, and Back-Arc Basins*, M. Talwani and W. C. Pitman III, eds., Maurice Ewing Series 1, Am. Geophys. Union, Washington, D.C., pp. 57–61.

Mitchell, A. H. G. (1976). Tectonic settings for emplacement of subduction-related magmas and associated mineral deposits, in Geol. Assoc. of Canada, *Spec. Paper 14*, pp. 3–21.

Miyamura, S., and S. Uyeda, eds. (1972). The crust and upper mantle of the Japanese area, *Part I—Geophysics*, Japanese National Committee for Upper Mantle Project, Geol. Survey of Japan, Kawasaki, 119 pp.

Montecchi, P. A. (1976). Some shallow tectonic consequences of subduction and their meaning to the hydrocarbon exploration ist in circum-Pacific energy and mineral resources, *Am. Assoc. Petrol. Geol. Mem.* 25:189–202.

Mulder, C. J., P. Lehner, and D. C. K. Allen (1975). Structural evolution of the Neogene salt basins in the eastern Mediterranean and the Red Sea, *Geol. Mijnb.* 54(3/4): 208–221.

Nairn, A. E. M., and F. G. Stehli (1975). *The Ocean Basins and Margins, Vol. 3, The Gulf of Mexico and the Caribbean*, Plenum Press, New York, 705 pp.

Office de la Recherche Scientifique et Technique Outre-Mer, Bureau de Recherches Géologiques et Minières, Institut Français du Pétrole, Inter-Union Commission on Geodynamics (1976). International symposium on geodynamics in South-West Pacific, Noumea, New Caledonia, August 27–September 2, 413 pp.

Riddihough, R. P., and R. D. Hyndman (1976). Canada's active western margin—the case for subduction, *Geosci. Can.* 3(4):269–278.

Ringwood, A. E. (1977). Petrogenesis in island arc systems, in *Island Arcs, Deep Sea Trenches, and Back-Arc Basins*, M. Talwani and W. C. Pitman III, eds., Maurice Ewing Series 1, Am. Geophys. Union, Washington, D.C., pp. 311–324.

Schermerhorn, L. J. G. (1976). Volcanism and metallogenesis, *Geol. Mijnb.* 55(3/4):205–210.

Scholl, D. W. (1974). Sedimentary sequences in the north Pacific trenches, in *The Geology of Continental Margins*, C. A. Burk and C. L. Drake, eds., Springer-Verlag, New York, 493–504.

Scholl, D. W., and M. S. Marlow (1974). Sedimentary sequence in modern Pacific trenches and the deformed circum-Pacific eugeosyncline, in *Modern and Ancient Geosynclinal Sedimentation*, R. H. Dott, Jr., and R. H. Shaver, eds., Soc. Econ. Palaeontol. Mineral. Special Publ. 19, pp. 193–211.

Scientists Aboard *Glomar Challenger*, Leg 57 (1978). Japan trench transected, *Geotimes* 24(4):16–20.

Seely, D. R. (1977). The significance of landward vergence and oblique structural trends on trench inner slopes, in *Island Arcs, Deep Sea Trenches, and Back-Arc Basins*, M. Talwani and W. C. Pitman III, eds., Maurice Ewing Series 1, Am. Geophys. Union, Washington, D.C., pp. 187–198.

Seely, D. R., and W. R. Dickinson (1977). Structure and stratigraphy of forearc regions, *Am. Assoc. Petrol. Geol. Continuing Educ. Notes Series 5*.

Shouldice, D. H. (1971). Geology of the western Canadian continental shelf, *Bull. Can. Petrol. Geol.* 19:415–436.

Sillitoe, R. H. (1976). Andean mineralization: a model for the metallogeny of convergent plate margins, in Geol. Assoc. of Can., *Spec. Paper 14*, pp. 59–100.

Spense, W. (1977). The Aleutian Arc: tectonic blocks, episodic subduction, strain diffusion, and magma generation, *J. Geophys. Res.* 82(2):213–230.

Strong, D. F., ed. (1976). Metallogeny and plate tectonics, Geol. Assoc. of Can., *Spec. Paper 14*, 660 pp.

Sugimura, A., and S. Uyeda (1973). *Island Arcs: Japan and Its Environs*, Elsevier, New York, 247 pp.

Sutton, G. H., M. H. Manghnani, R. Moberly, and E. U. McAfee, eds. (1976). *The Geophysics of the Pacific Ocean Basin and Its Margin, A Volume in Honor of George P. Woollard, Geophys. Monogr. 19*, Am. Geophys. Union, Washington, D.C., 480 pp.

Sykes, L. R. (1970). Seismicity of the Indian Ocean and a possible nascent island arc between Ceylon and Australia, *J. Geophys. Res.* 75:5041.

Talwani, M., and W. C. Pitman III, eds. (1977). *Island Arcs, Deep Sea Trenches and Back-Arc Basins*, Maurice Ewing Series 1, Am. Geophys. Union, Washington, D.C., 470 pp.

Toksöz, M. N., and P. Bird (1977). Formation and evolution of marginal basins and continental plateaus, in *Island Arcs, Deep Sea Trenches and Back-Arc Basins*, M. Talwani and W. C. Pitman III, eds., Maurice Ewing Series 1, Am. Geophys. Union, Washington, D.C., pp. 379–393.

Turcotte, D. L., W. F. Haxby, and J. R. Ockendon (1977). Lithospheric instabilities, in *Island Arcs, Deep Sea Trenches, and Back-Arc Basins*, M. Talwani and W. C. Pitman III, eds., Maurice Ewing Series 1, Am. Geophys. Union, Washington, D.C., pp. 63–69.

Vedder, J. G., L. A. Beyer, A. Junger, C. W. Moore, A. E. Roberts, J. C. Taylor, and H. C. Wagner (1974). A preliminary report on the Continental Borderland of Southern California, U.S. Geol. Survey Miscellaneous Field Studies Map MF-624.

Watts, A. B., and M. Talwani (1974). Gravity anomalies seaward of deep sea trenches and their tectonic implications, *Geophys. J.* 36:57–90.

Young, I. F., and R. L. Chase (1966). Gravity and seismic reflection profiles over the sandspit fault, Queen Charlotte Islands, British Columbia, B.C. Dept. Mines Pet. Resources Geol. Fieldwork.

Young, I. F., and R. L. Chase (1977). Marine geological-geophysical study: southwestern Hecate Strait, British Columbia, from Report of Activities, Part A, *Geol. Surv. Can. Pap.* 77-1A, pp. 315–318.

9 | Mobile Fold Belts and Ancient Continental Margins

INTRODUCTION

In the great mountain ranges of the world, geologists can actually see the product of processes that occurred deep in the earth. Mobile fold belts contain remnants of ancient continental margins, and large uplifts in these belts allow us to view the product of earlier deep crustal processes related to subduction. Present subduction occurs on active margins and profoundly influences wide areas within the adjacent continent.

We all stand in awe before high mountains, yet, there are even higher topographic features hidden under the oceans. In addition, the basement underlying episutural sedimentary basins is commonly 6–9 km deep, and a very large portion of the megasuture is underlain by such deep basins (Figure 8.6). These basins were formed during and in conjunction with Tertiary subduction processes and form an integral part of the history of mobile belts. Because they were discussed previously, we will not dwell on them any longer. Instead, we will concentrate on the fold belts themselves.

In approaching this subject, we would first like to discuss some crustal aspects of mobile fold belts as we see them today. Then, we will comment on the reconstruction of mobile fold belts. (Such reconstructions, in fact, ultimately depict ancient continental margins.) Finally, we will select examples to show the relevance of studying outcropping, old "geosynclinal" rocks and compare those rocks from today's continental margins.

THE "BASEMENT" OF FOLD BELTS

Typical fold belts display outer zones with folded and thrusted sediments and an inner, dominantly igneous zone. Rocks in the inner igneous zone include fragments of oceanic crust; fragments of rigid continental crust that has been mobilized and metamorphosed during the formation of the fold belt; intrusions; and more or less metamorphic, deformed sedimentary and volcanic sequences.

Ignoring sediments, volcanics, and instrusions, the basement of folded belts consists of the following types (for example, see Figure 9.1):

1. Oceanic basement involved in the subduction process. These are the ophiolites (an association of ultramafic rocks, gabbros, sheeted dykes, pillow lavas, and the overlying deep-sea sediments, i.e., radiolarian rocks). Scientists have not determined whether ophiolites are fragments of true oceanic crust or whether they represent marginal sea crust. Reliable geochemical criteria to differentiate between the two are lacking.

2. Downward continuation of rigid continental sialic lithosphere of the continental cratons under the outer margin of the folded belt in A-subduction zones.

3. Rigid continental basement that experienced internal uplifts but not remobilization. (Rocks yield radiometric ages that are indicative of the age of an old basement.)

4. Remobilized sialic crust (characterized by flow folds

in orthogneisses and their metasedimentary cover). These rocks are widespread in many folded belts (e.g., Shuswap complex of the Canadian Cordillera, Pennine nappes of the Alps). Pervasive thermal events in these areas mobilized the old sialic basement, caused partial melting of basement rocks, and reset the radiometric clocks. This is reflected by young radiometric age determinations that often overprint older basement ages. By field work, geologists have been able to correlate such young radiometric ages with main deformation phases in the external zone of the folded belt.

From this, it is evident that mountain ranges are formed in a nonrigid manner that involves flow-folding of large portions of sialic crust. So, strictly speaking, a theory of *rigid*-plate tectonics is not applicable to observations in mountain ranges. Viewed over longer geological time spans (say, of the order of 10 million years), rigid-plate tectonics can be directly applied to most areas that are outside our megasuture but not to the genesis of the structures inside that megasuture.

If sialic crust is really "softened" in the manner suggested, it would be important to determine the begin-

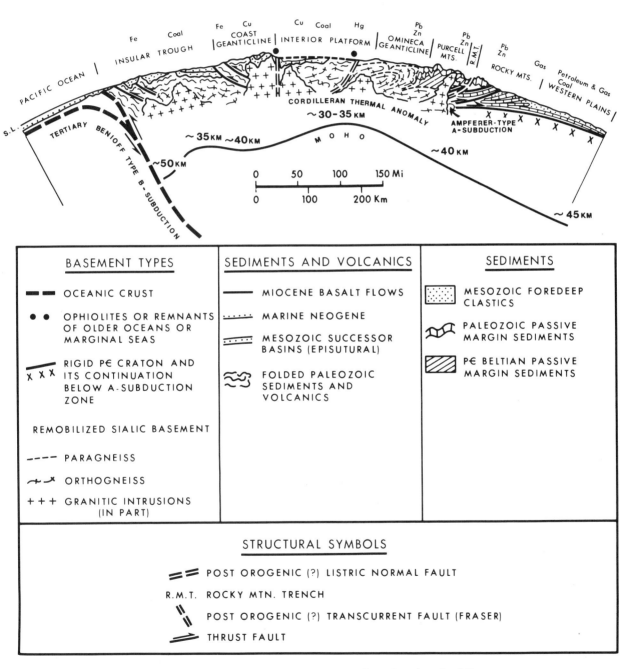

FIGURE 9.1 Schematic section across Southern Canadian Cordillera.

ning and duration of the softening process. Mountain-building (orogenic) processes typically stretch over some 80 million to 100 million years. But there is debate whether the process is more or less continuous, or whether worldwide folding phases occur. The softening of the basement appears to begin early and to continue intermittently throughout the mountain-building process. Cooling follows the thermal "softening" event and is reflected by K–Ar radiometric ages of igneous rocks. Potassium–argon dates suggest uplift and often correspond with phases of intensive clastic deposition in the adjacent foredeeps.

Can "softening" of a sialic crust be detected today by geophysical means? Long-distance refraction surveys in the Alps and Apennines have been interpreted to suggest seismic-velocity reversals in the crust. (See Angenheister

et al., 1972; Giese *et al.*, 1976.) Some geophysicists question the existence of these velocity reversals. Even if they do exist, different geological interpretations of them can be made. A high-velocity lower crust and mantle may have been thrust over higher (low-velocity) crust, or they may simply indicate hotter rocks.

The uncertainty of the interpretations, and the fact that the debate is directly related to the mode of deformation of continental crust adjacent to an active subduction zone, puts a high premium on crustal studies in Cenozoic–Mesozoic mountain ranges. Most of these measurements will have to be made on land, but some could be done in the sea (e.g., Timor, Eastern Mediterranean).

Modern reflection seismic techniques, supplemented (when needed) by the judicious use of refraction work, should provide significant new insights into the nature of

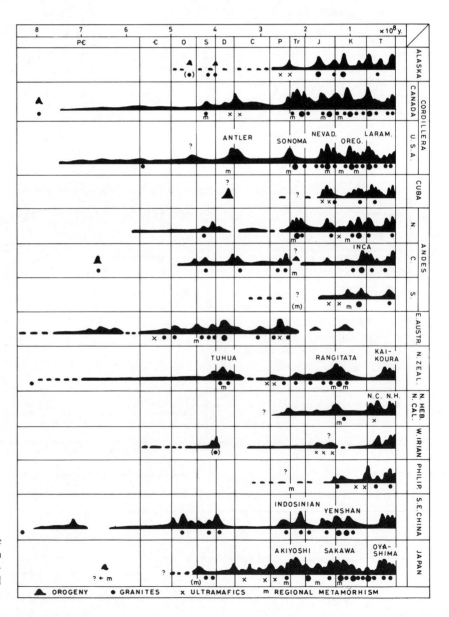

FIGURE 9.2 Diagram showing orogenic events in the circum-Pacific area. [From Matsumoto (1977). Reproduced by permission of the National Research Council of Canada.]

the lower crust and mantle of folded belts. The initial success of the COCORP program should encourage aggressive planning in this area. All work will have to be supported by conventional geophysical surveys (gravity and magnetics) and detailed surface and subsurface geology. Such work is an indispensable preamble for a program of continental drilling in the Cenozoic–Mesozoic folded belts. Not unlike IPOD-DSDP drilling, the scientific significance of a continental drilling program is critically dependent on good reconnaissance combined with first-class, detailed site surveys.

RECONSTRUCTIONS OF FOLDED BELTS

In a recent paper, Matsumoto (1977) has compiled evidence suggesting that the circum-Pacific has been the site of intermittent subduction since the mid-Paleozoic. The phenomena plotted on Figure 9.2 are all directly related to subduction processes. Combined with Figure 5.1, the breakup of Pangea, we visualize a long-lasting subduction regime always hugging the continents that today surround the Pacific.

In order to follow the subduction process back in time, methods of reconstructing ancient folded belts have to be devised. Some of the genetic types of folded belts that are recognized are as follows:

(a) Some folded belts are due to the *alternating creation and collapse of marginal seas and backarc basins.* This concept has been developed by E. Scheibner for the Tasmanides of Australia (see Figure 9.3). That marginal basins have a short life span is suggested by the overwhelming majority of today's basins, which are only Tertiary in age. Few, if any, older marginal basins are preserved and intact in ancient mountain ranges.

(b) Other folded belts, such as the Alps and the Himalayas, are thought to result from the *collision of two continents,* in which one continent (e.g., Australia) rides on a lithospheric plate being subducted under another continent. The two continents collide. What first was B-subduction turns into A-subduction, as the subducted continent is forced under the overriding plate. Because of its buoyancy, the "subductee" cannot go far, but the process is responsible for producing high mountains and forming foredeeps such as the oil-rich Middle East Basin.

(c) Some folded belts *capture small continents,* the vicissitudes of subduction being such as to virtually surround a continent by folded belts. An excellent example of this is the Szechuan Basin in China, where an old Pre-

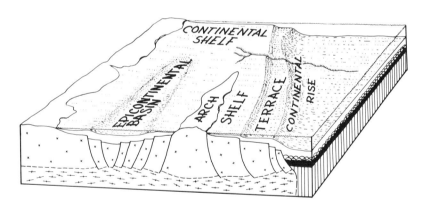

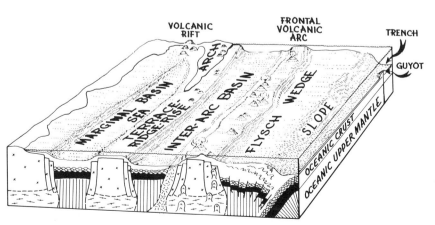

FIGURE 9.3   Block diagrams illustrating a reconstruction of the basic tectonic units of the Tasman fold belt of New South Wales. [From Scheibner (1976). From the explanatory notes on the tectonic map of New South Wales, reprinted by permission of the Geological Survey of New South Wales, Australia.]

cambrian platform is surrounded on all sides by Mesozoic and Paleozoic fold belts.

(d) Lastly, there are fold belts with obvious *strike-slip* deformation, such as southern California, Alaska, New Zealand, the mountains of Afghanistan and Pakistan, and the Caribbean borderlands of South America.

Without reconstructions of folded belts, we are not likely to understand our active margins. Three approaches offer some possibilities of understanding the development of mobile belts:

1. *Paleomagnetic work* has helped to determine the paths of drifting continents. For the Paleozoic, there is no pristine—that is, neither subducted nor obducted—ocean crust known. Therefore, the positions of continents cannot be reconstructed by matching magnetic stripes. However, paleomagnetic methods can be used on well-dated, synchronous rocks of any age in mountain ranges to resolve three questions:

(a) Do arcuate-fold trends reflect primary structures preceding the folding, or is the arcuate nature due to the folding processes?

(b) What rotations did any portion of a folded belt undergo with respect to an adjacent stable craton?

(c) Were specific areas in the folded belt in the same latitudes brought in later by strike-slip movements?

2. *Reflection seismic work in subduction zones* frequently allows us to map the continuation of the top of the underthrust crust below the folded belt. This is particularly true for A-subduction zones, where continental crust slips under the folded belt. Petroleum geologists have mapped and explored such areas in great detail, and reconstructions were worked out and published many years ago. It is true that such reconstructions cannot be made with any assurance in B-subduction zones, because there are no deep wells penetrating the accretionary wedges we see on reflection seismic data. Reconstructions for

B-subduction zones will have to wait for the results of current and future IPOD-DSDP drilling.

3. *Scientific drilling on continental margins* could, in a number of cases, be undertaken more safely and would be cheaper *on land.* IPOD drilling is, by definition, confined to the oceans. Yet, some of the problems addressed by IPOD conceivably could be solved on land. The reports from the Ghost Ranch Workshop (Shoemaker, 1975) and the recommendations of the Panel on Continental Drilling of the Federal Coordinating Council on Science, Engineering, and Technology Committee on Solid Earth Sciences ignored this aspect of continental drilling. We recommend that a special panel study the potential of continental-margin drilling *on land.*

Sections across active margins and their associated subduction zones are strikingly barren of deep-seismic information in areas underlain by islands. The reasons for this are that seismic data in these areas are difficult to obtain; they require a different type of logistics; and commonly, geophysicists anticipate that results in this type of terrain will be unsatisfactory. However, considering the often nondescript nature of the "accretionary wedge" on marine seismic lines (see Figure 9.5), it would appear to be just as good, but cheaper, to calibrate such melange complexes on land. A similar case for drilling on land could be made for a better determination of the rocks underlying the volcanoes of island arcs, but here again, any work would have to be preceded by the acquisition of adequate geophysical data.

Examples of specific areas that might be attacked are most islands on island arcs (e.g., Barbados, Kodiak, Nias Island in Indonesia) or areas where marine subduction zones come on land (e.g., the west coast of Colombia, the Olympic Peninsula of Washington, or the Nicoya Peninsula of Costa Rica). These areas are probably the most difficult to cover with reflection seismic techniques, but the data for marine accretionary wedges are not very good

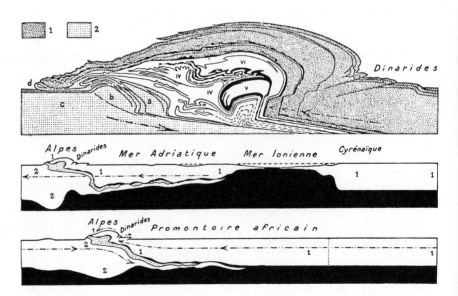

FIGURE 9.4 Alpine collision. (After Argand, 1924.) The top of the figure is a detailed drawing of the left-hand side of the center section of the figure. 1, Africa; 2, Eurasia; I–VI, Pennine remobilized basement folds; a, b, c, rigid basement of Alpine foreland.

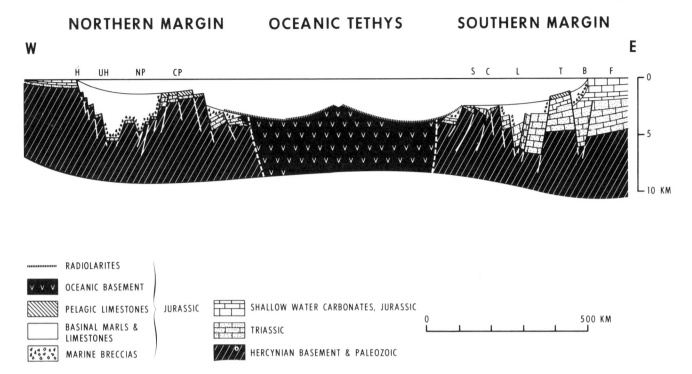

FIGURE 9.5   Reconstruction of Alpine Geosyncline at the end of the Jurassic. [Redrawn after Laubscher and Bernoulli (1977).]

either. A deep well into the no-data zone could give some solid information where none is obtainable by other methods.

GEOLOGICAL FIELD OBSERVATIONS ON ANCIENT MARGINS

For decades, geologists went to the mountains to study outcrops. Theories of mountain building were derived largely from such work. As early as 1857, James Hall observed in the Appalachians that intensely folded sediments were several times thicker than formations of the same age in undisturbed regions. This led to the development of the geosynclinal theory of mountain building—that no mountain range could form without thick sediments being deposited first. Toward the end of the last century, European geologists decided that deep-sea sediments outcropped in the folded belts. Later, ophiolites and the overlying radiolarian rocks were recognized as originating the deep sea. Theories developed that postulated deep oceans in the Alpine geosyncline, flanked by an African shelf and a European shelf. Much later, following a suggestion by Drake *et al.* (1959), geosynclines were compared with continental margins.

Through the years, many scientists have tried to reconstruct the original sites of deposition of sediments found in folded belts. Elaborate drawings illustrated many types of subsiding basins filled with thick sediments and vol-

canics (the geosynclines), but no one really knew what a simple, nondeformed basin looked like. Only in the last decades have reflection seismic lines revealed the anatomy of such basins and their regional setting. Geophysical data obtained on continental margins, drilling data from industry, and DSDP data from the deep oceans have been most helpful.

We now study stratigraphic sections in mountain ranges by comparing them with their analogs on continental margins and in the deep oceans.

SUTURE ZONES—BOUNDARIES OF LITHOSPHERIC PLATES

In 1924, Argand held an advanced concept of the Alpine–Himalayan ranges. He considered them to be the product of a continental collision. He pointed out that stratigraphic sequences in the northern Alps belonged to the Eurasian shelf, that those in the southern Alps belonged to the African shelf, and, in between, he recognized oceanic sediments. Figure 9.4 dramatically illustrates how he perceived the collision. A modern drawing of the Alpine geosyncline was made by Laubscher and Bernoulli (1977). In their illustration (Figure 9.5) we can clearly perceive the rifted edges of Europe and Africa and the opening Tethys in the middle.

Geologists now can recognize the deep-water sequences and their associated ophiolites caught in narrow suture zones that separate the colliding lithospheric plates. Such suture zones are often characterized by

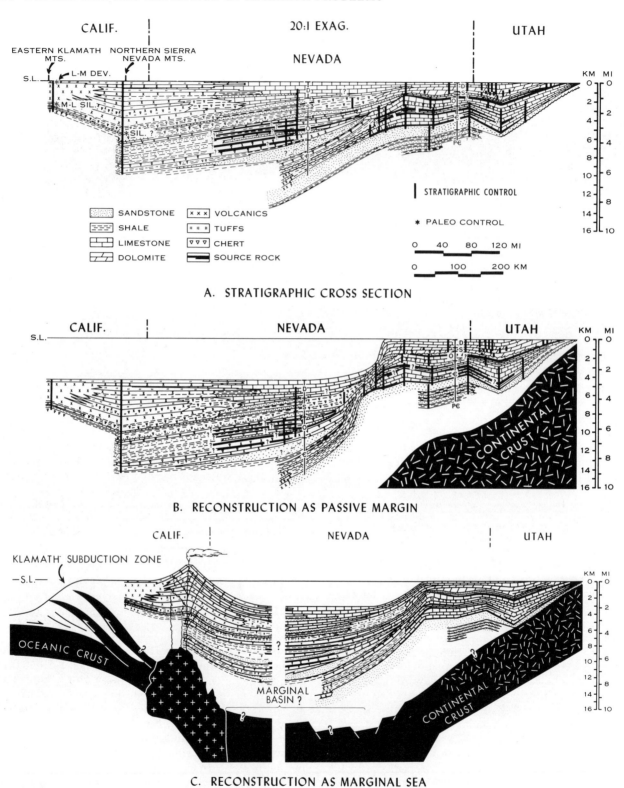

FIGURE 9.6 Great Basin (three sections). A, Stratigraphic cross section; B, reconstruction as passive margin; C, reconstruction as marginal sea.

high-pressure, low-temperature mineral assemblages (blueschists), pelagic sediments, ultrabasics, associated gabbros, and pillow lavas. These rocks appear to be mixed into an incoherent, chaotic mass—the melange. Fine exposures of melange sequences can be seen in California, Washington, Oregon, and Alaska.

Dating radiolaria and other microfossils in cherts and pelagic sediments that overlie the ophiolites can establish the time of ophiolite formation and helps to bracket the time that these units were structurally emplaced; that is, the time of subduction.

BORDERLANDS

Geologists working in mountain belts often try to explain the presence of volcanic or plutonic rocks and the presence of extracratonic clastic sources by postulating ancient volcanic island arcs or remnant magmatic arcs that rifted. Microcontinental blocks involving rifted continental margins that drifted offshore with marginal ocean basins forming behind them also form important borderland terranes. Identification of these various classes of borderlands is essential in order to understand the tectonic history of many formerly active continental margins that now are preserved *in* continents (Churkin and Eberlein, 1977). Today, only Alaska, California, and the northeast margin of Canada have offshore borderlands.

In the geological record, however, offshore borderlands appear to have been active along the margins of the primitive North American continent, including the Arctic. The western Pacific margin appears to be a good current analog for much of the Paleozoic history of the Cordilleran fold belt, the Appalachian fold belt, and for the Innuitian fold belt along the Arctic margin.

Some of these ancient borderlands may be completely foreign plates or blocks that collided with the North American continent. Some scientists hypothesize that they came from Asia, from parts of Europe, and so forth. Other borderlands presumably were emplaced by strike-slip movement along an ancient continental margin.

Studies of modern-day borderlands that contribute sediment to basins adjoining a continental margin reveal criteria for identifying modes of sediment transport and directions to sediment sources. Similarly, volcanic eruptions distribute ash and other forms of volcanic detritus by either wind or ocean currents. The distance that variously sized materials can be transported from volcanic arcs and the mode of pyroclastic accumulation bear on the search for the location of volcanic centers within or adjacent to old arc terranes.

Borderlands and volcanic arcs have long been thought to be stepping-stones or centers of evolution for shallow-water faunas that apparently move from island to island, thus distributing themselves over wide parts of major ocean basins. The role of islands and borderlands is of great interest to students of paleogeography, in that such entities may have aided faunal migration. Volcanic arcs, and particularly borderlands, may have acted as barriers to fossil distribution, so that shallow-water organisms may have been prevented from moving across deep marginal seas or across major ocean basins.

THE WESTERN CORDILLERA OF THE UNITED STATES

The geology of the Basin-and-Range Province of western Utah and Nevada is highly complex. The stratigraphic record extends from the Precambrian to the Tertiary (Stewart *et al.*, 1977). The structural geologist sees thrust faults, strike-slip faults, and normal faults in addition to folds and large slump structures. Geophysicists recognize an attenuated crust and areas of high heat flow.

The geological record is incomplete, yet we know enough to realize that we are looking at the complex evolution of a continental margin. Figures 9.6(A)–9.6(C) give some interpretations. Figure 9.6(A) shows a simplified, stratigraphic cross section of the lower Paleozoic. These beds were underlain by great thicknesses of Proterozoic sediments, and the section, therefore, shows only a short interval of the area's evolution. Figures 9.6(B) and 9.6(C) show two different arrangements of the same stratigraphic data, suggesting a passive-margin arrangement and an island-arc arrangement. The passive-margin arrangement can be rejected, because it places the thick volcanic sequences of the Sierra Nevada in an uncommon basinal position for which there is no recent analog. Figure 9.6(C) is more satisfactory but poses a problem because the passive margin on the right is known to extend all the way from the northern Yukon to southern Nevada. We also know that this margin subsided from the late Proterozoic until the end of the Pennsylvanian, more than 700 million years. Passive margins do not commonly persist that long; nor are they expected to be that continuous in marginal-basin settings.

Figure 9.7 shows Churkin's (1974a) "time-and-motion" drawings of the area. He suggests that the Great Basin was a site of marginal basins opening and closing behind volcanic arcs during much of Paleozoic time. A long process of sedimentation and deformation followed throughout the Mesozoic, modifying the originally oceanic crust to continental crust. In the Cenozoic, after at least 40 m.y. of quiescence and stable conditions, substantial crustal and upper-mantle changes occurred. These are recorded by elevation of the entire region, crustal extension resulting in basin-and-range faulting, extensive volcanism, high heat flow, and a low-velocity mantle. These phenomena are superimposed on the inherited subcontinental crust that developed from an oceanic origin in Paleozoic time and that possibly retained some of its thin and layered characteristics. The present anomalous crust in the Great Basin represents an accretion of oceanic geosynclinal material to a Precambrian continental nucleus. This accretion apparently was an intermediate step in the process of converting oceanic crust to a stable continental landmass or craton.

These interpretations are speculative. To place them on

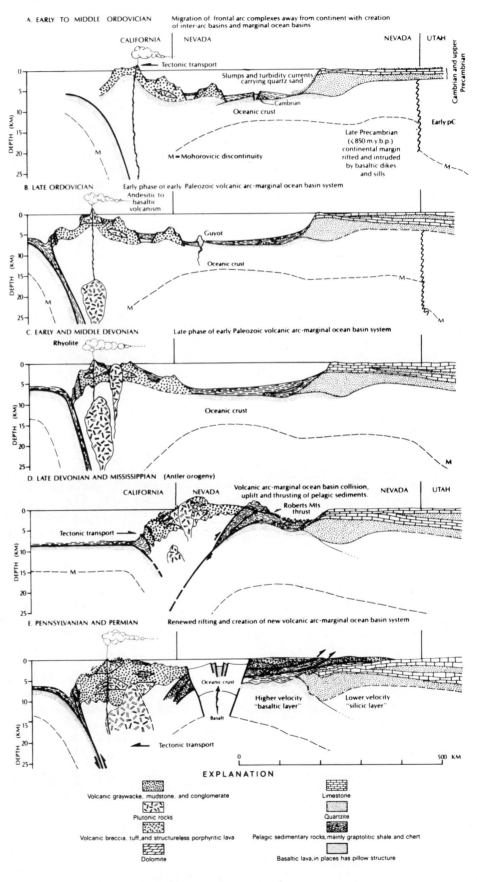

FIGURE 9.7 Preliminary reconstruction of the major stages in the Paleozoic development of the western margin of North America. [From Churkin (1974a). Reprinted by permission of the Society of Economic Paleontologists and Mineralogists.]

firmer ground, we must establish criteria to determine the water depth and physiographic setting in which a sedimentary rock was deposited. To accomplish this, fossil assemblages must be interpreted in terms of the site of burial.

UNCONFORMITIES

Perhaps one of the most unsettling discoveries marine geologists have made is the widespread occurrence of continental slope and deepwater unconformities. In the past, stratigraphers commonly believed that unconformities represented emersion from the seas and subaerial exposure. Some studied the paleogeomorphology of such unconformities in considerable detail. Much of their work is still valid. But unconformities occurring in and under the oceans add an entirely new dimension to traditional stratigraphy.

Most slopes are unstable. Sediments are not easily preserved there. The inherent instability of slopes is further accentuated by the presence of deep geostrophic (contour) currents. We know very little about sedimentation (or nonsedimentation) on continental slopes. We need to identify fossil slopes in mountain ranges, but it is difficult to recognize and reconstruct them.

Seismic stratigraphy is making a welcome impact on the solution of these problems. Modern multichannel seismic-reflection data are providing exciting stratigraphic details on modern and ancient slopes in sedimentary basins. To put outcropping rock sequences in the correct perspective, seismic observations must be merged with lithologic and paleontologic studies done by surface geologists.

DEEP-SEA TURBIDITES

Deep-sea turbidites (sediments and rocks deposited from density currents) have been identified by deep-sea drilling and dredging activities. It is important to trace these turbidites back to their source. This leads to the study of sediment dispersal patterns and their relation to submarine canyons and deep-sea fan deposits (Figure 3.6). Criteria that allow us to identify coarse-grained turbidites as channel deposits in deep-sea fans or as nearshore deposits can also help us to identify similar ancient rocks. Sedimentary structures and fossils, such as bedding-plane features, sizes of clasts, orientations of clasts, and infaunas (organisms that live in sediments rather than on them) can be used to differentiate between a coarse-grained, conglomeratic, channel deposit and a nearshore conglomerate that is interlayered with deeper marine sediments.

Continental-rise deposits that accumulated at the base of an ancient continental slope are coalesced units of detritus that traveled down the slope through submarine canyons. Criteria for separating channel deposits from intrachannel deposits are needed. Efficient physical depth markers are also needed to calibrate facies in mountain belts. Paleoecological studies may be valuable in deter-

mining depth indicators and give clues to the character and pattern of water circulation.

Allochthonous debris-flow deposits involving carbonate detritus from reef or bank margins are known. Criteria by which to recognize allochthonous breccias and conglomerates have been developed. Much less is known, however, of noncarbonate detritus that has no associated, obviously displaced faunas.

We are at a stage where the transport and deposition of coarse-grained submarine debris flows need to be studied in modern sedimentary environments, in flume and other laboratory studies, and compared with ancient deposits. We need to know how the *processes* affect the geometric pattern of the deposit in turbidity flows and sediment fluidization. Examples of this type of study are the work of Ingersoll (1976) and papers in Stanley and Kelling (1978).

Good places to conduct field studies of ancient examples are the following:

Upper Paleozoic of the Innuitian fold belt in Arctic Canada
Lower Paleozoic of Yukon Territory
Middle Paleozoic of Alaska
Middle Paleozoic of Basin-and-Range Province
Cretaceous of Mexico and southern California
Tertiary of California and Oregon

STUDY THE OLDEST ROCKS IN OUTCROPPING MIOGEOSYNCLINAL SEQUENCES

"Miogeosyncline" was the name for the realm in which thick sequences of nonvolcanic rocks were deposited. Today, this is the "ancient passive margin." Although many geologists claim that such passive margins outcrop in mountain ranges, it is not certain that the outcropping sequence was, in fact, part of a continental margin. Furthermore, the outcrops are often discontinuous, badly deformed, and not easily compared with a good multichannel seismic line.

Nevertheless, it would be useful to take a concentrated look at the deepest portion of geosynclines that may be ancient passive margins. Fieldwork would help to distinguish the rifting from the drifting phase. One mountain range already has been studied in this manner—the Alps (see Figure 9.5). Others that might be investigated include the lower Proterozoics of southeast British Columbia. They contain widespread turbidite sequences that are riddled with sills of basic intrusive rocks, and they appear to be on an early continental rise. This may be an example of an outcropping ocean–continent boundary.

REFERENCES AND BIBLIOGRAPHY

Angenheister, G., H. Bögel, H. Gebrande, P. Giese, R. Schmidt-Thome, and W. Zeil (1972). Recent investigations of surficial and deeper crustal structure of the Eastern and Southern Alps, *Geol. Rundsch. 61*:349–395.

Argand, E. (1924). La tectonique del'Asie, Compt. Rend. IIIe Cong. Geol. Liège, Imprimerie Vaillant-Carmanne.

Burke, K., J. R. Dewey, and W. S. F. Kidd (1977). World distribution of sutures; the sites of former oceans, *Tectonophysics* 40(1/2):69–99.

Churkin, M., Jr. (1974a). Paleozoic marginal ocean basin—volcanic arc systems in the Cordilleran foldbelt, in *Modern and Ancient Geosynclinal Sedimentation*, R. H. Dott, Jr., and R. H. Shaver, eds., *Soc. Econ. Paleontol. Mineral. Spec. Pub. 19*, May.

Churkin, M., Jr. (1974b). Deep-sea drilling for landlubber geologists; the Southwest Pacific, and accordion plate tectonics analog for the Cordilleran geosyncline, *Geology* 2(7): 339–342.

Churkin, M., Jr., and G. D. Eberlein (1977). Ancient borderland terranes of the North American Cordillera: Correlation and microplate tectonics, *Geol. Soc. Am. Bull.* 88:769–786.

Dewey, J. F. (1977). Suture zone complexities: A review, *Tectonophysics* 40:53–67.

Dickinson, W. R., ed. (1974). Tectonics and sedimentation, *Soc. Econ. Paleontol. Mineral. Spec. Pub. 22*, 204 pp.

Drake, C. L., M. Ewing, and G. H. Sutton (1959). Continental margins and geosynclines; The East Coast of North America north of Cape Hatteras, in *Physics and Chemistry of the Earth*, Vol. 3, L. H. Ahrens, F. Press, K. Rankama, and S. K. Runcorn, eds., Pergamon Press, Elmsford, N.Y., pp. 110–198.

Giese, P., C. Prodehl, and A. Stein (1976). *Explosion Seismology in Central Europe*, Springer-Verlag, Berlin, 429 pp.

Ingersoll, R. (1976). Evolution of the Late Cretaceous fore-arc basin of northern and central California, Stanford U., PhD thesis, unpublished.

Laubscher, H., and D. Bernoulli (1977). Mediterranean and Tethys, in *The Ocean Basins and Margins IV: Mediterranean*, A. E. M. Nairn, F. G. Stehli, and W. Kanes, eds., Plenum, New York, pp. 1–28.

Matsumoto, T. (1977). Timing of geological events in the circum-Pacific region, *Can. J. Earth Sci. 14*:551–561.

Scheibner, E. (1976). Explanatory notes on the tectonic map of New South Wales, Department of Mines, Geol. Surv. of New South Wales, Australia, 283 pp.

Shoemaker, E. M. (1975). Continental drilling: Report of the workshop on continental drilling, Ghost Ranch, Abiquiu, New Mexico, Carnegie Institution of Washington, Washington, D.C., 56 pp.

Sillitoe, R. H. (1977). Metallogeny of an Andean-type continental margin in South Korea: Implications for opening of the Japan Sea, in *Island Arcs, Deep Sea Trenches and Back-Arc Basins*, M. Talwani and W. C. Pitman III, eds. Maurice Ewing Series 1, Am. Geophys. Union, Washington, D.C., pp. 303–310.

Stanley, D. J., and G. Kelling, eds. (1978). *Sedimentation in Submarine Canyons, Fans, and Trenches*, Dowden, Hutchison and Ross, Inc., 416 pp.

Stewart, J. H., C. H. Stevens, and A. E. Fritsche (1977). *Paleozoic Paleogeography of the Western United States*, Pacific Section of Soc. of Econ. Paleontol. and Mineral., Los Angeles, Calif., 502 pp.

10 | Seismicity and the Deep Structure of Continental Margins

A. INTRODUCTION

Most of the earth's seismic activity is concentrated along the margins of continents (Figure 10.1). Destructive or potentially destructive earthquakes occurring at shallow depths within these margins account for well over three fourths of the earth's total seismic-energy production. Earthquakes are evidence of the continuing violent processes that have shaped continents over eons of geological history.

The theory of plate tectonics has provided an explanation for why so many of the great earthquakes are concentrated in narrow zones: most continental margins are, or have been at some time in the past, the boundaries between major lithospheric plates engaged in geologically rapid motion relative to each other. These motions give rise to the stresses that cause large earthquakes and that continually modify the face of our planet.

To fully understand the dynamics of destructive earthquakes and the fundamental mechanisms that cause them, we need to increase our knowledge about the structures and dynamics of continental margins. An adequate understanding of these problems will fail to emerge if we confine our attention to the upper levels of the crust; our studies must extend deeply into the earth's mantle.

Besides being directly relevant to dynamical questions, the information gleaned from studies of the deep structure of margins is critical to any theory of the evolution of continents. The nature of the subsurface transition from the oceanic to the relatively stable continental environment is not at all well understood. Geophysical data indicate that profound differences between oceanic and continental crusts extend well into the upper mantle. At these depths, the transition zone includes not only today's continental margins but nearly all tectonically active regions of the continents—which generally are regions that were continental margins at some time in the relatively recent geological past.

B. SEISMICITY OF THE MARGINS

Global networks of seismometers are capable of locating most potentially destructive earthquakes (magnitude > 4.5 on the Richter scale) with a precision of about ⊥ 25 km. However, much can be learned about the tectonics of seismically active regions by monitoring smaller events and locating them with increased precision. That kind of work usually requires local or regional networks of seismometers that are separated by tens to several hundreds of kilometers. Accurate location also requires good azimuthal distribution of stations around the source region. For continental-margin seismicity, this coverage cannot be achieved with land-based stations alone. They must be augmented by seismometers properly sited in the marine environment. Recent advances in the technology

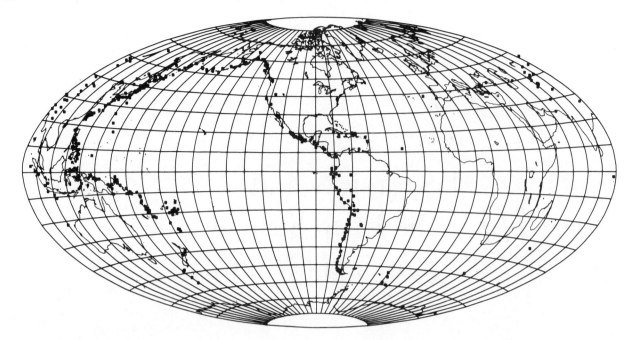

FIGURE 10.1 The distribution of earthquakes of magnitude 7.5 or greater, 1897–1974. (Courtesy of the National Oceanographic and Atmospheric Administration.)

of ocean-bottom seismometers (OBS) at a number of universities and oceanographic institutions now permit such seismic investigations to be done routinely.

Ocean-bottom seismometers are miniaturized, self-contained seismic stations capable of withstanding the harsh environment of the deep ocean. In some OBS designs, incoming signals activate the recording device. Magnetic tape or digital memory-buffers are used to obtain complete recording of the event and some sample of the preceding noise activity. This mode permits the long recording times (1 month or more) necessary to sample the microseismic activity adequately. In more advanced OBS packages, data are logged in digital format. Such logging increases the dynamic range and greatly facilitates data analysis. At the end of the recording period, the OBS's are recalled to the surface by acoustic command or released from the bottom at a preset time. Figure 10.2 illustrates some microearthquake data collected by an array of OBS's on the East Pacific Rise near the mouth of the Gulf of California.

OBS technology is quite new and is being used to study continental-margin seismicity and structure. More instrumental development is needed, especially to increase the frequency range (bandwidth) and low-frequency sensitivity, if full use is to be made of OBS capabilities to study shear and surface waves from seismic events of larger magnitudes.

Using ocean-bottom seismometers in conjunction with land-based seismic arrays will considerably increase our knowledge about both the major and minor seismicity of continental margins. Specific questions that should be addressed include:

1. What are the spatial and temporal relations between minor seismicity and the occurrence of large earthquakes?

2. Is minor seismicity of active margins concentrated along major faults or distributed within large volumes of the lithosphere?

3. What is the nature of the seismic gap occurring between trenches and island arcs in some subduction zones?

4. What is the minor seismicity of passive margins? What features along passive margins are seismically active?

C. TECTONIC STRESSES ALONG THE MARGINS

At present, little is known about the state of stress within the earth, despite the obvious importance of stress in earthquake processes. Especially little is known about absolute stress magnitudes. For example, there is no consensus among geologists or geophysicists about the average absolute-stress fields prior to seismic faulting; estimates vary by more than an order of magnitude. On the one hand, for earthquakes, seismologists routinely measure stress drops (i.e., the difference between initial and final stress) that are generally less than 100 bars. Stress drops have been used as measures of absolute stress. On the other hand, experiments in rock mechanics suggest that, at lithostatic pressures appropriate for the middle to lower crust—about 3–6 kbar—absolute stresses of several kilobars are necessary for fracture of crustal rocks. Until this question is resolved, it will be impossible to

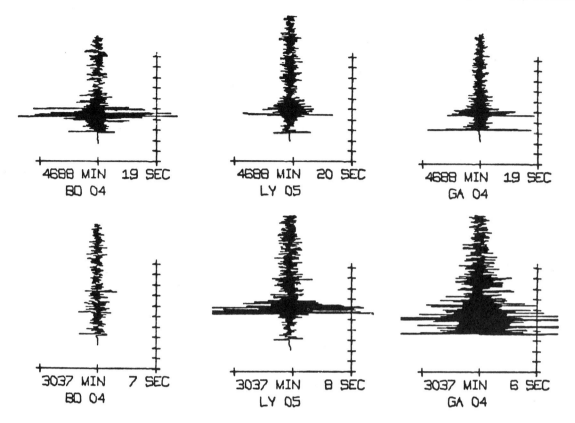

FIGURE 10.2 Fracture zone microearthquakes recorded by ocean-bottom seismometers near the mouth of the Gulf of California. (Courtesy of T. H. Jordan, Scripps Institution of Oceanography.)

understand adequately earthquake dynamics or theories about the mechanism that drives the plates.

Studies of stress fields along active margins should be a cornerstone for any research program on earthquake dynamics. Estimates have been made on convergent active margins from seismic measurements (i.e., stress drop and stress-tensor orientation), but the techniques need further development. In particular, computational schemes for properly handling surface tractions during the rupture process (important for large events) need to be devised.

For stress-field studies, data collected by global networks of digitally recording, broad-bandwidth seismometers will be critical. At present, three such networks are deployed or being deployed (Figure 10.3):

1. High-Gain Long-Period (HGLP) Network;
2. Seismic Research Observatory (SRO) Network; and
3. International Deployment of Accelerometers (IDA) Network.

In addition, the U.S. Geological Survey is proposing to upgrade 15 to 20 World-Wide Standard Network Seismograph (WWSSN) stations to seven-channel, digital-recording capability. These networks will provide the high-quality data base necessary to study continental-margin earthquakes, and they deserve strong federal support.*

For the study of small earthquakes, e.g., the aftershock sequences of large events, local networks, including ocean-bottom seismometers, are necessary.

Ocean-bottom topography outboard from the trenches is influenced by the stress regime of convergent active margins. This topography places independent constraints on the model of stress regime of those margins. Several scientists, using elastic plate models, have estimated that the stresses within the descending slab necessary to support the topographic elevation of the so-called "outer rise" are of the order of several kilobars. These are values similar to those of the strength of rocks studied under laboratory conditions, but the latter values are an order of magnitude (or more) greater than stress-drop estimates derived from seismology. Such values, however, are sensitive to assumptions based on the elastic-plate model, and further studies employing nonlinear plastic rheologies should be attempted.

*For further information on these networks, consult Committee on Seismology (1977) and Panel on Seismograph Networks, Committee on Seismology (1977).

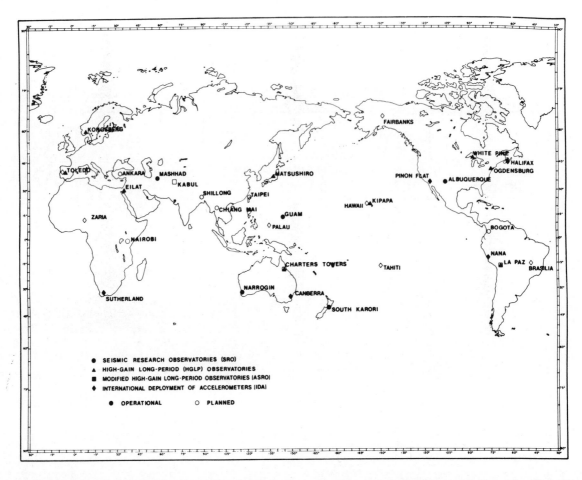

FIGURE 10.3   The distribution of digitally recording seismic stations. (Panel on Seismograph Networks, Committee on Seismology, 1977.)

To model earthquake dynamics, it is necessary to know not only the ambient stress field but also the nature of temporal changes in the stress and strain fields of seismically active areas, especially on a time scale of one to several tens of years. Measurements of these changes require (a) precise geodetic techniques that can sense very small vertical and horizontal displacements and (b) the development of new stress-monitoring devices. Recently developed satellite-ranging and lunar-laser-ranging methods and the techniques of long-baseline interferometry are promising. Temporal changes also can be monitored by ultrastable strainmeters and gravimeters based on laser and cryogenic technologies. In studying U.S. margins, more program coordination among the responsible federal agencies should be achieved.

The direct measurement of stresses at shallow depths by downhole hydraulic fracturing has improved our knowledge of continental stress fields. This technique is fairly expensive, even on land, and it is still imperfect. Nevertheless, the feasibility of collecting such measurements should be examined.

Information on stress and strain fields studies is needed to answer the following specific questions:

1. How far into the continental–island-arc region does the deformation associated with oceanic-plate subduction extend?
2. Which portions of the subduction zone are under relative tension, and which are in compression?
3. Do the different arcuate structures of subduction zones correspond with different conditions of stress within the volcanic arc?
4. What determines the equilibrium position of the overriding plates with respect to the surface of the underthrust lithosphere?
5. How are the orogenic forces in convergence regions transmitted from the subduction zone to the mountains on the overriding plates?
6. How important, deep, and extensive are the high-friction regions that are presumed to heat up the upper surface of the subducted slab to produce magmas?
7. What is the stress field in marginal basins?

8. What are the strengths of rocks in continental-margin fault zones, and how are these strengths affected by fluid content?

D. CONTINENT–OCEAN TRANSITION

The continental margins are regions of transition between two very different crustal types—the thin, simatic crust of the oceans and the thick, sialic crust of the continents. The difference between these two regions is not confined to the crust: profound variations extend to an as-yet undetermined depth into the mantle. Furthermore, the margins are regarded as the sites of continental accretion. Continents are the most spectacular surface manifestation of the long-term chemical evolution of the earth.

The nature of the continent–ocean transition is a fundamental problem confronting modern earth science. Probing this transition zone is not an easy task. By its very nature, the continental margin is laterally heterogeneous on many scales. The techniques for mapping this heterogeneity must have high lateral and vertical resolution and must be capable of great depth penetration. No one geophysical technique can provide this resolution; a combination of techniques is required.

The highest-resolution methods are seismic surveys that use compressional waves and artificial sources. These can be broadly classified into two types—reflection and refraction.

Reflection profiling with artificial sources has been extensively developed in the search for petroleum, and its capabilities of using advanced multichannel recording systems are detailed elsewhere in this report (see pp. 151–153). High resolution and increased signal-to-noise ratio are achieved by stacking procedures that take full advantage of data redundancy. The multichannel seismic-reflection method seems to be ideally suited to study the upper 10 km of the crust. By two-ship, constant-offset profiling, recording can be extended to even greater depths, perhaps allowing the entire crustal column and even the uppermost mantle to be sampled. Further development of the reflection-profiling method deserves high priority.

The seismic-refraction technique involves a greater separation between the source and receiver. This method has the advantage that very deep penetration (200 km or more) can be achieved with artificial sources. The development of reliable ocean-bottom seismometers has reawakened the interest of the scientific community in the use of refraction techniques with which to study continental-margin structure.

The geometry of refraction profiling with OBS's is illustrated in Figure 10.4. A ship firing explosive charges steams away from the OBS, and the seismic energy reflected at depth is recorded for later recovery. To conserve magnetic tape, the shots are fired and recorded at prearranged intervals. Recording in digital format is preferred, since it facilitates subsequent data processing. To maximize the usefulness of the information so obtained, sophisticated analyses are used to interpret the data. The most successful of these techniques has been the time-domain modeling of the arrivals by the computation of synthetic seismograms. The use of amplitude and waveform information greatly enhances the resolution of the refraction methods.

Further information can be obtained by shooting to an array of seismometers instead of to a single sensor. Frequency-wavenumber processing can then be employed to extract the horizontal phase velocities of various arrivals. This kind of processing further constrains the class of acceptable velocity models, especially in regions characterized by large vertical velocity gradients. The use of OBS arrays can also provide constraints on the lateral structure beneath the array. An example of the application of frequency–wavenumber analysis to data from an OBS array is illustrated in Figure 10.5.

The advantages and disadvantages of reflection and refraction profiling are complementary. The reflection technique has high spatial resolution but limited depth of

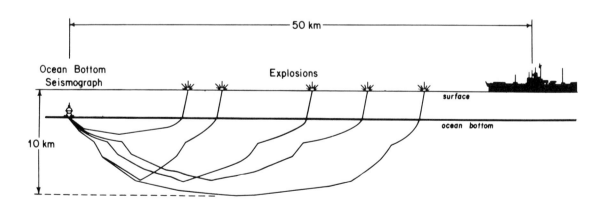

FIGURE 10.4 The geometry generally used in refraction profiling with ocean-bottom seismometers. The separation between sources and receivers can be extended to 1500 km, allowing penetration to depths exceeding 200 km. (Courtesy of T. H. Jordan, Scripps Institution of Oceanography.)

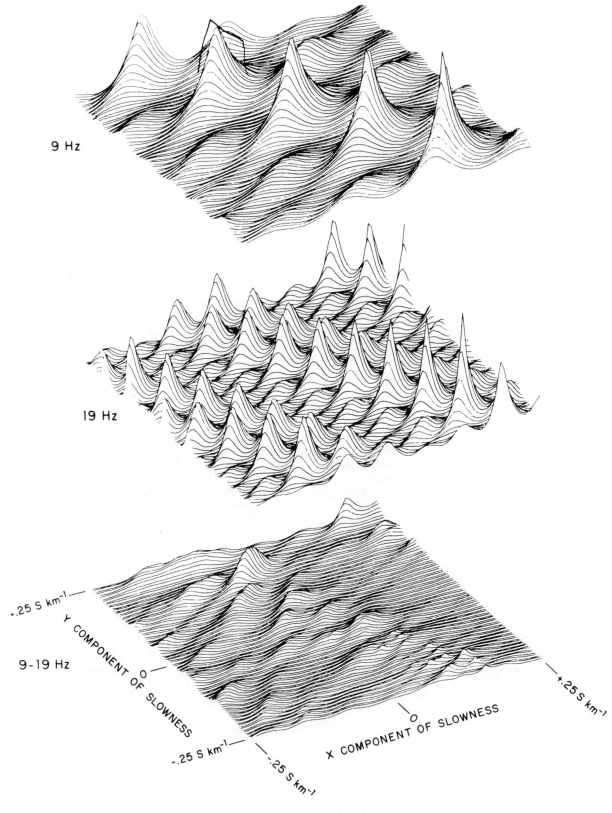

FIGURE 10.5 Examples of spatial filters applied to digitally recorded data from an ocean-bottom seismometer array used to isolate the direction of an incoming signal. (Courtesy of L. Dorman, Scripps Institution of Oceanography.)

penetration and limited ability to resolve compressional velocities. The refraction technique can achieve deep penetration and good vertical velocity resolution but involves extensive horizontal averaging of structure.

Neither of these techniques has yet been very useful for shear-velocity studies. Shear velocities are critical parameters for determining the composition and state of crustal and mantle rocks, as shear velocities are particularly sensitive to the temperature and fluid content of materials. Artificial sources of seismic energy are inefficient in exciting shear waves, especially in the marine environment, and methods that rely on artificial sources have so far provided very little shear-velocity information. There is some hope that the study of converted phases (i.e., *P*-energy converted to *S*-energy at sharp interfaces) can be used to derive shear-velocity profiles. Converted phases have been identified on both reflection and refraction profiles of oceanic crust. Because of the importance of shear-wave information to continental-margin studies, efforts to develop means of obtaining data on shear waves should be encouraged.

However, the precise delineation of shear-velocity structure will undoubtedly require the use of nature's own seismic sources—earthquakes. Shear-wave travel times and surface-wave velocities are particularly useful in modeling shear-velocity structures. The development of broader-band OBS packages would permit these waves

to be recorded directly on the margins of the continents and should be given high priority.

For studies of the continent–ocean transition, information from global seismic networks is again critical. Seismic surface waves and free oscillations are sensitive to lateral variations at great depths, and advanced theoretical techniques for modeling these variations on a global scale are rapidly being developed. Body waves other than first-arriving *P*- and *S*-waves are proving to be quite useful in describing these deep lateral variations. For example, Figure 10.6 illustrates the large-scale variations in shear-velocity associated with the passive continental margin bordering northern Siberia as expressed in the travel times of shear waves reflected in multiples from the core–mantle interface (multiple *ScS* waves). These data were obtained from the WWSSN. The data emerging from the more advanced digitally recording networks will allow signals such as these to be processed in a routine fashion and will considerably expand the data base.

Structural studies of the continental margins should be directed toward the following questions:

1. What are the properties and the lateral variations of the major discontinuities (sediment–rock interface, Mohorovičić discontinuity, upper-mantle discontinuities)? What are the variations in physical properties of the lithologic units between these discontinuities?

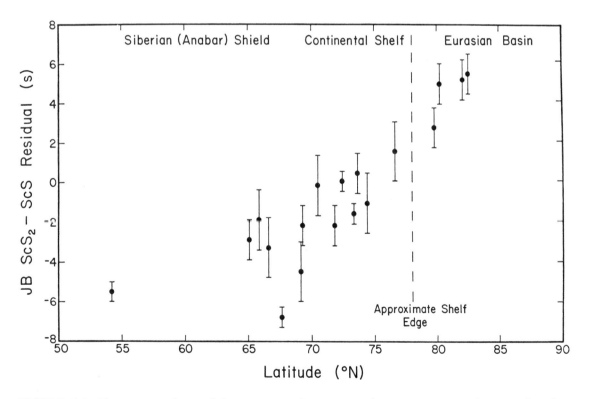

FIGURE 10.6 The variation of vertical shear-wave travel times across the passive continental margin of Northern Siberia. The large magnitude of the residuals on the monotonic trend observed in this figure is indicative of a broad continent–ocean transition zone extending to great depth. (Courtesy of S. A. Sipkin and T. H. Jordan, Scripps Institution of Oceanography.)

2. What is the configuration of the Gutenberg low-velocity zone beneath the margins? Do other low-velocity zones exist within the crust and upper mantle? If so, what are the velocities within these zones, and what is their lateral variation?

3. Are the seismic velocities in these regions anisotropic? If so, what is the orientation and magnitude of the anisotropy? What is the relation between anisotropy and ambient tectonic stress?

4. What is the attenuation structure of the margins?

5. To what depth into the mantle do the lateral variations associated with the continent–ocean transition extend?

E. THE FATE OF SUBDUCTED LITHOSPHERE

Beneath convergent active margins, oceanic lithosphere is thrust deeply into the mantle beneath the continents. Any theory of crustal and mantle dynamics must explain the eventual fate of this material. At present, the data are not sufficient for definitive answers to the following questions:

1. Does the lithospheric slab retain its identity below the Benioff Zone? If so, how deeply into the mantle does it extend before it reaches thermal equilibrium?

2. What are the dynamics of phase changes within the slab? Are any upper-mantle discontinuities elevated or depressed within the slab?

3. What mechanical factors determine the angle of subduction? What processes are responsible for lateral and vertical changes in this angle?

4. What is the mechanism of deep-focus earthquakes? What factors determine the magnitude of seismic-energy release in the deep-focus zones?

5. What are the variations of rheological properties across the slab–mantle interface? To what extent is frictional heating important in determining the thermal regime of the slab?

6. What is the fate of volatiles and sediments subducted with the slab?

7. What is the relation between Benioff Zone seismicity and volcanic activity?

Seismology provides powerful tools for answering these questions. Teleseismic studies of Benioff Zone earthquakes can yield information about the nature of these earthquakes and the physical properties of the slab. Some information about deep-focus events has been interpreted as indicating that mechanisms involving changes of volume (phase changes) may be responsible for initiating these earthquakes. Other studies of deep-focus events have shown that the high seismic velocities associated with the lower-temperature slab extend below the earthquake zone, perhaps far into the lower mantle. If these conclusions are substantiated, then we will have direct evidence that the lower mantle is participating in a convection process that drives the plates, a critical constraint on geodynamics models.

Structural studies of the lithospheric slab and studies of Benioff Zone seismicity will make significant progress with the help of ocean-bottom seismometers. With these instruments, signals from deep-focus earthquakes can be recorded seaward of trenches, providing information complementary to that obtained with land-based networks.

F. RECOMMENDATIONS

Effective monitoring of continental-margin seismicity, studies of earthquake dynamics, and investigations of the deep structure of continental margins depend critically on the collection of high-quality, broad-band seismic data from both global and local networks of instruments. This Panel strongly supports the recommendations contained in the recent National Research Council report by the Panel on Seismographic Networks of the Committee on Seismology, and specifically, with regard to the global networks:

Recommendation 1. Stable funding . . . should be established to assure . . . continuing operation of the WWSSN as a basic research facility for U.S. investigators.

Recommendation 2. Stable funding should be established . . . to continue operation, maintenance, and improvement of the [digitally recording] HGLP; SRO; ASRO; and IDA seismograph stations. Further, [a number of] WWSSN stations should be upgraded to include digital recording capability. . . . Facilities for the organization, storage, retrieval, and distribution of digital data from the above observatories [should be provided].

The study of the structure and seismicity of the continental margins on both local and regional scales requires the use of ocean-bottom seismometers. Therefore:

Recommendation 3. A strong national program in ocean-bottom seismology needs to be established and maintained. Elements of this program should include sufficient funds for instrument development.

REFERENCES AND BIBLIOGRAPHY

Bullen, K. E. (1963). *An Introduction to the Theory of Seismology*, 3rd ed., Cambridge U. Press, Cambridge, England, 381 pp.

Committee on Seismology (1977). *Trends and Opportunities in Seismology*, a report based on a workshop held at Pacific Grove, Calif., Jan. 3–9, 1976, Assembly of Mathematical and Physical Sciences, National Research Council, National Academy of Sciences, Washington, D.C., 158 pp.

Gutenberg, B., and C. F. Richter (1954). *Seismicity of the Earth (and Associated Phenomena)*, Princeton U. Press, Princeton, N.J.

Panel on Seismograph Networks, Committee on Seismology (1977). *Global Earthquake Monitoring: Its Uses, Potentials, and Support Requirements*, Assembly of Mathematical and Physical Sciences, National Research Council, National Academy of Sciences, Washington, D.C.

Richter, C. F. (1958). *Elementary Seismology*, W. H. Freeman, San Francisco, Calif.

II | RESEARCH TOOLS AND METHODS

11 | Geophysical Methods

A. INDUSTRIAL AND ACADEMIC GEOPHYSICS: STATE OF THE ART

INTRODUCTION AND SYNTHESIS

Traditionally, academic geophysicists have developed methods for measuring and interpreting significant physical properties of the subsurface and improved the resolution of current methods. The domain of their investigation extends across continental margins to the deep oceans and through the crustal layers. The recent development of ocean-bottom seismometers (OBS) and the measurements of heat flow and electromagnetic properties result from this motivation.

The exploration geophysicist concentrates mainly on sediments and predictions to optimize exploratory drilling. Increasingly, attention is shifting from the definition of structures to the more demanding task of predicting lithology and subtle variations of lithology and fluid content. This is accomplished by acquisition and processing methods that improve signal-to-noise ratios over extended frequency ranges (Dobrin, 1976; Telford *et al.*, 1975).

Seismic resolution has been improved to the extent that the subtle variations in physical properties that distin-

guish a gas-filled sandstone reservoir from an adjacent water-filled sand can be recognized under optimum conditions. Because other lithologies have wave velocities and densities similar to those of gas sands, the so-called direct-detection methods are ambiguous, particularly in undrilled basins. This limitation leads geophysicists to attempt to measure additional physical rock properties, such as shear-wave velocities and attenuation.

An independent approach to the resolution of lithology is through the methods of seismic stratigraphy, wherein reflection patterns have been observed to be diagnostic of different depositional situations (see pp. 61–64).

The geological problems of the continental margin will tax the resolution and depth penetration of geophysical methods as they are now used in either academe or industry. Judicious combinations of academic and industrial techniques could be fashioned to solve some of these problems. For example, in seeking crustal resolution with seismic common-depth-point (CDP) shooting, bandwidth, multiplicity, and shot-to-detector distances will need to be changed *vis-à-vis* common industrial applications to the sediments. In another example, the very high resolution of industrial seismic direct-detection methods can be applied to discriminate between a wide range of lithologies. This capability would be extremely important in reconstructing the detailed evolution of sedimentation on both passive and active margins.

Geophysical interpretations involve many variables

For details on geophysical methods, see Appendixes B.1–B.7.

151

with different degrees of uncertainty and have to be done over and over again in order to increase the probability of making correct predictions. Both new facts and new concepts require repeated scrutiny. The academic community pioneered the integration of several kinds of geophysical measurements. Looking ahead, gravity and magnetic modeling on the margins can produce even more powerful results for future applications if it incorporates the potential for geometric resolution and depth penetration that industrial CDP-reflection concepts and techniques have to offer.

Through the Deep Sea Drilling Project (DSDP) the academic community acquired the ability to test hypotheses promptly that otherwise might remain idle speculations. The learning process in the earth sciences has been greatly accelerated because many problems now can be subjected to a relatively simple iteration: model building–testing by drilling–corrective feedbacks. Be-

cause of its power and expense, the drill should be used only to test carefully selected, high-priority hypotheses.

Corollary with drilling should be complete in-hole logging so that the full information potential of a borehole can be thoroughly exploited. Because they link subsurface and surface geophysical methods, logs multiply the significance of both. All factors must be properly assessed in any attempt to develop cost-effective programs for solving geophysical problems of continental margins.

ACQUIRING THE DATA

An industrial marine geophysical vessel (e.g., the R/V *Hollis Hedberg*) and its exploration system described in Appendix C is complex and costly. Its system has five components—seismic, gravity, magnetics, geochemistry, and onboard processing capabilities—of which the geochemistry and onboard processing can be regarded as

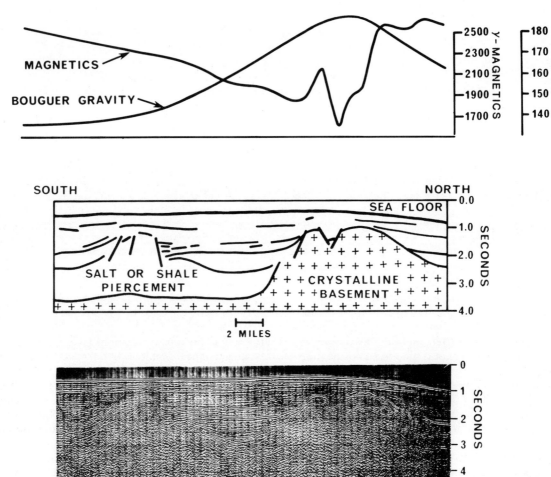

FIGURE 11.1 Integrated multisensor interpretation. On the seismic section (bottom panel), two large intrusive bodies are evident on the left and right sides of the profile, but there is insufficient resolution to differentiate their origins. By adding potential data (top panel), it is seen that the large positive gravity anomaly and sharp magnetic expression indicate the right-side feature to be crystalline basement. From a gravity low and smooth magnetics for the left-side feature, it may be deduced that it is a shale or salt piercement. (Courtesy of E. S. Driver, Gulf Science and Technology Company.)

partially experimental at this stage. Although this ship and its multicomponent system are fashioned to help solve problems in hydrocarbon exploration, the potential of this approach should be carefully considered by both academic and government geologists and geophysicists.

Onboard processing is justified basically by the fact that seismic noise factors on the shelf and slope often render single-channel monitors uninterpretable, and, when surveys fail to realize their exploration objective, extreme waste results. Conventional industrial exploration in remote or hostile environments involves (1) reconnaissance geophysics, (2) detailed geophysics, and (3) drilling in season. Each step in that sequence is separated by about a year's interval. In essence, the explorer is blind with respect to the seismic reconnaissance details until playback at a land-based processing center (usually months after recording). Onboard processing makes it possible to identify and detail interesting anomalies in one survey, accelerating the whole campaign by as much as a year— perhaps more.

Thanks to their participation in the DSDP, academic geophysicists are now also under pressure to guide a very costly drilling program accurately, efficiently, and promptly. Much of their geophysical work will require very deep resolution of shelves and slopes. Seismic noise problems will increase as the reflection signal weakens with depth. Lacking high-quality onboard processing, the academic geophysicist, too, is essentially blind with respect to the detailed resolution of multifold CDP.

Evaluating all the factors—ship time, transit to the survey area, the requirements of the drilling program, and mobilizing scientists—against the objectives of the geophysical survey, a certain degree of onboard processing will no doubt prove to be the most cost-effective way for the academic geophysicist to ensure against inadequate expeditions.

SEISMIC PROCESSING

By adapting digital recording and computer processing to seismic interpretation, geophysicists have produced a major breakthrough in information recovery. Paper B.3 on seismic data processing (see Appendix B) explains the concepts and methods involved. The digital revolution has provided the interpreter with amplitude, attenuation, and precise velocity information in a practical way.

Inversion of a seismic trace to an approximation of a velocity log has long been a goal of the geophysicist. By combining a velocity curve with a reflection-coefficient function, a pseudo-velocity log can be produced. As a practical development, this is an outgrowth of constantly improving accuracy in amplitude preservation, velocity extraction, and noise suppression.

The success of sophisticated processing schemes depends on the quality and bandwidth of the original recordings. No single display of a seismic section can present all of its information content. The processing geophysicist is therefore motivated to maximize the

amount and variety of the seismic information to give to the interpreter in a simple format. Color displays serve this purpose. The color spectrum can be modulated by amplitude, frequency, polarity, or velocity. These results can then be superimposed on the conventional black-and-white variable amplitude or variable density plot. Plotting dipping reflections (migration) and three-dimensional methods will have pertinent applications to unraveling the complex structures of active margins.

INTEGRATED MULTISENSOR INTERPRETATIONS

Within recent years, a major breakthrough has been made in improving the accuracy of shipboard gravity meters. A key factor was gyrostabilization, which made it possible to reduce those errors introduced by wave action and cross-coupling. One-half milligal variations in the earth's gravity field can now be detected in the background noise (caused by the ship's acceleration) of up to 100,000 mgal.

Industry is developing vertical and horizontal magnetic gradiometers. The academic community has developed effective deep-tow magnetometers. Industry's methods of interpreting multisensor data have been developed, in large degree, from work done by the academic community. Figure 11.1 vividly demonstrates the power of integrated interpretation.

LITHOLOGIC RESOLUTION

The section on seismic stratigraphy (pp. 61–64) and two papers in Appendix B (B.5 and B.6) reflect the intense effort directed at resolving lithology from seismic data. (See also Payton, 1977.)

At best, P-wave (compressional-wave) seismic exploration provides only P-wave velocity and density. Because many lithologies and variations of lithology and fluid content have similar P-wave velocities and densities, the interpretation is ambiguous. Measuring seismic shear waves (S-waves) is another way to explore physical properties of rocks.

Seismic stratigraphy is an independent interpretive method. It depends on the ability to recognize unconformities and reflection patterns that are diagnostic of different depositional environments.

LOGGING AND THE RELATIONSHIP OF THE BOREHOLE TO SURFACE GEOPHYSICS

Papers on this subject are included in Appendix B.7 to emphasize the synergistic effects of a link between rocks below the surface and surface geophysics.

Synthetic seismograms (artificial seismic-reflection records manufactured from velocity-log data) express the ideal subsurface response. They permit discrimination between signal and noise on recorded seismograms, make it possible to identify reflection markers, and provide an indication of the bandwidth necessary in recording significant impedance contrasts.

Seismic lithologic studies are most effective when the lithology in question or its variation can be modeled from well logs (records of physical measurements as a function of depth in a borehole).

Logs of shear-wave velocity, density, magnetic properties, and electrical resistivity can provide links to the corresponding surface method in a fashion analogous to synthetic seismograms. All methods of direct detection, whether they be measurements of physical properties or observations of diagnostic patterns, are considerably more effective when they are calibrated by borehole data.

ANTICIPATED FUTURE DEVELOPMENTS

"Advances in Geophysical Methods for the Detection of Hydrocarbons," (Appendix B.1) delivers an interesting chronicle of the evolution of seismic developments from concept through instrumentation to applications. In the immediate future, modifications in recording instruments and methods of processing will be required in order to bias the system in favor of high frequencies. Eventually, computers may perform many of the interpretive tasks that geologists and geophysicists now do.

COMMON-DEPTH-POINT APPLICATIONS TO PROBLEMS OF CONTINENTAL MARGINS

Appendix B.2, pp. 233–237, gives examples of how industrial CDP methods may need to be modified in order to most cogently reveal the deep structure and lithology of continental margins.

CURRENT GEOPHYSICS CAPABILITIES IN ACADEME

Geophysicists in the academic community work with less-advanced equipment than do their industrial colleagues, particularly in terms of navigation apparatus for multichannel seismic-reflection studies and concomitant onboard and postcruise data processing. For navigation, most university ships rely on integrated data from satellite fixes, pit logs, and gyrocompasses. These have provided, by and large, positioning accuracies sufficient for most research projects. University vessels can, of course, be outfitted with additional positioning equipment if the need arises. Industry developed multichannel seismic-reflection equipment and techniques; only in recent years have a few university groups begun to use such equipment. High initial equipment costs, operating costs, and data-processing costs have thrust university marine geophysics behind industry. The University of Texas (at Galveston) and Lamont-Doherty Geological Observatory of Columbia University have 24-channel systems. Woods Hole Oceanographic Institution has a 6-channel system, which is currently being expanded to 12 channels. Scripps Institution of Oceanography of the University of California has a 12-channel system. These systems, although lacking some of the resolving capabilities of the newer industrial systems, are, nevertheless, powerful tools in the hands of the academic community.

The multichannel capability adds a new dimension to studies of the genesis and development of continental margins. It also permits researchers to interpret and present their seismic findings within a framework that is compatible with industrial data. Because the industrial effort is primarily focused on shallower margin waters and university scientists more often work in deeper waters, the ability to relate the data sets to each other is vital to gain a more complete understanding of the whole continental-margin complex.

Other geophysical techniques are employed in essentially similar ways by both academic and industrial groups. Some of these techniques are marine gravity- and magnetic-field measurement and interpretation; single-channel seismic-reflection profiling; and the use of expendable radio sonobuoys for wide-angle reflection and refraction measurements. These methods were developed largely within the academic community, with substantial help from military R&D efforts. Such techniques will continue to be valuable elements in continental-margin studies.

During the past few years, several academic groups have developed and used ocean-bottom instrumentation, which, while primarily geared for research in seismology, can also be used for electromagnetic studies. Ocean-bottom seismometers and ocean-bottom hydrophones (OBH) are now available in substantial numbers. These instruments can be deployed on the seafloor in various arrays, and they are particularly well suited for seismic-refraction measurements in areas as structurally complex as continental margins. These instruments also provide a means for detailed seismicity studies that can provide important insights into the tectonic development of margins.

An expanding spread-reflection technique that uses multichannel equipment and that is available to both industrial and academic groups has considerable potential for continental-margin studies. In this technique, one ship deploys a multichannel receiving array and a second ship acts as source ship. The two ships steam in opposite directions away from a common position and record a CDP profile with large moveout dimensions. The profile can be expanded into a refraction profile in which the refraction paths are centered on the CDP. Two important features of this technique are (1) the reflection and refraction profiles, having been centered on a common point, offer strong mutual support in analysis, and (2) the large range of moveout angles in the reflection profile may permit good observations of reflected transformed shear waves, thus presenting the possibility of determining a v_p/v_s ratio as a function of depth in the section, which ratio could be a most useful diagnostic parameter.

ELECTROMAGNETIC INDUCTION STUDIES

Electromagnetic (EM) induction studies can determine the electrical conductivity structure beneath continental margins. This structure can be related to sedimentary basins in the upper crust or to temperature and/or compositional structure in the mantle. Because of the probable complications of interpreting two- (or even three-) dimensional structures with limited array deployments, the continental margins have been purposely avoided in the past few years. The "noise" due to ocean currents on the oceanic electrical field is also probably significantly larger in shallow water.

Nevertheless, it should be noted that the first successful oceanic EM work (Cox *et al.*, 1971) was the investigation of the deep structure at a continent–ocean boundary. Similar work on the Atlantic margin (Cochrane and Hyndman, 1974) led to the inference that substantial conductivity anomalies exist in the continental crust, and the petroleum companies are becoming more interested in higher-frequency studies to determine the character of sedimentary basins. As more reliable and less expensive instruments are developed, the application of these techniques for continental-margin studies should be enhanced, probably within the next decade.

HEAT-FLOW MEASUREMENTS ON CONTINENTAL MARGINS

Determination of accurate values of geothermal flux on and across continental margins would satisfy several important scientific objectives, depending on the type of margin. Measurements on active margins and relatively young (≤ 60 m.y.) passive margins would serve to elucidate the transient tectonic processes that formed them. On older passive margins, such measurements would help to answer questions about the vertical distribution of heat sources beneath continents and oceans, using the "edge effect," or horizontal discontinuity, of heat sources at such margins.

Unfortunately, accurate heat-flow measurements are very few, or lacking, on continental margins, particularly passive margins. A significant obstacle to making heat-flow measurements in these areas has been the inability, using standard techniques, to sufficiently penetrate temperature probes into the rather firm sediments of the shallow areas. Overcoming the penetration obstacle might be accomplished by vibracoring or, if that fails, by shallow drilling. With that provision, scientists would be able to measure gradients and heat flux.

The relatively shallow water on shelves precludes standard oceanic techniques; commercial borings are seldom measured accurately enough. On continental rises, rapidly deposited and thick sediments may significantly depress the surface geothermal flux from its equilibrium value. These perturbations can be reduced or eliminated analytically with information on sediment thicknesses and deposition rates.

Ocean techniques for measuring heat flow have been used with some success for active margins, particularly along the island-arc region extending from the Kurils to Southeast Asia. We know of some systematic variations in heat flow, such as normal heat flow on the oceanic-plate side, lower values in the trenches and/or the lower landward trench slope, and higher values on the island-arc side of the trenches. However, individual measurements show considerable variability within this pattern, and the same pattern does not always exist in all regions. Different trenches and island arcs show significantly different mean heat flow. We do not know what causes these differences. They may result from varying tectonic styles. Other geological and geophysical studies near the same locations as the heat-flow measurements will help us to understand them.

For both young and old passive margins, the best distribution of measurements would be on transverse profiles across the margins, extending into the continental and ocean structures proper. Such profiles across young margins of differing ages would help to clarify the tectonics of initial rifting of continents and the associated transient heat-transfer processes. Heat-flow measurements would be more valuable for tectonics if the elevations of both continental and oceanic basement were determined along the profiles. We could so derive a sensitive measure of the excess (or deficiency) of heat in the lithosphere. Similar profiles across older passive margins may be used to deduce the distribution of heat sources in the upper lithosphere of the adjacent continent and ocean structures. Here, deep geophysical information will be needed to determine the sharpness of the structural transition.

To date, few heat-flow measurements have been made on continental slopes or shelves. That is largely because seasonal fluctuations in temperatures of near-bottom waters are sufficiently high to potentially mask thermal gradients in the upper few meters of sediment. We need probes tens of meters long—perhaps as long as 100 m—to eliminate this problem fully. With the current technologies available, unfortunately, this means making use of holes drilled by DSDP or by oil-exploration efforts, neither of which has provided sufficient holes or interest to allow a wide-scale study of upper-margin heat flow.

B. RELATIONSHIP OF THE BOREHOLE TO SURFACE GEOPHYSICS

Because boreholes serve as crucial links between the subsurface domain and surface geophysics, and as laboratories for improving techniques, no opportunity should be lost to exploit a borehole for the purpose of measuring the physical properties of rocks.

BOREHOLE GEOPHYSICS

Geophysical fields are measured at or near the earth's surface. The geophysicist interprets these fields in terms of variations in rock properties in three-dimensional subsurface space.

Exploration geophysicists have an advantage over their solid-earth colleagues, because their interpretation of the subsurface is frequently tested by the drill, and boreholes enable them to measure directly rock properties previously inferred from surface measurements. With both cause and effect available, the explorationist can fully use and frequently guide technological developments in exploration geophysics.

The borehole not only reveals subsurface data, it also permits measurements to be made in its vicinity. This exploration technique is especially valuable when the targets are small or when the seafloor and/or subsurface are too complex for unambiguous interpretation of surface-bound data.

WELL LOGS

Measuring rock properties with tools lowered into the borehole is a well-developed technology. All physical properties of rocks of interest to the geophysicist can be measured with these tools (Schlumberger-Doll Research Centers, 1972a, 1972b, 1974; Pirson, 1970). The measurements are usually presented in the form of well logs, in which the measured property is displayed as a function of depth along the borehole.

The elastic constants as a function of depth are obtained from sonic logs (well logs of travel times for acoustic waves). There are two types of sonic logs: interval-time logs and full-wave logs. The interval-time log measures the transit time of the first arrival from a dilatational source to two detectors sufficiently separated from the source to ensure an *in situ* measure of the compressional velocity. Shear velocity can be measured directly in some cases by proper recording procedures, or it can be derived from the tube-wave velocity. The full-wave sonic log, which is a recording of the entire wave train registered by a detector, facilitates shear-wave identification and measurement.

The interval range over which interval times are measured must be chosen to ensure that *in situ* measurements of rock properties are not affected by the presence of the borehole. The interval within the borehole can vary from inches to the total depth of the borehole. The smallest interval is always chosen in order to obtain maximum resolution. The magnitude of the interval will dictate the frequency content of the source-generated signal, the nature of the source itself, and its location. For intervals in the inch range, the source is ultrasonic, operating in the high kilohertz range. The source and detectors are mounted on pads pressed against the borehole wall and on a sonde centered along the borehole axis. To remove the effect of borehole caliper changes, two sources straddling a quartet of receivers provide a reversed interval-time measure. This "borehole-compensated tool" is available for intervals of less than 9 feet. Beyond this separation, a single source to multiple receivers must be used. For intervals beyond 30 feet, the normal seismic-reflection frequency band is employed; namely, 5–10 Hz. The source can be placed either in the borehole or at the surface. Inhole sources such as a gun perforator have been used, but surface sources are more common. Currently, a nonexplosive, repeatable source, such as an air gun operating at the surface, has been found to be most suitable. A dynamic downhole detector clamped to the borehole wall is preferable to a hydrophone exposed to the borehole fluid. The accuracy of measurements is improved when the source signature and downhole detector signal are digitally recorded. Shear waves can also be measured by this basic system if the source and detector polarization are chosen properly, e.g., a shear vibrator source at the surface and horizontally polarized detectors clamped to the borehole wall at depth.

The anelastic properties of the subsurface can also be measured by using the transmitted signal if proper attention is given to signal strength and downhole detector coupling.

A rock-density log is obtained by using the gamma–gamma borehole tool and/or the borehole gravity meter. The gamma–gamma tool has shallow penetration into the rock surrounding the borehole, while the gravity meter averages the rock density between the downhole observation stations and, consequently, averages the densities of rock further removed from the borehole. Rock resistivity can be measured by a variety of tools. The most suitable instrument for geophysicists is the deep-induction tool.

MODELING

Knowing something about the distribution of rock properties in a vertical profile or sometimes even in three dimensions allows the geophysicist (1) to compute theoretical effects of the observed model (inverse problems) or (2) to compare theoretical computations with observed data (the forward problem). The latter gives the geophysicist a measure of the expected signal, a means to determine the resolving power of this signal, and an opportunity to evaluate methods (such as modifications to the source signature) by which to improve this resolving power, data recording, or data processing. The forward problem also provides a method by which geophysicists can separate signal from noise and with which they can devise methods to reduce noise.

EXPLORATION IN BOREHOLES

The borehole gives access to the subsurface and permits geophysical fields to be measured vertically into the earth as well as horizontally at the earth's surface. The fields are corrected for the known vertical variation in rock properties. Anomaly fields are interpreted in terms of anomalous distributions in rock properties laterally from the borehole or below the total depth of the borehole.

The ultra-long-spaced electric log (ULSEL) has been effective in profiling the flanks of salt domes. Density anomalies near boreholes were detected by borehole gravimeters. Gravity measurements at the earth's surface did not pick up these anomalies. Vertical seismic arrays in boreholes have not been used extensively in the West, but for Russia, Gal'perin (1974) documents their usefulness.

C. EFFICIENT USE OF SCIENTIFIC PERSONNEL AND SURVEY EQUIPMENT IN OBTAINING OBJECTIVES

A serious effort to provide a better understanding of continental margins in the fashion contemplated by this Panel requires a special planning and management system for personnel and equipment in order to achieve selected objectives. In Figure 11.2, the major components of such a system are arranged around a circle in a time sequence. The central circle represents system management, which is shown as controlling critical decisions between component phases. The system is depicted dynamically, which means that feedback loops occur among the components to ensure self-correction. This feedback also exists among the problem-solving campaigns that are carried on concurrently. Ensuring and facilitating such feedback should be a major function of project management.

In making the decisions required for transition between phases of exploration, a principle of cost-effectiveness may be useful. Figure 11.3 portrays a

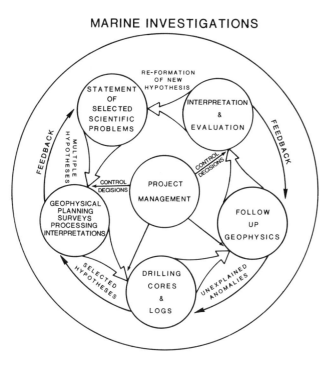

MARINE INVESTIGATIONS

FIGURE 11.2 A conceptual model that visualizes scientific problem solving on continental margins as a dynamic process. Five component phases are shown in time sequence. Transition from one phase to the next requires a management decision recognizing the scientific results of the preceding phase plus a monetary commitment to undertake the next phase. Presumably, the process reduces the number of tenable hypotheses or postulated geological models. Feedback provides for self-correction. An unexpected anomaly encountered at any phase may necessitate return to an earlier phase for more data before progressing forward. (Courtesy of E. S. Driver, Gulf Science and Technology Company.)

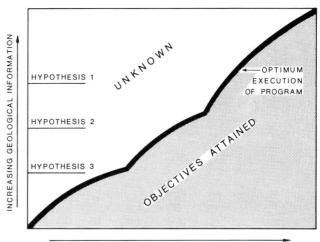

FIGURE 11.3 A decision-making process in marine investigations. The darkly shaded curve entitled "Optimum Execution of Program" represents the desired relationship between resources expended and gathering factual data (horizontal scale) and the resultant increase in geological information (vertical scale). If a geophysical or drilling company fails to provide enough data to test postulated hypotheses adequately, waste occurs (the white field). When marine investigations are carried out with inadequate instrumentation (when better is available), money and scientific talent are wasted. (Courtesy of E. S. Driver, Gulf Science and Technology Company.)

hypothetical curve relating increasing effort in geophysics and drilling to a corresponding increase in geological information. Such a curve will differ, depending on the areas and problems. One of the main functions of earth-science investigators is to define such a curve for their areas of investigation and for the various parts of the world in which they work.

Once established for a problem, this relationship curve helps the manager of resources to avoid waste in accomplishing scientific objectives. It is wasteful to fall short of testing critical hypotheses because the effort was not quite sufficient to penetrate a threshold leading to the resolution of the problem. On the curve in Figure 11.3, waste is represented by any combination of hypotheses with inadequate effort.

On the vertical scale, three hypotheses are ranked according to the relative amount of geological information needed to test them. The geophysical or drilling program should be designed with some latitude of problem resolution and depth penetration with an awareness of the remaining "unknown."

One issue that emerges from this type of analysis is the optimum tradeoff between the number of U.S. marine research vessels and the completeness of their sensing and navigational equipment. In our judgment, the overall scientific goals on the continental margins would be more effectively accomplished by fewer, but more completely equipped, vessels. This is particularly true now that increasing emphasis is being placed on deeper drilling.

Figure 11.4 attempts to translate the effectiveness of marine geophysical expenditures as they relate to the resultant geological information. Obviously, any such generalization is both subjective and area-dependent. Nevertheless, it is the best current judgment based on over 10 years of experience.

The vertical scale is predicted on the need to understand continental margins where the sedimentary sections are thick and complexly structured. This is expressed in the conclusion that deep lithological and structural resolution is worth approximately three times as much as shallow resolution at this point in the history of accumulating marine geological information. It also stresses the critical nature of definitive deeper data as they relate to drilling costs and safety. The horizontal scale is much more factual, as it is based on current cost information.

The inferences to be derived from the graph are:

1. It is very costly to staff and to position an oceanographic vessel accurately, even with the most basic geophysical instrumentation; and

2. Having made this commitment, the incremental cost to add adequate seismic sensors represents a good value in terms of the additional geological information that can be obtained.

The decision as to when to use 96 channels instead of 48, or when to use $2X$ air guns instead of X depends on the magnitude of the impedance contrasts one is seeking and the noise background.

Figure 11.5 compares 6-fold and 48-fold stack processing of identical raw data and migrated versions of the same data. This figure gives a fair indication of the improvement in effective bandwidth and signal-to-noise

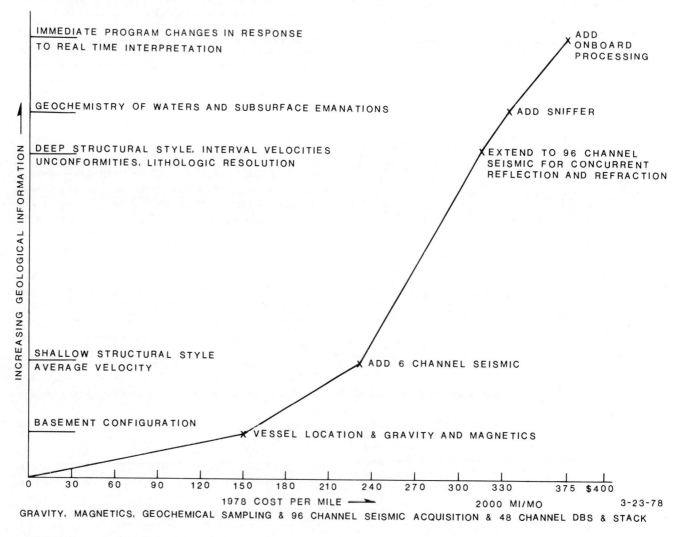

FIGURE 11.4 Cost effectiveness of marine geophysics. (Courtesy of E. S. Driver, Gulf Science and Technology Company.)

ratio resulting from 48-fold stacking. All deconvolution and display factors are the same for 6-fold and 48-fold. Note the improved overall reflection continuity on the 48-fold stack and, specifically, the following:

(a) There is a question whether the event at 1.4 sec, Shot Point 450, correlates on the 6-fold with the event at 1.280 sec or the event at 1.370 sec, Shot Point 510. Looking at the corresponding events at Shot Point 510 on the 48-fold, one can see that the event at 1.370 is suppressed relative to the event at 1.280. From this, it may be concluded that the lower event may have been mostly multiple energy and, as such, was more effectively suppressed by the high fold of stack. Proper migration of the right-dipping event at 1.300 sec, Shot Point 490, will no doubt show it to be above the previously mentioned events.

(b) Similar migration would resolve the apparent contradiction of interfering events about 2.0 sec at Shot Point 190.

(c) The right-dipping fault between 1.0 and 1.5 sec under Shot Points 160 to 200 is sharply defined on the 48-fold but might be missed on the 6-fold.

There are many other similar instances of improved structural and lithologic resolution attributable to the increased fold of stacking. However, some caution is needed in making these comparisons—the neophyte may pass from a clearly defined event on the 48-fold to its faint counterpart on the 6-fold and erroneously conclude that the 48-fold is superfluous. The critical question is whether neophytes would have recognized the event on the 6-fold alone and whether, on that basis alone, they could defend any of their geological conclusions and predictions with sufficient authority to justify the large expenditure of drilling a well.

Stacking is no panacea, but it is extremely powerful in multiple suppression, and it is an essential technique for extracting velocity information. Flowers (1976) discusses some of the advantages and pitfalls related to modern seismic reflection. For some spectacular high-resolution color displays, see also Taner and Sheriff (1977).

D. RECOMMENDATIONS

In light of several geological problems that require geophysical measurements in order to be better understood, we make the following recommendations:

1. *No opportunity should be lost to exploit boreholes for the purpose of:*

(a) *measuring the physical properties of the rocks;*
(b) *forming a link between the subsurface and surface geophysics; and*
(c) *serving as a laboratory to improve techniques.*

For example, we recommend that the wells drilled by the Continental Offshore Stratigraphic Test Group (COST), a consortium of petroleum companies, be used to calibrate the effectiveness of the v_p/v_s ratios derived from expanded spreads as well as to extend downhole measurements and instrumentation.

2. *Long, multisensor (reflection, refraction, potential, and electrical methods) geophysical traverses across strike should be recorded from cratonic provinces over orogenic belts and across both active and passive continental margins.* Some complementary traverses should be recorded along strike. All traverses should tie boreholes in the most effective way. Special in-line and three-dimensional shooting techniques may be required to achieve resolution beneath detachment surfaces. This kind of work and associated LANDSAT studies would form a bridge between land and marine geology—a neglected bridge.

3. *More long-distance refraction measurements should be made on continental margins.* Such measurements should be made jointly with deep-penetration reflection and other geophysical techniques that use surface and bottom seismometers. These measurements should be made in conjunction with similar projects on land.

4. Many vessels in the U.S. oceanographic fleet are inadequately instrumented for the resolution and deep penetration needed to solve significant geological problems on continental margins. Most of the fleet's research ships lack advanced multichannel seismic-reflection systems. Efficiency in achieving primary geological objectives requires up-to-date instrumentation. Using modern geophysical equipment saves dollars. *When a budget is limited, it is preferable to have a smaller number of well-instrumented vessels employed in marine geology and geophysics than to have a larger number of inadequately instrumented ships. We therefore recommend that two modern geophysical research vessels be outfitted to work on the U.S. East Coast and West Coast, respectively (for details, see p. 187).*

5. Broader-band seismic-reflection and -refraction systems should be used. These enhance resolution and improve measurements of amplitudes, velocity, and attenuation. Greater multiplicity of recordings is necessary to improve signal-to-noise ratios over this extended bandwidth. This could involve improved sources, bottom shooting, and recording techniques. Digital recording is recommended for dynamic range and to take optimum advantage of processing programs. Seismic stratigraphy and lithologic resolution would be enhanced.

6. *A sea-bottom source of shear waves should be developed.* To do this, industry's activities will need to be coordinated with those of government and academe.

7. *A standard seismic test tape or test procedure should be adopted to specify the actual response of the various systems employed and to facilitate their integration.* Correct polarity and understanding of phase response are essential for lithologic deductions. We rec-

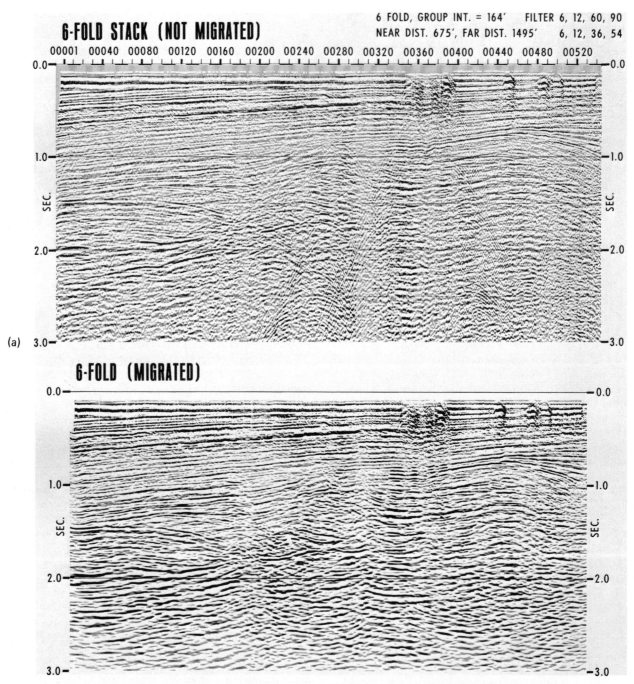

FIGURE 11.5 A comparison between 6-fold [(a) above] and 48-fold [(b) opposite] stack processing of the same raw data. The bottom panels illustrate the same data, migrated. (Courtesy of E. S. Driver, Gulf Science and Technology Company.)

ommend that the Society of Exploration Geophysicists be approached to determine whether any effort of this sort is under way and, if not, whether such an endeavor could be undertaken.

Present marine geophysical techniques are important in any study of continental margins, but problems arise when attempts are made to merge information collected by one institution with that collected by another. This is well illustrated by discrepancies observed between the interpretation of multichannel and single-channel seismic-reflection data within the same area. Differences in navigational accuracy and the corrections applied to raw data (e.g., diurnal corrections applied to raw magnetic data) often make it difficult to compile data from different

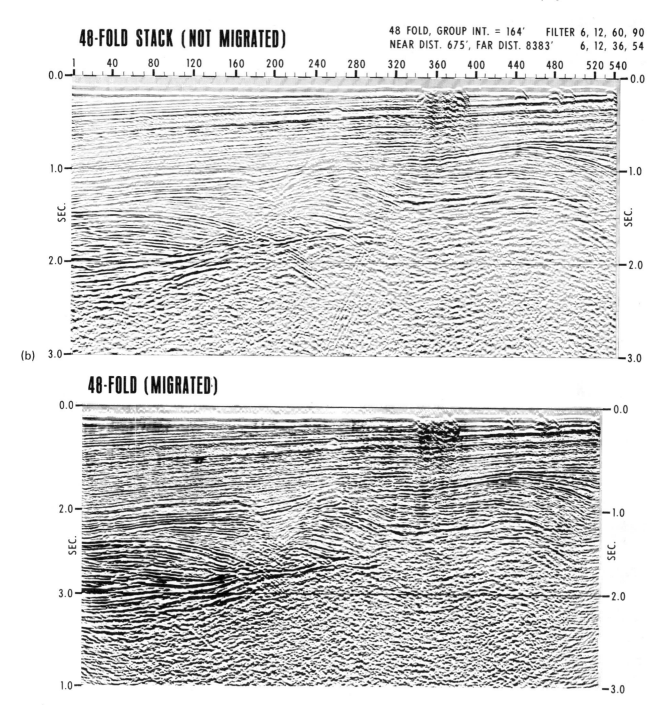

48-FOLD STACK (NOT MIGRATED)

48 FOLD, GROUP INT. = 164'   FILTER 6, 12, 60, 90
NEAR DIST. 675', FAR DIST. 8383'   6, 12, 36, 54

(b)

48-FOLD (MIGRATED)

sources. Examination of margins thought to be of different ages or formed in different ways would be extremely valuable if the same geophysical and geological techniques and instruments were used.

8. *Since they supply additional descriptions of subsurface rocks, magnetotelluric, electrical, and heat-flow measurements are recommended.*

9. *On the East Coast, the National Magnetic Anomaly Map should be extended offshore to include the M-sequence of magnetic stripes.* We endorse the current work being done by the Naval Research Laboratory on the U.S. East Coast.

REFERENCES AND BIBLIOGRAPHY

Buhl, P., P. L. Stoffa, T. K. Kan, C. Windisch, J. Ewing, and M. Talwani (1977). A critical angle reflection experiment to map the M-discontinuity: preliminary data analysis (abstract), *EOS, Trans. Am. Geophys. Union* 58(6):510.

Cochrane, N. A., and R. D. Hyndman, 1974. Magnetotelluric and magnetovariational studies in Atlantic Canada, *Geophys. J. R. Astron. Soc.* 39:(2)385–406.

Committee for a National Magnetic Anomaly Map (1976). National Magnetic Anomaly Map, a report of the *National Magnetic Anomaly Map Workshop* held at Golden, Colorado, Feb. 17–19.

Cox, G. S., J. H. Filloux, and J. C. Larsen (1971). Electromagnetic studies of ocean currents and electrical conductivity below the ocean floor, in *The Sea*, Vol. IV, Part 1, John Wiley and Sons, New York, pp. 637–693.

Dobrin, M. B. (1976). *Introduction to Geophysical Prospecting*, 3rd ed., McGraw-Hill Book Co., New York, 630 pp.

Edwards, R. N., and J. P. Greenhouse (1975). Geomagnetic variations in the eastern United States: Evidence for a highly conducting lower crust? *Science* 188:726–728 (May 16).

Filloux, J. H. (1977). Ocean-floor magnetotelluric sounding over North Central Pacific, *Nature* 269(5626):297–301.

Fitch, A. A. (1976). *Seismic Reflection Interpretation, Geoexploration Monographs Series 1*, No. 8, Gebruder Borntraeger, Berlin.

Flowers, B. S. (1976). Overview of exploration geophysics: Recent breakthroughs and challenging new problems in circum-Pacific energy and mineral resources, *Am. Assoc. Petrol. Geol. Mem.* 25:203–210.

Gal'perin, E. I. (1974). Vertical seismic profiling, *Soc. Explor. Geophys. Spec. Pub. 12.*

Herron, T. J., W. J. Ludwig, P. L. Stoffa, T. K. Kan, and P. Buhl (1977). Structure of the ridge crest of the East Pacific rise from multichannel seismic reflection data (abstract), *EOS, Trans. Am. Geophys. Union* 58(6):511.

Hyndman, R. D., and N. A. Cochrane (1971). Electrical conductivity structure by geomagnetic induction at the continental margin of Atlantic Canada, *Geophys. J. R. Astron. Soc.* 25:425–446.

Kan, T. K., P. L. Stoffa, P. Buhl, T. J. Herron, and M. Truchan (1977). Wave equation migration: a powerful tool to improve deep sea multi-channel seismic reflection data (abstract), *EOS, Trans. Am. Geophys. Union* 58(6):510.

Ocean Sciences Board (1976). *Multichannel Seismic Reflection System Needs of the U.S. Academic Community*, Assembly of Mathematical and Physical Sciences, National Research Council, National Academy of Sciences, Washington, D.C., 30 pp.

Payton, C. E., ed. (1977). Seismic Stratigraphy—Applications to Hydrocarbon Exploration, *Am. Assoc. Petrol. Geol. Mem. 25.*

Pirson, S. J. (1970). *Geologic Well Log Analysis*, Gulf Publishing Co., Houston, Texas, 370 pp.

Schlumberger-Doll Research Center (1972a). The essentials of log interpretation practice, Schlumberger Technical Services, Paris.

Schlumberger-Doll Research Center (1972b). *Log Interpretation*, Vol. I, Schlumberger, Ltd., New York.

Schlumberger-Doll Research Center (1974). *Log Interpretation*, Vol. II, Schlumberger, Ltd., New York.

Sheriff, R. E. (1973). *Encyclopedic Dictionary of Exploration Geophysics*, Soc. Expl. Geophysicists, Tulsa, Okla.

Sheriff, R. E., and T. A. Lauhoff (1977). Marine geophysical exploration: The state of the art, *IEEE Trans. Geosci. Electron. GE-15*(2):67–73.

Talwani, M., C. C. Windisch, P. L. Stoffa, P. Buhl, and R. E. Houtz (1977). Multichannel seismic study in the Venezuela Basin and the Curacao Ridge, in *Island Arcs, Deep Sea Trenches and Back-Arc Basins*, M. Talwani and W. C. Pitman III, eds., Maurice Ewing Series 1, Am. Geophys. Union, Washington, D.C., pp. 83–98.

Taner, M. T., and R. E. Sheriff (1977). Application of amplitude, frequency and other attributes to stratigraphic and hydrocarbon determination, in Seismic Stratigraphy—Applications to Hydrocarbon Exploration, *Am. Assoc. Petrol. Geol. Mem. 25.*

Telford, W. M., L. P. Geldart, R. E. Sheriff, and D. A. Keys (1975). *Applied Geophysics*, Cambridge U. Press, Cambridge, England.

Windisch, C., P. L. Stoffa, P. Buhl, M. Talwani, J. Ewing, T. K. Kan, and S. Murauchi (1977). A critical angle reflection experiment to map the M-discontinuity: The field experiment (abstract), *EOS, Trans. Am. Geophys. Union*, 58(6):510.

12 | The Application of Remote Sensing

A. LANDSAT IMAGES

LANDSAT 1 (launched in July 1972) and LANDSAT 2 (launched in January 1975) use multispectral scanners to provide repetitive coverage of the earth's surface resources and environmental conditions. LANDSAT imagery gives ground resolution of 80 m and has provided valuable information about the extent and duration of changes in surface conditions caused by man or by nature. In addition to a multispectral scanner similar to that on the earlier satellites, LANDSAT C will also carry a thermal infrared scanner and a panchromatic return-beam vidicontube that can give surface images with a 40-m ground resolution. The National Aeronautics and Space Administration (NASA) plans to launch LANDSAT D early in the 1980's. That satellite will carry a thematic mapper that will provide multispectral images of the earth's surface with a 30 m spatial resolution (CORSPERS, 1976).

SEASAT-A, NASA's ocean-dynamics satellite, is planned for launch in 1978. Its synthetic-aperture imaging radar, with ground resolution of about 25 m, should be of great value in coastal studies. It is to be hoped that the radar will be activated over the land extension of continental margins.*

LANDSAT images are obviously important for coastal-zone management, because they offer a unique synoptic view of marine and coastal systems and shallow nearshore bathymetry. LANDSAT images adjunct to the coastal studies advocated on pp. 35–37 are useful to obtain a worldwide overview of sedimentation patterns in the coastal zone. In addition to the intrinsic interest of such studies, these studies will provide models of reservoir distributions in subsurface.

Suspended sediments frequently provide natural water-mass tracers that can be used to map nearshore and estuarine current circulation. Repeated imaging of the same area should help to study seasonal variations in these circulation patterns. So, LANDSAT and SEASAT data will be of particular relevance to a program in shelf-sediment dynamics.

B. SATELLITE ALTIMETRY

The figure of the earth or geoid can be determined from terrestrial gravity data and from observations of artificial satellites. The geoid in oceanic areas, however, corresponds to the mean sea surface to within a meter or two, so recent developments in satellite altimetry (McGoogan et al., 1975; Leitao and McGoogan, 1975) are currently of much geophysical interest.

Satellite radar-altimeters measure the distance be-

*The SEASAT-A spacecraft failed in orbit on October 9, 1978, as a result of a massive short circuit in its electrical power system.

163

tween the ocean surface and the altimeter, which, when subtracted from the calculated height of the altimeter above the reference ellipsoid, gives the geoid undulation. A satellite altimeter was in use during the SKYLAB and GEOS-3 missions. The altimeter transmitted a radar pulse downward and received the pulse reflected from the sea surface. The altimeter measured the sea-surface height over a "footprint" of about 14 × 14 km². The accuracy of data from the SKYLAB and GEOS-3 mission launched in April 1975 is about 1 m in the global mode and 50 cm in the intensive mode. Forthcoming missions such as SEASAT-A will have an accuracy of about 10 cm, which will give extremely important information on ocean currents and circulation.

The first experiment done during the SKYLAB mission revealed that the well-known gravity low over the Puerto Rico Trench was reflected as a depression in the sea surface. J. Marsh at the Goddard Space Flight Center (GSFC) Geodynamics Branch is currently constructing a gravimetric 5° × 5° geoid in the northwestern Atlantic Ocean and a geoid profile across the Blake Plateau and the southeast U.S. continental margin. This profile agrees with the geoid derived from satellite tracking data to within about 2 m. SEASAT-A will carry an altimeter with a 2-km footprint and 10-cm accuracy.

These altimeter measurements may prove to be a most useful tool in outlining the *broad* structure of an Atlantic-type margin (see Figure 12.1). The geoid resolves features with wavelengths longer than about 30 km and therefore excludes short-wavelength features that correlate closely with topography in free-air gravity-anomaly profiles. Currently, two features of much interest are the

broad sedimentary basins underlying the continental rise and the transition zones between oceanic and continental crust.

C. SATELLITE GRAVIMETRIC GEOID

The broad features of the global geoid can be determined by precise satellite tracking, and this has been done with increasing detail and precision since the early days of the space age. At GSFC, a series of gravimetric geoids, based on satellite and surface gravity data, has been constructed, the latest being GEM-10 (Goddard Earth Model-10). Such geoids show the long-wavelength features of the geoid. They are important to the study of continental margins. First, they provide information about mass distribution in the upper mantle, subject, of course, to the ambiguity inherent in all gravity investigations. Second, they provide reference fields for surface investigations or, as previously mentioned, for satellite altimetry measurements.

The latest geoid reveals generally good correlation between gravity and Pacific-type continental margins. Atlantic-type margins, on the other hand, have much less pronounced gravity signatures. However, there are exceptions to these relations for both types of margin, and these exceptions may prove more interesting.

D. MAGSAT AND CRUSTAL MAGNETISM INVESTIGATIONS

The GSFC Geophysics Branch, in conjunction with the U.S. Geological Survey, is carrying out investigations related to crustal magnetism as measured by satelliteborne magnetometers. Regan *et al.* (1975) demonstrated that crustal magnetism could be mapped on a global basis from satellites. Therefore, a satellite (MAGSAT) carrying both vector and scalar magnetometers at about 350 × 500 km² altitude will be launched in late 1979 specifically for crustal anomaly studies and to update field models. It will have a 97° inclination and about a 6-month lifetime, sufficient to produce a near-global magnetic map. The accuracy of scalar magnetometer measurements will be 3γ in orbit, and the accuracy of the vector magnetometer will be 6γ in each component.

Satellite magnetic-field measurements can also reveal information on crustal structure at depths above the Curie isotherm. This information is specifically relevant to continental margins. For example, Regan *et al.* (1975) provide regional magnetic reference fields that can serve as background for aerial or surface magnetic surveys. The relation between regional magnetic anomalies and continental margins must have some relationship to the respective structure and composition. Also, at least in some areas, such as Australia, the satellite magnetic map appears to outline ophiolite belts, which are generally considered to be former continental margins.

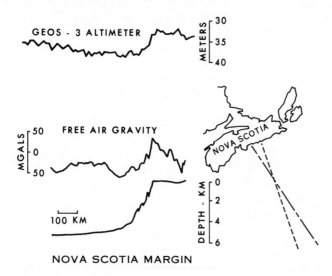

FIGURE 12.1 GEOS-3 satellite altimeter profile across continental margin off Nova Scotia. The margin is associated with a step in the geoid of about 6 m. (Courtesy of Lamont-Doherty Geological Observatory of Columbia University and the National Aeronautics and Space Administration.)

E. MONITORING PLATE MOTION AND DEFORMATION

Classical geodesy and earlier satellite geodesy are evolving rapidly with space technology, which is providing new and extremely accurate ways of measuring crustal motion and deformation. These techniques involve laser ranging to-and-from satellite (SLR) and to-and-from laser reflectors (LLR) now in place on the moon, as well as very-long-baseline microwave interferometry (VLBI). The latter technique involves the use of large radio telescopes to receive and record signals from extremely distant quasar sources. It is expected that in the near future, accuracies of about 5 cm will be achieved over baseline distances exceeding 3000 km.

If successful, these techniques will permit scientists actually to measure motions between plates and within plates. This direct test of the plate-tectonics hypothesis would not fail to impact continental-margin studies, because it would provide a direct means to relate plate motions to seismicity and volcanicity on active margins. Several experiments are currently in progress:

1. The Pacific Plate Motion Experiment (PPME) will use large, fixed radio telescopes in California, Texas, Alaska, Massachusetts, and eventually Hawaii to measure the motion of the Pacific plate relative to the North American plate and to determine deformation within the North American plate itself.

2. The Astronomical Radio Interferometric Earth Surveying (ARIES) experiment will use relatively small portable antennas in conjunction with the Goldstone (California) antenna to measure distances of a few hundred kilometers in California to accuracies of, eventually, 5 cm.

In addition to this, an experiment involving SLR is being carried out along the San Andreas Fault, which can be considered part of the continental margin in a broad sense; so that anything learned about the fault will be relevant to continental margins in general. The San Andreas Fault Experiment (SAFE), being carried out by the GSFC Geodynamics Branch, is a project using laser tracking systems to measure the relative motion of the North American and Pacific plates along the San Andreas fault system over a very long baseline—about 1,000 km. One tracking station is at Otay Mountain, near San Diego, and the other at Quincy, California. The technique used is to bounce laser pulses off retroreflectors on the Satellite LAGEOS and other satellites equipped with cubed corners in order to obtain distance measurements accurate to 5–10 cm. Preliminary measurements suggest right-lateral

movement of about 9 cm/year between Quincy and Otay Mountain. The project is planned to run through 1981. Other stations may be established aound the Gulf of California.

REFERENCES AND BIBLIOGRAPHY

Anon. (1976). National Aeronautics and Space Administration briefing on Earth and Ocean Dynamics Program (EODAP), June 4, available from NASA.

CORSPERS (1976). *Resource and Environmental Surveys from Space with the Thematic Mapper in the 1980's*, a report by the Committee on Remote Sensing Programs for Earth Resource Surveys, Commission on Natural Resources, National Research Council, National Academy of Sciences, Washington, D.C.

CORSPERS (1977). *Microwave Remote Sensing from Space for Earth Resource Surveys*, a report by the Committee on Remote Sensing Programs for Earth Resource Surveys, Commission on Natural Resources, National Research Council, National Academy of Sciences, Washington, D.C., 148 pp.

Leitao, C. D., and J. T. McGoogan (1975). SKYLAB radar altimeter: Short wave-length perturbations detected in ocean surface profiles, *Science 186*:1208–1209.

Lowman, P. D., Jr. (1976). Geoscience applications of space technology, 1975–2000, a report available from the Technical Information Division, Code 250, Goddard Space Flight Center, Greenbelt, Md.

Marsh, J. G., and S. Vincent (1974). Global detailed computation and model analysis, *Geophysical Surveys 1*, D. Reidel Publ. Co., Dordrecht, Holland, pp. 481–511.

McGoogan, J. T., C. D. Leitao, and W. T. Wells (1975). Summary of SKYLAB S-193 altimeter altitude results, NASA technical memorandum, *Rep. No. NASA Tm X-69355*, 323 pp.

NASA (1976). Outlook for space, a report to the NASA Administrator, by the Outlook for Space Study Group, available from NASA Scientific and Tech. Info. Office, Washington, D.C.

Otterman, J., P. D. Lowman, and V. V. Salomonson (1976). *Surveying Earth Resources by Remote Sensing from Satellites*, Geophys. Surveys 2, D. Reidel Publ. Co., Dordrecht, Holland, pp. 431–467.

Regan, R. D., J. C. Cain, and W. M. Davies (1975). A global magnetic anomaly map, *J. Geophys. Res. 80*:794–802.

Short, N. M., P. D. Lowman, Jr., S. C. Freden, and W. A. Finch, Jr. (1976). *Mission to Earth: Landsat Views the World*, NASA SP-360, NASA Scientific and Technical Information Office, Washington, D.C.

Vonbun, F. O. (1976). Earth and ocean dynamics satellites and systems—An overview, *Proceedings of the XXVI International Astronautical Congress*, L. G. Napolitano, ed., Pergamon Press, New York.

Williams, R. S., Jr., and W. D. Carter, eds. (1976). ERTS: A new window on our planet, *USGS/Prof. Paper 929*, U.S. Government Printing Office, Washington, D.C.

13 | The Role of Drilling for Scientific Purposes

Drilling reveals the truth, and in geology, truth is hard and expensive to come by. Petroleum geologists know this best, for they often see their predictions evaporate as dry holes are abandoned. The experience bruises the ego, is sobering, but delightfully humanizes a soul. It would be salubrious to scientists in academe and government to more intimately share similar experiences. The Deep Sea Drilling Project (DSDP) gave many scientists their first major opportunity to test their scientific predictions by the drill.

Offshore drilling evolved in the mid-1920's from extensions of nearby onshore exploration to the waters of the Gulf of Maracaibo and off California. In the 1930's, rigs were mounted on barges in the swamps of Louisiana. Finally, in 1947, Continental, Union, Shell, and Superior Oil companies formed the CUSS group in California to develop new offshore drilling techniques. The "CUSS 1," constructed in 1956, became the first true floating drilling rig.

In connection with the Mohole Project,* CUSS 1, using four temporary dynamic-positioning units, drilled a one-bit hole with 560-ft (171-m) penetration in 11,672 ft (3558 m) of water in offshore California. (Dynamic positioning "anchors" the ship to a beacon on the seafloor by means of acoustic and electronic devices rather than by mooring lines.) That hole marked the first use of dynamic positioning. Also, that hole reached a depth that was not exceeded until the *Glomar Challenger* finally drilled a deeper one.

Industry has developed a wide range of drilling rigs (Figure 13.1). Some 400 units should be available by 1978, about a third of them for operation in U.S. waters. In 1976, industry drilled 132 wells in water depths exceeding 600 ft (183 m). Fifty-eight wells have been drilled in waters deeper than 1000 ft (305 m). Esso Exploration holds the record; they drilled a well in 3460 ft (1055 m) of water in the Andaman Sea. The total depth of Esso's drill well exceeded 14,000 ft (4267 m). There are now about seven rigs, overall, that can drill in waters deeper than 3000 ft (914 m). Ten additional rigs are being constructed, some of them capable of drilling depths of about 6000 ft (1829 m). For example, the *Discoverer Seven Seas* drill ship (owned by Offshore International S.A.) (Figure 13.2) has an operating water depth of 250–600 ft (76–183 m) and a rated drilling depth of 25,000 ft (7620 m). A deepwater drill ship costs about $50 million to build and to outfit. Operating costs are about $100,000 per day.

Geer (1975) gives a capability forecast for industry in Figure 13.3. The solid line shows an estimate of the current and short-term extensions to mobile drilling capability (excepting the Arctic). The dashed line predicts future capability based on new programs. Geer mentions some additional problems:

*The Mohole Project was an effort designed to drill a hole into the earth's crust and through the Mohorovičić discontinuity. Project Mohole was cancelled in 1967.

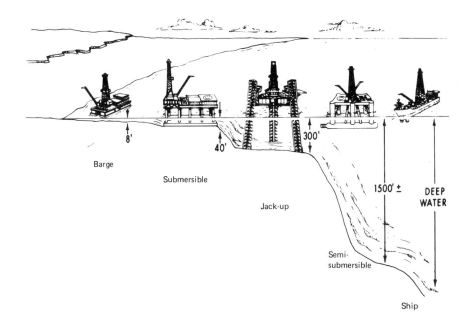

FIGURE 13.1 Family of drilling rigs. (Courtesy of R. L. Geer, Shell Oil Company.)

Drilling for oil and gas in deeper waters and the Arctic offshore will require innovative extensions of current drilling practices coupled with new technology. The petroleum industry has the development capability but unless the offshore prospects become available, the process of novel and innovative development will move slowly. The major delay in development of the new generation of floating rigs with deepwater drilling capability similar to the SEDCO 445 is the questionable need for this type of deepwater drilling equipment in view of the uncertain government policy, lease sale schedules, and economic climate.

The technical problems which confronted developers of these deepwater drill ships included: Reliable mooring and dynamic positioning systems of the vessel, well re-entry techniques and equipment, electro-hydraulic control systems for safe and reliable control of the subsea wellhead blowout preventor (BOP), and competent marine riser systems. A substantial amount of engineering will be required to optimize the systems and equipment for routine use down to about 3000 feet. Beyond this depth, additional research and development effort will be necessary to refine or replace the marine riser and the BOP control system.

. . . Drilling for oil in the Arctic offshore environment will be an especially difficult and expensive undertaking. Arctic offshore drilling structures must be able to resist very large loads imposed by moving ice sheets, or they must be able to move off location quickly.

To date, the search for petroleum in the offshore Arctic waters has focused on the shallow water, solid frozen, fast ice areas. Exploratory drilling operations have been conducted from artificial (gravel) islands in the Canadian Arctic. As operations move into deeper Arctic waters (beyond about 60 feet) the portion of the Arctic Ocean that is covered all or part of the time by moving pack ice is encountered. The industry is committed to developing operating capability in this regime as shown by the four proposed means [Figure 13.4 of the present report] for conducting exploratory drilling operations in Arctic waters. However, all of these mobile offshore Arctic rig concepts will be very expensive to build and operate. Costs may run 2–3 times normal offshore drilling costs. Therefore, it will be imperative that a competent

understanding of the Arctic offshore environments be developed and new technology evolved to fill the needs as the industry moves out from the shore.

Commercial offshore drilling mostly evaluates offshore prospects. There are, however, some cases in which drilling addresses problems that are less directly related to specific prospects. In the past, petroleum companies did shallow coring in offshore areas aimed at getting some information on shallow layers. Such data helped to interpret seismic data. The results of one such program were published by Lehner (1969).

More recently, the U.S. Geological Survey undertook a coring program to obtain reliable data for environmental appraisals and to help in evaluating offshore resources (Hathaway *et al.*, 1976). Geotechnical and engineering properties of sediments were measured, freshwater aquifers were defined, possible resources such as phosphates, sand, and gravel deposits were studied, and again, geophysical data were calibrated.

Since 1975, a consortium of 31 petroleum companies —the Continental Offshore Stratigraphic Test (COST) group—has been drilling a number of wells on U.S. continental margins. These tests are aimed at getting data for stratigraphic information in areas coming up for offshore sales. Government regulations stipulate that all geological information on such wells be published 60 days after such leases are made. COST wells usually are drilled in positions best suited for calibration of stratigraphic sequences. They often are quite deep. For instance, the COST B-2 well extends to a total depth of about 16,000 ft (4877 m). The scientific value of such information is immense, but it must not be overlooked that these wells are primarily aimed at getting general stratigraphic information in sale areas (Scholle, 1977). They are not oriented

FIGURE 13.2   *Discoverer Seven Seas* drill ship. (Courtesy of Offshore International, S.A.)

toward solving basic scientific problems, e.g., the drilling into the marginal high of the Atlantic or testing differences in diagenesis between adjacent but differing carbonate facies.

Let us briefly review the DSDP. In 1964, four major oceanographic institutions got together and formed the Joint Oceanographic Institutions for Deep Earth Sam-

pling (JOIDES) consortium. A national program for coring ocean sediments was designed. The *Glomar Challenger*, capable of drilling in water depths up to 20,000 ft (6096 m) with a penetration of about 2500 ft (762 m), was constructed and launched in spring 1968. Over 400 holes were drilled since that time (Figure 13.5). Drilling in 10,000–15,000 ft (3048–4572 m) became routine; the

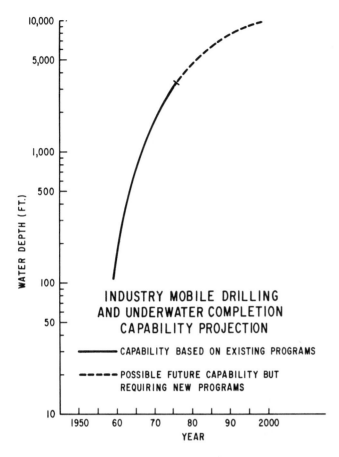

FIGURE 13.3 Industry mobile drilling and underwater completion capability projection. (Courtesy of R. L. Geer, Shell Oil Company.)

amount to $137 million, including some $19 million in foreign contributions that became available in the late IPOD phase of the DSDP program. Contributing countries are the Soviet Union, Germany, the United Kingdom, France, and Japan.

Some of the major achievements of the DSDP include the following:

Confirmation of seafloor spreading;
Calibration and verification of magnetic chronology;
Age of the ocean crust is less than 200 million years;
Organic-rich shales in deep-ocean basins (mainly the Atlantic);
Red Sea brines related to spreading center;
Changing geochemistry of the oceans;
Changing sediment-distribution patterns;
Large-scale vertical movements;
A unifying view of global stratigraphy.

The JOIDES Ad Hoc Subcommittee on the Future of Scientific Ocean Drilling (FUSOD) held a conference in March 1977, at Woods Hole, Massachusetts. For this conference, many working papers were prepared and reviewed. Large portions of these papers have been used for this report. The conference participants recommended a new, 10-year, deep-ocean drilling program using *Glomar Challenger* technology as well as the *Glomar Explorer*—equipped with a riser and well-control system. Program costs are estimated to total $450 million for 10 years (including *Glomar Explorer* conversion).

In the oil business, drilling marks the end of a long chain of exploration activities. These often begin with geological, gravity, and magnetic reconnaissance surveys. Then, a seismic-reflection reconnaissance grid is made and subsequently detailed by additional surveys to define the prospect and locate a preferred drilling site. During and after drilling, the well is logged (see p. 156) to obtain petrophysical data and to help relate rock properties to the distribution of seismic reflectors.

deepest water depth being 20,483 ft (6243 m). The deepest penetration below the ocean floor was 5709 ft (1740 m). This deep a bore was possible because a reliable hole re-entry capability was developed. Re-entry means that the drill string or other instrumentation can be guided back into an existing drilled hole at once or after returning from a port call. In 1976, one hole drilled in over 14,000 ft (4267 m) of water was re-entered nine times.

The *Glomar Challenger* has no riser system* and no blowout-prevention mechanisms. Therefore, drilling sites are carefully scrutinized for any safety risk. Several independent panels review drilling-site proposals. IPOD drilling is scheduled to cease in 1979, and DSDP will be phased out by 1980. The total DSDP costs by that time will

---

*A marine riser is a large-diameter tube (407–508 mm) installed between the floating drilling rig and a seafloor well-head. The functions of the riser and the well-control system on the drilling vessels are (1) to allow drilling fluids to circulate from the hole, through the drill pipe and back to the ship, and (2) to control pressure and borehole sealing at the seafloor, through the well-control system, in areas of potential hydrocarbon traps.

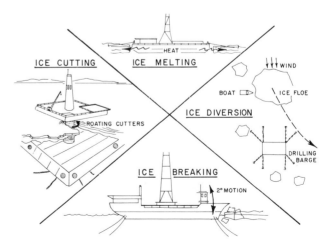

FIGURE 13.4 Mobile drilling concepts for offshore Arctic waters. (Courtesy of R. L. Geer, Shell Oil Company.)

With some justification, the DSDP program was criticized by industry's earth scientists for inadequate preparatory work and inadequate logging. In fact, most holes were drilled on the basis of poor to very poor seismic data. Much better data could have been obtained (at a higher cost, of course) using seismic contractors. Moreover, there are few cases in which regional seismic lines of decent quality illustrate the geological context within which the holes were located. Finally, few holes were logged, and these were not logged completely. The impression all this gives is that, in the euphoria of getting exciting data, and given the momentum of the operation, quality was often sacrificed for the sake of quantity. A small amount of progress in the right direction is being made in the current IPOD program.

In the future, adequate scientific preparation for these holes is a must. The high costs, coupled with fiscal responsibility, simply demand proper preparation for drilling. In emphatic agreement, we would like to quote from the report of the JOIDES Ad-Hoc Subcommittee of the Future of Scientific Ocean Drilling:

It cannot be overemphasized that the hole is not the experiment, it is the ground truth that translates geophysical parameters into geological reality. Just as drilling an exploration well in search of hydrocarbons in a new area follows a long program of regional and site specific geological and geophysical investigations, so must a scientific well follow a well integrated program of broad scale problem definition and small scale site examination. Although the drill is a very powerful tool for helping to solve some fundamental scientific problems, it cannot solve, or even define, the problems on its own. An investment in a drilling vessel without a commensurate investment in scientific investigations on an appropriate scale could not be supported in a good conscience. In fact, the committee feels so strongly about this subject that it recommends the drilling program envisaged in this report *only if adequate funding is assured for scientific studies for*

(i) *broad scale problem definition*
(ii) *small scale site examination and preparation*
(iii) *sample analysis*
(iv) *interpretation and synthesis*

to be carried out in conjunction with a drilling program.

Not only is it necessary to provide funding for these scientific activities, it is essential to do so in a manner that the funds so earmarked cannot be diverted for drilling activities. Otherwise, there is always the danger that contingencies arising in connection with an expensive drilling program will swallow the funds

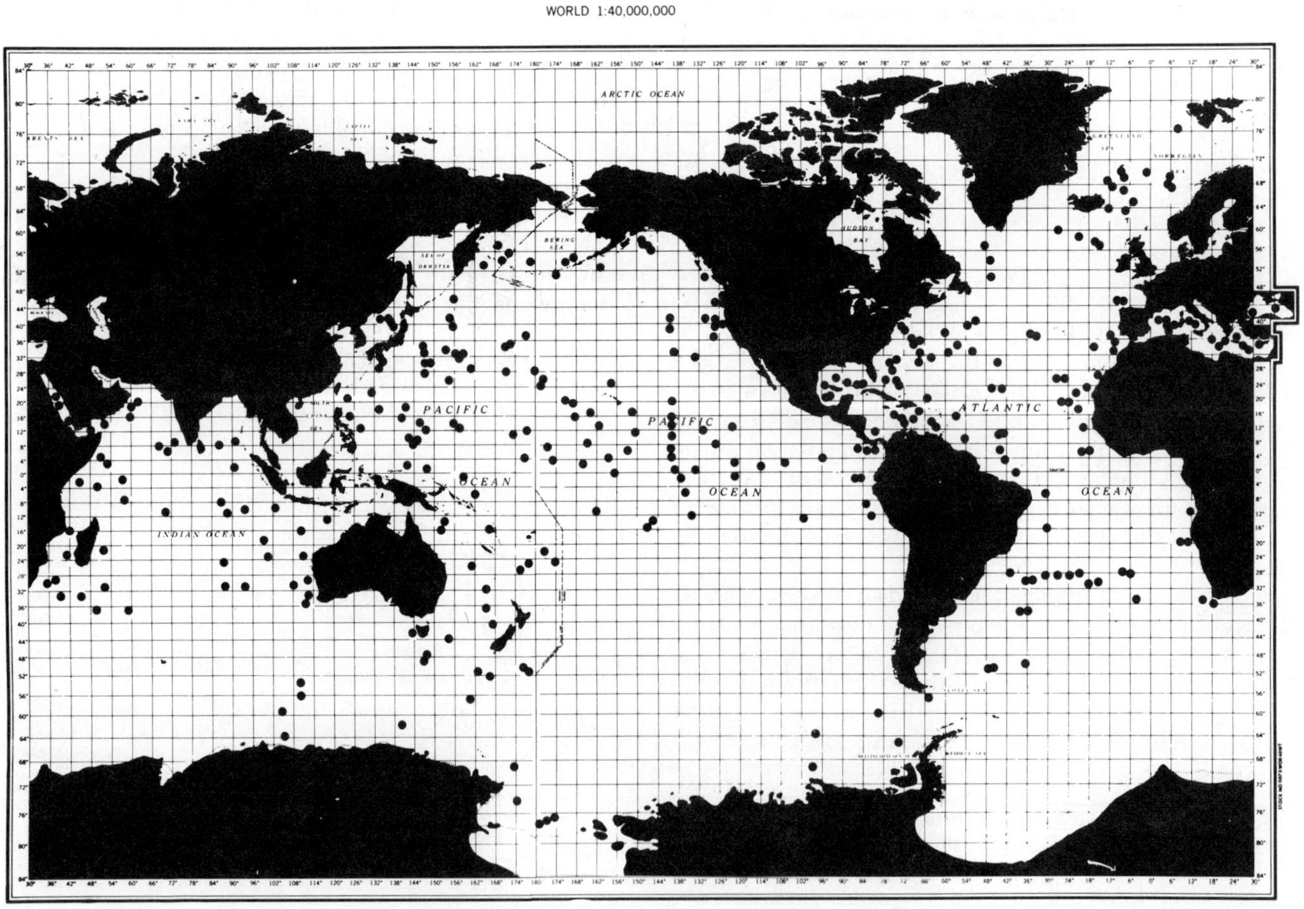

WORLD 1:40,000,000

FIGURE 13.5 Location of DSDP holes. (Courtesy of the Deep Sea Drilling Project.)

needed for geophysical studies, site location surveys, sample studies, interpretation and syntheses, and thereby completely frustrate the overall aim of the scientific program.

The position of this Panel with regard to drilling for scientific purposes is elaborated in Part III, Chapters 15 and 16 (see, particularly, pp. 188–190).

REFERENCES AND BIBLIOGRAPHY

Ad Hoc Committee on Technology of Drilling for Energy Resources (1976). *Drilling for Energy Resources*, National Research Council, National Academy of Sciences, Washington, D.C.

Anon. (1976). *Glomar Explorer*'s many technical innovations. *Ocean Industry*, December, pp. 67–73.

FCCSET Committee on Solid Earth Sciences (1977). *Continental Drilling:* Recommendations of the Panel on Continental Drilling of the Federal Coordinating Council on Science, Engineering, and Technology, Washington, D.C.

Geer, R. L. (1973). Offshore drilling and production technology—Where do we stand and where are we headed? Third Annual Meeting, Div. of Prod., API, Denver Hilton, Apr. 9–11.

Geer, R. L. (1975). National needs, current capabilities and engineering research requirements, Offshore Petrol. Ind., National Technical Information Service (NTIS) No. PB 249110, prepared for Committee on Seafloor Engineering, Marine Board, National Research Council, Washington, D.C.

Hathaway, J. C., J. S. Schlee, C. W. Poag, P. C. Valentine, E. G. A. Weed, M. H. Bothner, F. A. Kohout, F. T. Manheim, R. Schoen, R. E. Miller, and D. M. Schultz (1976). Preliminary summary of the 1976 Atlantic Margin Coring Project of the U.S. Geol. Surv., November, available from the U.S. Geological Survey, Washington, D.C.

JOIDES Subcommittee (1977). *Report on the Future of Scientific Ocean Drilling*, based on a conference held March 7–11, 1977, at the Swope Center, Marine Biol. Lab., Woods Hole, Mass., available from the JOIDES Office, U. of Washington, Seattle, Wash., 92 pp.

Lehner, P. (1969). Salt tectonics and Pleistocene stratigraphy on the continental slope of the northern Gulf of Mexico, *Bull. Am. Assoc. Petrol. Geol.* 53:2431–2479.

O'Brien, T. B. (1976). Drilling costs: A current appraisal of a major problem, *World Oil*, October, pp. 75–78.

Peterson, M. N. A., F. C. MacTernan, and S. Serocki (resource person) (1977). Abstract of a report to JOIDES on advanced drilling concept studies by the Deep Sea Drilling Project, Feb. 25, available from the JOIDES Office, U. of Washington, Seattle, Wash.

Planning Committee of the Joint Oceanographic Institutions for Deep Sea Sampling (JOIDES) for the Deep Sea Drilling Project (DSDP) (1977). A proposal for research in the International Program of Ocean Drilling (IPOD) by deep-sea drilling in the Atlantic Ocean during the period 1979–1981, available from the JOIDES Office, U. of Washington, Seattle, Wash.

Scholle, P. A., ed. (1977). Geological studies on the COST No. B-2 well, U.S. Mid-Atlantic Outer Continental Shelf Area, *Geol. Surv. Circ. 750*, 71 pp.

Snyder, L. J. (1977). Deepwater drilling has its own technology, *Offshore*, May, pp. 315–322.

14 | The Role of Ships and Submersibles for Continental-Margin Research

THE U.S. OCEANOGRAPHIC RESEARCH FLEET

The U.S. research fleet consists of the following:

(a) The academic-institutions-based fleet. This consists of 28* ships plus the deep-sea submersible, *Alvin*. Of the fleet, 20 ships were constructed with federal funds—ten by the National Science Foundation (NSF) and ten by the U.S. Navy. The scheduling of the fleet is coordinated through the University National Oceanographic Laboratory System (UNOLS). UNOLS is a unique organization, developed as a planning and coordinating mechanism for oceanographic facilities, and is a joint effort by the academic community and the federal funding agencies, principally the NSF, the Office of Naval Research (ONR), the National Oceanographic and Atmospheric Administration (NOAA), the Department of Energy (DOE, formerly ERDA—the Energy Research and Development Administration), the Environmental Protection Agency (EPA), and the U.S. Geological Survey (USGS). Support for operation of this fleet is provided principally by NSF (68 percent), with ONR, USGS, DOE, the Bureau of Land Management (BLM), and other federal agencies contributing most of the balance of a total annual budget of about $22 million. Approximately half of the fleet is engaged primarily in "blue-water" oceanography; the remainder, in coastal

*Plus one under construction to replace an old ship retitled in Fall 1976.

and continental-shelf research. Areas covered include geochemical, physical, biological, and geological–geophysical research.

Both UNOLS and NSF (in coordination with other agencies) are currently developing fleet plans for the coming 10–15 years.

The current distribution of ship size and preliminary projected fleet size is as follows:

General Purpose	Actual 1977	Preliminary UNOLS Projections 1990†
Small coastal (under 99 ft)	9	16
Large coastal (100–149 ft)	4	8
Intermediate (150–199 ft)	9	11
Large (200–240 ft)	5	5
Very large (240 ft +)	2	4
Specialized		
G&G large/intermediate	0	2
Arctic intermediate	0	1
TOTAL SHIPS	29	47
Deep submersible	1	2
Submersible tender	1	2
Manned ocean station	0	1
Shallow habitat	0	1

†UNOLS preliminary projections may change somewhat with further iterations of their plan.

172

The academic fleet, as indicated in this table, is mixed in ship size, distribution, age (not indicated in the table), and age and sophistication of onboard equipment. The ships range in age from the 202-ft R/V *Vema*, built in 1923 to the 177-ft R/V *Endeavor*, built in 1976. Most of the large ships of the academic fleet were built in the mid-1960's, whereas six of the intermediate-size ships were constructed in this decade. In marine geological–geophysical research, a few of the vessels, such as the R/V *Ida Green* of the University of Texas (not under UNOLS) and the R/V *Conrad* of the Lamont-Doherty Geological Observatory of Columbia University (part of UNOLS), are equipped to conduct 24-channel multichannel seismic work. The University of Texas research ships (R/V *Ida Green* and R/V *Longhorn*) are not part of UNOLS and operate at an annual cost of $0.5 million. The bulk of the fleet, however, uses the standard single-channel reflection system, with air gun–sparker sound source, sonobuoy refraction, 3.5-kHz echo-sounder, gravimeter, magnetometer, and coring and dredging systems. Scripps Institution of Oceanography has a 12-channel multichannel system, and Woods Hole Oceanographic Institution has a 6-channel system. Navigation systems aboard these vessels range from satellite to Loran systems. The sharp rise in the cost of operating ships since 1970 has not been fully matched by an increase in the ship-operating funds for support of the academic fleet. Consequently, one or several fleet ships have operated considerably less than full time or have been laid up entirely, each year for the past 4–5 years. A major cause of this has been the steady decline in ONR ship support, only partially offset by ship use by other federal agencies such as the Bureau of Land Management (BLM).

(b) The NOAA fleet is a fleet of 23 research vessels ranging in size from the 303-ft *Oceanographer* to vessels under 100 ft in length. Of this fleet, 13 ships are involved in hydrographic–geological–geophysical–physical–chemical research. The rest of the fleet is involved in biological (largely fisheries) research. The nonbiological research is concentrated primarily in mapping, charting, and physical oceanographic research around the continental margins of the United States, extending into the oceanic regions out to 200 miles. Capital ships of the NOAA fleet, such as the *Oceanographer*, *Discoverer*, *Researcher*, and *Surveyor*, are used for open-ocean research and for environmental studies in hardship areas such as the oceanic areas surrounding Alaska and in the North Pacific. The *Discoverer* and *Surveyor* are working exclusively on BLM work. The cost of operating this fleet is about $34 million/year.

(c) The U.S. Navy research fleet. The U.S. Navy carries out limited geological and geophysical research work along the continental margins. Much of the Navy's research is done aboard research vessels based at the individual laboratories or operated by the Oceanographer of the Navy through the Naval Oceanographic Office or NORDA at Bay St. Louis, Mississippi. The research fleet has eight vessels; they are engaged primarily in physical

oceanographic and acoustical research work. Very little marine geophysical or geological work is carried out. Two of these vessels, the USNS *Chavenet* and the USNS *Harkness*, are 395 ft in length and displace 4200 tons. The other vessels range in size from 200 to 280 ft in length. The cost of operating this fleet is about $38 million/year.

(d) The industry research fleet. This is the largest and the most highly specialized fleet in existence for the study of continental margins. Typically, the vessels are used for oil prospecting along the world's continental shelves; are of the order of 150–200 ft in length; and are equipped with long, multichannel acoustic arrays and complete onboard seismic-processing systems. Contracting companies offering their services are too numerous to list in this summary. Several petroleum companies operate seismic-prospecting vessels of their own. An outstanding example of these ships is Gulf's *Hollis Hedberg* (see Appendix C).

With the present budget, the academic community will not be able to equip or update the capabilities of its research ships to be competitive with the capabilities of commercial ships such as the R/V *Hollis Hedberg* (see recommendation on p. 187).

(e) The submersible research fleet. (For a summary, see Heirtzler and Grassle, 1976.)

DSRV *Alvin*: The *Alvin* is the nation's most versatile and widely used research submersible. Its depth capability is 12,000 ft (actual) and 14,000 ft (theoretical). The *Alvin* is used for most continental-slope as well as deep-ocean (primarily ridge-crest) studies. Together with her mother ship, the catamaran, *Lulu*, the *Alvin* has worked in both Pacific and Atlantic waters, although the bulk of her dives have been concentrated along the Atlantic margin of North America. With an underway endurance of about 10 hours under water, the *Alvin* has served largely as a precise bottom-sampling (biology, geology, chemistry) platform as well as a vehicle for extending the scientist's eye onto the ocean floor. The latter approach has been extremely successful when the *Alvin* worked closely with detailed surface-ship-generated bottom-photo surveys and multibeam sonar dive-site surveys. Support for the *Alvin* and the *Lulu* operations is provided by NSF, NOAA, and ONR. It amounts to $1.5 million annually. Shiptime for the *Alvin* is apportioned by a UNOLS review group, which examines competing requests.

DSRV *Seacliff* and DSRV *Turtle*: These are two 6000-ft-depth submersibles operated by the U.S. Navy and based in San Diego, California. Both vehicles occasionally have been made available through ONR to academic investigators and have largely been used in the mid-Pacific, West Coast, and Panama Basin regions for geological and biological studies. Both vessels are to be converted to a 20,000-ft depth capability within the next 5 years, thus replacing the bathyscaph, *Trieste*, for deep-ocean basin operations. The ONR plans to continue using the two modified submersibles for ONR-sponsored research projects.

The Nuclear Research Submarine NR-1 is operated by

the U.S. Navy out of New London, Connecticut. The *NR-1* has been used extensively by ONR investigators for the study of sedimentation processes on the shelf and slope of the eastern seaboard of the United States and also in the Caribbean and the North Atlantic. Although its depth is limited, the *NR-1* has considerable bottom-endurance time, limited only by human considerations. Therefore, it is especially valuable in the long-term continuous observation of bottom shelf-processes.

Shallow-water (less than 1000-ft-depth capabilities) submersibles are frequently leased by U.S. government agencies as well as by academic institutions for shelf studies.

FIGURE 14.1 The R/V *Alcoa Seaprobe*. (Courtesy of Alcoa Marine Corporation–Ocean Search.)

In a review of a book (Geyer, 1977) on submersibles and their uses in oceanography and ocean engineering, Ballard (1978) says: "... The submersible is a precision tool, not an exploratory one. It is the tool of last resort, used only after the proper homework has been conducted by conventional surface-ship techniques." Ballard believes that part of Geyer's book "should have concentrated on what is needed for submersible programs to carry out more effective work, such as improved support-ship facilities, the development and use of advanced navigational and data-logging systems, and the development of better scientific instrumentation within the submersible for use *in situ*." He also believes that Geyer's book should have placed greater emphasis "on the other ingredients needed to conduct a successful submersible program, such as detailed predive bathymetric reconnaissance, followed by surface-ship programs using deep-towed geophysical instrument systems and unmanned photographic studies." Ballard's observations and philosophy about uses and methods of using submersibles parallel those of this Panel.

(f) *U.S. Geological Survey Fleet.* The USGS currently operates one ship specifically configured for geological studies of the U.S. continental margins, the R/V *Lee*, 208 ft in length, displacing 1297 tons. The *Lee* is equipped with a 24-channel multichannel seismic, air-gun sound-source system. Onboard data processing with high-density tape allows real-time, high-data volume analysis. Coupled to the multichannel system is a shallow-penetration, high-resolution seismic system that consists of a minisparker 3.5-kHz echo sounder and a uniboom array. Gravity and magnetic data are collected under way. Navigation is by satellite and Loran. The 180-ft-long R/V *Sea Sounder*, a chartered ship, has a high-resolution seismic system, Vibracoring equipment, and TV and sonar systems for the study of sediment bottom-transport processes. The cost of operating the fleet is $2.2 million annually.

(g) *Industry.* Worldwide, there are some 16 cable-controlled, unmanned submersibles commercially available. Of these, five have been built in the United States. These vehicles are used mainly for inspecting offshore structures. Because they cost less and are less hazardous, they have replaced manned submersibles for some limited tasks. Chapman and Westwood (1977) have summarized the current capabilities. Little is known about the potential of these vehicles as research tools in shallow waters.

(h) An unusual newcomer on the scene is the dynamically positioned *Alcoa Seaprobe* (see Figure 14.1). This vessel combines the capacity to recover 200-ton payloads from 6000 ft below the surface of the water with the ability to search, core, drill, and sample deposits in water depths up to 18,000 ft. Instruments are at the end of drill pipe and include sensor systems (sonar, TV, still camera, and target illumination) and recovery devices such as grappling claws and coring tools. The potential of this vessel as a research tool merits an in-depth evaluation.

BUDGET SUMMARY FOR RESEARCH FLEET OPERATIONS

Cost of ship operations exclusive of scientific costs:

Organization	Number of Ships	Annual Cost ($ Millions)
Academic research fleet	28	22
Submersible *Alvin/Lulu*	1	1.5
NOAA fleet	23	34
U.S. Navy research fleet	13	38
U.S. Geological Survey research fleet	2	2.2
TOTAL (U.S. Government/Academe)	67	97.7

As stated on page 187 of this report, this Panel believes that much could be achieved with the construction or purchase of two modern geophysical research vessels. The details of the recommendation are spelled out in our conclusions on page 187. In addition, academic research and government survey vessels (i.e., NOAA or USGS) currently operating along U.S. continental margins should collect a full range of underway data such as those on gravity, magnetics, bathymetry. These data, if not processed and used by the institution, should be made available to the community through NOAA's Environmental Data Service. This is particularly important, since, while the cost and accuracy of data gathering have decreased, ship-operating costs have risen astronomically.

REFERENCES AND BIBLIOGRAPHY

Ballard, R. A. (1978). Submersibles and their use, *Am. Assoc. Petrol. Geol. Bull.* 62(5):875.

Chapman, R., and J. Westwood (1977). Unmanned submersibles—operational requirements, *Offshore Services*, April, pp. 26–40.

Geyer, R. A., ed. (1977). *Submersibles and Their Use in Oceanography and Ocean Engineering*, Elsevier Scientific Publishing Co., New York, 369 pp.

Heirtzler, J. R., and J. F. Grassle (1976). Deep-sea research by manned submersibles, *Science 194*:294–299.

National Oceanic and Atmospheric Administration (1977). Manned Undersea Science and Technology—Fiscal Year 1976 report, Rockville, Md.

University National Oceanographic Laboratory System (1975). Preliminary report UNOLS long-range planning meeting, May 1, UNOLS Office, Woods Hole Oceanographic Institution, Woods Hole, Mass.

University National Oceanographic Laboratory System's Advisory Council (1978). On the orderly replacement of the academic research fleet, a report of the UNOLS Advisory Council, UNOLS Office, Woods Hole Oceanographic Institution, Woods Hole, Mass., 15 pp.

Vadus, J. R. (1975). International review of manned submersibles and habitats, presented at Atlantic International Search and Rescue Seminar—LANTSAR '75, April 22–25, New York, published by the National Oceanic and Atmospheric Administration, available from the National Technical Information Service, Order No. PB 24642879WO.

Vadus, J. R. (1976). International status and utilization of undersea vehicles 1976, prepared for Inter-Ocean '76 Conference, June 15–19, Düsseldorf, published by the National Oceanic and Atmospheric Administration, available from the National Technical Information Service.

III | CONCLUSIONS AND RECOMMENDATIONS

15 | General Conclusions

CONCERNING THE ROLES OF ACADEME, INDUSTRY, AND GOVERNMENT

In Appendix A, we review the relative roles played by industry, government, and academe in continental-margin research. Conclusions based on that review can be only judgments. We were not prepared to conduct a study of statistics regarding financial support and manpower levels devoted to continental-margin research. Such statistics are not easy to obtain for several reasons. (1) What constitutes research is not quite clear. The difference between "surveying," "exploration," and "research" is poorly defined and mostly a matter of opinion. (2) We are including in our purview of continental margins the landward continuation of subsea margins. Thus, we would have to isolate solid-earth continental-margin statistics from general oceanographic and earth-science budgets of various agencies. (3) We did not have the time or expertise to distinguish between, compare, and verify the multitudinous funding statistics submitted to us.

Joining the lament of the academic oceanographic institutions over the leveling-out in funding would be easy. It also would be tempting to concur with Senator Hollings, who, in a recent speech* pointed to the lack of

*Remarks by the Honorable Ernest F. Hollings, U.S. Senator from South Carolina, on the occasion of the First Annual Doherty Lecture sponsored by the Center for Oceans Law and Policy of the University of Virginia, held in Washington, D.C., at the National Academy of Sciences, May 11, 1977, *Marine Technol. Soc. J. 11* (No. 3), 30–33 (1977).

coordination suggested by the existence of 21 federal ocean activities scattered among six department and five agencies. He referred to the 1960's as a "decade of ocean rhetoric."

As far as the earth sciences are concerned, the last 15 years were probably the most glorious and fruitful years of this century. Much of the work was done by U.S. oceanographic institutions, and that work was amply funded by the federal government. The excitement of plate tectonics has not yet dwindled, to wit, Project FAMOUS, or the recent exploration of the Galapagos Ridge. Today, we see continental margins in a completely different light, and much of the content of this report was simply inconceivable in the early 1960's.

All this was made possible by a funding structure that, in large part, still exists today (except for the relatively recent shift in funding emphasis from the Office of Naval Research to the National Science Foundation). Whether the success was despite the funding scene or because of it can be debated. We believe that the great diversity of funding goals was very helpful in balancing various views. That balance could have been upset had the funding structure for research been more centralized.

We believe that the diffuse distribution of the federal oceanographic effort among many agencies has not inhibited good research. The spread of the effort and funding among several mission-oriented agencies has supported a greater variety of approaches, thus averting some of the blunders characteristic of overcentralization.

Massive duplication in the earth sciences is largely avoided with the help of various advisory mechanisms, peer review of proposals, overlapping membership on committees, and, in some cases, with the help of our professional societies.

Probably the single factor most responsible for the past success and cohesion of the earth-sciences effort was the personal relationship between geological and geophysical oceanographers. Most of these scientists began their careers in one of three major U.S. oceanographic institutes (Woods Hole Oceanographic Institution, Scripps Institution of Oceanography, and Lamont-Doherty Geological Observatory). An interdisciplinary mode of research advocated by great scientific leaders like Maurice Ewing created lasting personal bonds and a strong network of communications that extends well beyond academic oceanographic institutions.

Coordination of information concerning plate tectonics has been an example of communications at its best. On a global scale, the model was based on magnetic-survey data collected by the Navy and academic oceanographic institutions and on magnetic field-reversal studies produced by the U.S. Geological Survey. The field-reversal studies were correlated with the marine data by two British geophysicists working at Princeton, an institution not involved in the "big science" push for more data. Plate tectonics also has roots in seismological data from the World-Wide Standard Seismograph Network (WWSSN), formerly coordinated by the National Oceanographic and Atmospheric Administration and now by the U.S. Geological Survey, with substantial funding support from the National Science Foundation and from temporary stations operated by academic institutions. Studies of the earth's interior and its properties by scientists in academe and government provided clues about the nature of movements at depth.

A great number of scientists from all sectors of the community made substantial contributions to the development of plate-tectonics models of continental margins. Industry, with advanced seismic-reflection techniques, has provided greatly improved seismic cross sections of continental-margin structure. The same techniques led to detailed seismic stratigraphy and to models of worldwide sea-level change that are making it possible for scientists to establish relationships between continental and oceanographic tectonic events. Academe made major contributions, among which are data indicating that the differences between continents and oceans extend to great depths beneath the earth's surface. Government geoscientists provided both onshore and offshore data that clarified effects of plate-tectonics processes and related them to problems of resources and natural hazards.

Each of these communities was (and is) motivated by different factors—economic, intellectual, environmental, political, or other—but each contributed to the model's development. Each group responds to short-term missions or objectives; each must have the physical and intellectual resources required to cope with long-term needs. Bridling activities of any one of these groups in the interest of economy retards the advance of both scientific and practical objectives.

Are we losing momentum? For the solid-earth sciences, the answer is a qualified yes. Let us focus on the most sensitive areas.

Federal money for oceanographic research has increased substantially in recent years, but most new funds have gone for research in mission-oriented government agencies. Funding for basic oceanographic research done by scientists in academic institutions has barely kept up with inflation.

The consequences of level funding of academic institutions will, in our view, be felt later. Deterioration and lack of new development of equipment, fewer young researchers, and the inability to attack major problems will negatively affect the vitality of these institutions—a vitality critically important to the conceptual understanding of continental margins.

It has been impossible for us to document that academic institutions were unable to obtain funds for work on continental margins because of increased industry and government research and exploration activities on margins. But it is fair to say that only a small fraction of total oceanographic funds available to academe in the past years was spent on U.S. continental-margin research. This is partly because academic institutions preferred to do research abroad and in deeper waters and partly because there was an impression that academic scientists might duplicate industry and government efforts.

Real environmental concerns have to be met to obtain approvals, particularly for U.S. margin research. Penetration of the seafloor by drilling, jetting, or even, in some cases, long piston coring requires permits from the U.S. Geological Survey. Seismic work using explosive sound sources requires special permits because of the danger of killing fish. Obstructions and sometimes excessive delays in securing these permits have discouraged some investigators from attempting research projects that must use these methods on U.S. continental shelves. Scientists who want to do the work have to devote an inordinate amount of time making arrangements to obtain the permits before proceeding.

It appears to be increasingly burdensome for academic scientists to obtain funds for work on continental margins. Ocean scientists must devote substantially more time to proposal writing and administration to get the same amount of dollars as in the past. Dollars that could more productively be spent on research are thus being diverted into procedural channels.

The Law of the Sea Conference has led to increased unilateral claims of exclusive economic zones, which effectively cover all continental shelves and many continental margins of the world (Figure 3). Sanctioning and continuing this trend may subject research on continental margins to lengthy international negotiations. Here again, funds will be sidetracked from research into administration.

In a nutshell, nationalism is shifting the emphasis from concern with oceanography as a science searching for principles and understanding to oceanography as a tool with which to accomplish practical goals (e.g., resource evaluation, platform stability, energy sources).

The roles and organization of government and industry for exploration on U.S. continental margins are subjects of a current political debate that is well outside the scope of our terms of reference and thus are discussed in only the most general terms in this report. We do make a case for strengthening the role of academic institutions in doing basic research on continental margins.

We think it sufficient to give just two of the best reasons for leaving a significant part of basic research in the care of our academic institutions: (1) Academic institutions are our best assurance for a pluralistic approach to elucidating difficult scientific problems with which we are concerned. (2) Government and industry rely on academic institutions to train modern scientists and researchers, and such training is best accomplished in a high-quality research environment.

PURE, APPLIED, AND DUPLICATION OF RESEARCH

Science planning is still plagued by the difficulty of separating "pure" science (or basic research) from "applied" science. Some take the view that "asking for the sake of asking questions" constitutes pure (or basic) science. By contrast, applied science would be that activity that would lead to some practical benefit. Activities in the earth sciences on continental margins are demonstrably both basic and applied in character.

Solid-earth science involves a range of activities extending from conceptual models that help to unravel the dynamics of the earth to straightforward prospecting for resources. One person's research is another person's survey and somebody else's exploration. Consider a multichannel reconnaissance line on a continental margin. If an oil company or seismic contractor shoots that line, it is viewed as exploration (or applied research). If the same line is shot by a government agency, e.g., the U.S. Geological Survey, it is, quite properly, considered to be a mission-oriented survey (applied research). Now, if the same line is shot by an oceanographic institution, its work financed by, say, the National Science Foundation, this line becomes classified as research (pure, basic). An informed (but skeptical) taxpayer, hearing that the same line in the same area was shot by all three groups, and not understanding the different uses of the information to be gained, no doubt would despair over duplication of effort and inefficiency and vigorously insist on greater coordination.

In reality, however, the matter is not so simple a case as duplication or lack of coordinated efforts. (1) The parties concerned would shoot the same line differently, thus collecting data differently, and each party would use different processing technologies. Chances are that different timing of the three projects would reflect differing vintages of a quickly evolving technology. (2) The line probably would not be located in precisely the same place. (3) Most important, all parties are looking at entirely different aspects of the problem and are using different techniques tailored to their particular problem.

Despite these differences, some overlap will usually occur. This is in keeping with the competitive spirit of research on continental margins. Such duplication can best be minimized by dialogue between the working groups concerned. Larger and more centralized organizations would provide watchdog mechanisms aimed at controlling "unnecessary" duplication. But the costs of watchdogs and their administrative overhead would, in our view, vastly exceed the benefits derived from their activities. Furthermore, the demoralizing effect of large, monolithic institutions on research morale and quality is immeasurable.

The original charter of the International Decade of Ocean Exploration (IDOE) emphasized resource management as a major goal. Therefore, IDOE was often viewed as a program in applied research. This impression was reinforced by the fact that there was no significant activity on domestic continental margins in IDOE's Seabed Assessment Program, presumably because other government agencies were doing mission-oriented work on U.S. margins. Note also that, in regard to potential Law of the Sea definitions of and distinctions between pure and applied research, a precedent was set when a project as esoteric as Project FAMOUS was undertaken as part of a program directed at better management of marine mineral resources. With respect to the Law of the Sea negotiations, all solid-earth research can and—in the prevailing climate of distrust—probably will be construed as having some bearing on resources and their exploitation.

Continental-margin work encompasses the full range from purely fundamental, process-oriented studies to the highly practical delineation of resources. It would be a serious mistake for any government agency, industry, or academic group to attempt to monopolize any segment of this range. Yet, the need for efficiency in the expenditure of manpower and financial resources requires that, in a continuously refined process, there be some determination as to who should do what. This requires a continuous dialogue, since the abilities, interests, and orientations of the participating entities perennially change.

DOMESTIC VERSUS INTERNATIONAL PROGRAMS

As a consequence of the acceleration of the unilateral claims to 200-mile territorial seas and the debates concerning the Law of the Sea, it is probable that continental-margin research in the future will be subject to some form of consent or approval by the governments of the adjacent coastal states (see pp. 218–221).

Foreign coastal states can now contract commercially high-quality geophysical and geochemical surveys for their resource evaluation. Although some of these surveys are fairly expensive, most coastal states should be able to justify them as high-priority items. Also, many states probably will be able to get partial support for such work from the World Bank or other groups that assist developing nations. These developments may severely restrict academic oceanographic surveys on foreign continental margins, but there are ways to turn these difficulties into research opportunities.

Because of the nature of the commercial data-gathering process, in which the contractors perform the work, scientists in foreign nations do not have much of an opportunity to participate. Often, they resent this "black-box" approach. They prefer to participate in projects in which they can be involved in the data gathering and processing. This degree of involvement demands training and total participation of foreign scientists in those projects and, in some cases, even a transfer of hardware. Future U.S. research on foreign continental margins is likely to be more welcome if it is also associated with a transfer of technology and training of foreign nationals. The commercial and political consequences of this mode of research are hard to evaluate, but there can be little doubt that it will put the science on a much broader international base.

Oceanographic institutions from developed nations other than the United States will compete for the opportunity to do research on foreign continental margins. We need a convincing showcase. Therefore, it would be judicious to have a strong academic research program on our own domestic margins (1) to develop the most advanced research we can conceive and afford and (2) to give some credibility to the concept that basic research is a pathfinder as well as a support for more applied activities.

CONCERNING BIG SCIENCE

Much earth-science research in recent years has been done through multidisciplinary, multi-institutional, and often costly ("big science") projects. The most spectacular successes have come from the Deep Sea Drilling Project. Other examples are Project FAMOUS, some of the imaginative projects undertaken under the auspices of IDOE (such as the Nazca Plate Project), and the CLIMAP Project. The Atlantic Shelf Project of the U.S. Geological Survey could be viewed as a big-science undertaking. Lately, on land, the Consortium for Continental Reflection Profiling (COCORP) project is developing into big science.

The advantages of the large-scale approach are evident. It is unlikely that the problems of continental margins will be solved without such a large-scale approach. The stimulus these projects have given to many research activities has been pervasive and, in the main, beneficial.

There are, however, some drawbacks. An attempt should be made to optimize large-scale programs of the future.

1. Proposal procedures tend to be elaborate and very detailed in technical and practical matters, reflecting the desire of the funding agency and the sincere effort of the investigators to justify with some sense of responsibility the expenditure of large amounts of public money. Yet, with the pressures and logistics of designing large programs, the scientific objectives and the relations between the research program and the expenditures are often unclear. The true nature of the scientific question is often lost in long, generic statements of past accomplishments, justifications for equipment purchases, and much verbiage addressing the relevance of the program rather than the problem-solving itself. As a result, the task of reviewers has been difficult, and we feel that the usual peer-review process has been decreasingly effective in evaluating scientific merit and eliminating waste and inefficiency. A considerable number of the nation's best scientists are tied up in writing, selling, and reviewing proposals. Once a project is under way, the situation seldom improves; the same talent is then wasted on complex bookkeeping, reporting, and accountability requirements. We fully accept the need for accountability, but we suggest that, in the decision-making process between big and small science, this inherent waste continuously be considered and that a real effort be made on the part of the funding agencies to simplify the proposal and reporting procedures.

2. Big science can be noncreative; it often leads to massive, undigested data banks. There is a strong tendency, both among researchers and especially in funding agencies, to emphasize the planning and data-gathering phases of big projects at the expense of the follow-up activities (interpretation, impact evaluation, follow-up experiments, technology transfer). This tendency, aptly labeled by some *Flucht nach Vorne* (escape to the future) has prevented some major projects from reaching their full potential.

3. There is also the possibility that large-scale science may drain funds and certainly drains personnel from smaller efforts. The smaller, individual programs are often the ones that develop the true conceptual advances. The continuing health of U.S. science depends on active and attractive small-science programs.

The program suggested in this report is big science. However, the total program should be an agglomeration of both small-scale and large-scale efforts, which, to be most productive, must be coordinated. Too often in the past, artificial boundaries (e.g., political, disciplinary, area of investigation, funding) have been created that inhibit a total understanding of the system. While a major effort is required to solve a major problem such as the history and nature of continental margins, we should avoid training a generation of young scientists who see big, cooperative research programs as the only means to subsist, a generation of young scientists who are not prepared to do the smaller-scale individual creative work.

16 | Recommendations

Parts I and II of our report contain many recommendations printed in italics. These are made within the scientific context and are given no priorities. In this chapter, we establish priorities, using three different styles of presentation.

1. *High Priority:* In this class, recommendations and conclusions are in **boldface print.**
2. *Second Priority:* Recommendations that are either of secondary importance or subsets of the high-priority items are presented in *italics.*
3. *Lower Priority:* High- and second-priority recommendations are repeated in Parts I and II, where they are *italicized.* In addition to these, Parts I and II contain lower-priority items, also printed in *italics.* These are good but less urgent programs.

Portions of the following recommendations are included verbatim in the Introduction, Summary, and Principal Recommendations at the beginning of this report. Here in Part III, we elaborate on the recommendations, adding some of the important component recommendations and giving the reader significant details related to the recommendations. A program manager or researcher who wants still more detail is referred to specific sections in the main body of the report (Parts I and II) that are relevant to individual recommendations.

HIGH-PRIORITY RESEARCH FOR THE 1980'S

A SEDIMENT DYNAMICS PROGRAM—HIGH PRIORITY

We recommend that a coordinated program be developed to study sediment dynamics on continental shelves, slopes, rises, and marginal basins. This program would study the entrainment, transport, and deposition of continental-margin sediments.

This program relates to the increased use of large areas of the continental margins—particularly the shelves—for food resources, mineral resources, waste disposal, oil exploration and exploitation, and recreation. Current research efforts need coordination. For example, a multitude of federal projects are under way in the Mid-Atlantic Bay (between Cape Cod and Cape Hatteras). With some exceptions, these programs appear to function independently of each other. Agencies that do in-house research and that contract out parts of their research are the Army Corps of Engineers, the Environmental Protection Agency, the National Oceanic and Atmospheric Administration (NOAA), and the U.S. Geological Survey (USGS). Research is also contracted by the Bureau of Land Management (BLM) and the Department of Energy. Basic research grants are awarded by the National Science Foundation (NSF) and the Office of Naval Research (ONR).

The basic mechanisms of boundary flow (near the seafloor), entrainment, and transport of sediments are understood poorly or not at all. There are almost no observations of any kind upon which to base initial questions, and we lack proper instruments partly because we do not know what measurements to make. Problems of boundary flow and sediment transport must be attacked with teams of physical oceanographers, marine geologists, and fluid dynamicists, with electronic engineers to design and build the equipment.

One agency should coordinate the sediment dynamics program with other federal agencies. A basic research program should be funded by the National Science Foundation, and an applied academic program could be funded by the Sea Grant Program of NOAA. As a by-product of an IDOE-sponsored workshop,* an advisory committee on shelf-sediment dynamics has been formed to coordinate research in that special field. The larger problems of the margins should be examined by similar panels treating the special dynamics and conditions of slopes, marginal basins, and rises. These panels should be grouped in a national margin–sediment dynamics project on the order of the Geodynamics Project.

Funding of sediment dynamics research should be increased from its roughly estimated current level of $10 million/year† to about $18 million/year (in 1977 dollars) in the 1980's. The additional $8 million/year should include the following:

- $5 million/year for sediment dynamics, which would provide a sophisticated equipment pool (e.g., solid-state current meters, shelf stations, wave followers) and would permit initiation of four or five major regional cooperative studies of differing margin types.
- $1 million/year for benthic boundary layer studies other than those covered by sediment dynamics.
- $2 million/year for studies of sedimentation and erosion on continental slopes and rises.

Primary Components of the Sediment Dynamics Program

1. *We recommend a research program to examine the dynamics of contemporary sediment transport from rivers to the continental rise and to examine boundary-layer and water-column processes.*

We need to know the rates of these transport processes at various energy levels, particularly under storm conditions. Mass balances must be defined for the major transport systems, bedload, and suspended load. Mathematical models must be developed and field tested.

Another poorly understood problem is how particles

*D. S. Gorsline and D. J. P. Swift, eds. (1977). Shelf Sediment Dynamics: A National Overview. Report of a workshop held in Vail, Colorado, November 2–6, 1976. Available from the National Science Foundation.
†Crudely estimated combined expenditures of NSF, BLM, NOAA, and USGS.

fall through the water column and how they move on, in, and through the sediment column once they have settled from suspension. Sediment settling can best be studied through sediment-trap experiments. Several such experiments that emphasize sediment settling are under way on marine continental margins with BLM and NSF support.

2. *As part of the International Decade of Ocean Exploration Program on Shelf and Nearshore Dynamics of Sediments (SANDS), we recommend a concentrated interdisciplinary research effort on the benthic boundary layer.*

Study of the benthic boundary layer involves all the basic oceanographic disciplines (physical, biological, chemical, and geological) and will entail long-term, bottom-mounted stations to make and record such measurements. Some of the instruments (e.g., nephelometers, current meters) are available; others (e.g., recording respirometers, devices to measure sediment mixing and early diagenesis) should be developed. The magnitude of benthic boundary layer problems requires a large-scale involvement of many scientists and correspondingly high levels of financial support.

High-resolution data are necessary to understand the following problems:

- Water–sediment chemical interaction;
- Metabolism at the sediment–water interface;
- Organism–sediment relationships;
- Variation in water turbulence and stability;
- Velocity gradients, other physical and chemical parameters;
- Deposition, consolidation, and stability of the seabed (e.g., its susceptibility to mass movement, erosion, and scour).

3. *We recommend an intensive study of morphology and the deposition and erosion of surficial sediments on slopes and upper rises of both passive and active continental margins.*

We probably know less about this physiographic region than any other of the oceans. We would like to know whether sediments are currently accumulating on the slope and rise or if large portions are being eroded. While the USGS has undertaken such studies on the shelf, both the NSF and the ONR appear reluctant to sponsor such studies on the shelf or the slope. Documenting late Quaternary sequences and rates of accumulation will require coring the upper few-to-tens of meters of the sediment column together with detailed topographic studies. The mechanics of mass movement on slopes are of interest. Few *in situ* data are available to support theoretical models of these processes.

Present conditions on the slope and rise may differ substantially from past conditions. It will, therefore, become important to compare and contrast the present slope and upper-rise regime with former, older slopes as exhibited on seismic lines. Eventually, these regimes also will need coring and detailed stratigraphic, paleontologic, and engineering studies.

4. *We recommend that the nonmilitary sector of the U.S. Government develop a multibeam swath-mapping capability, particularly for studies on the slope, the upper rise, and some of the adjacent deep-ocean regions.*

To study sediment dynamics and submarine morphology, we need extremely accurate bathymetric maps. These are best obtained by using the multibeam echo sounder that was developed at great expense by the Navy (the so-called "Harris Array").

Swath-mapping instrumentation requires a sizable vessel, and since three of the NOAA ships—*Researcher, Discoverer,* and *Surveyor*—already have elements of the array mounted on their hulls, *we recommend installation of deep ocean swath-mapping capability on two of these ships.* The vessels should be made available for nongovernmental as well as governmental researchers working on domestic continental margins. One vessel should operate on the East Coast and in the Gulf of Mexico and the other on the West Coast.

Installation costs are on the order of $500,000/vessel.

Operating costs are on the order of $1,500,000/vessel/year.

Secondary Component of Sediment Dynamics Program

5. *Quaternary coastal studies. In addition to mission-oriented studies of the coastal zone, we recommend more research on Quaternary sedimentation in differing coastal settings.*

To put coastal geology in a broader context, we need numerous local studies to understand the effects of Quaternary sea-level changes on coastal sedimentation and erosion. *We recommend that coastal universities and state laboratories intensify their efforts in this field.* We hope that studies of different situations will reveal general principles helpful in formulating a strategy for the conservation of coastal habitats. The projects should be funded at $2 million/year.

6. *Sniffing techniques. We recommend that environmental and sediment dynamics studies make full use of commercially available sniffing techniques.* These techniques should be further developed to trace and map the distribution of nonhydrocarbon chemical components in waters overlying the seafloor.

7. *We recommend increased research aimed at the evaluation of differences in sediment distribution and sedimentary processes on modern continental shelves as compared with pre-Quaternary shelves.* Because this work is crucially dependent on drilling data, it will have to be coordinated with future drilling plans.

A PROGRAM FOR GEOTRAVERSES ON DOMESTIC CONTINENTAL MARGINS—HIGH PRIORITY (See also p. 189)

We recommend recording long, multisensor (reflection, refraction, potential, and electrical) geophysical and geological traverses across land and marine segments of U.S. continental margins. Traverses should extend across both active and passive margins into the continental interiors. General problems that are involved include the kinematic, dynamic, and thermal evolution of active and passive margins. Geotraverses over land and water are recommended for the East Coast, Gulf Coast, and West Coast, with a large proportion of the whole geotraverse effort devoted to Alaskan continental margins.

These geotraverses should be made across the strike and across existing boreholes in the most effective way. Special seismic three-dimensional shooting techniques may be required to achieve resolution beneath detachment surfaces. This kind of work plus associated LANDSAT studies would form a bridge between land and marine geology.

A general strategy for such traverses is as follows.

(a) Compile previous work; continue reconnaissance on land and at sea, with emphasis on multichannel seismic data and long refraction lines.

(b) Narrow down to regional studies on land and sea.

(c) Concentrate on one or more transect bands (which typically may be some 100 miles wide and some 400–600 miles long, half on land and half over water). Advantage should be taken of existing well control.

(d) Emplace long-term monitoring devices on transects of seismically active margins. Temporal observation of seismic events, the stress–strain field, fluid distribution, and crustal displacements all will assist in developing models of continental margins.

(e) If necessary, complete the study with deep-drilling programs on land and in water, using current drilling capabilities on shelves and, eventually, deep riser-drilling (e.g., *Glomar Explorer*) for slopes and rises.

The costs of a transect will vary with accessibility, length, and complexity of the geological problems involved. For one typical land–sea transect, we estimate the costs to be in the following approximate range:

Geology and Geophysics	Costs ($ Millions)
Marine geophysics	3.0 to 5.0
Land geophysics	1.0 to 2.0
Marine geological and geochemical work	1.0
Land geological and geochemical work (including helicopter support as needed)	1.0 to 2.0
TOTAL cost for one net geotraverse (without drilling)	6.0 to 9.0

It is important that coastal academic institutions concerned with onshore geology, oceanographic institutions, government agencies, and industry operate together in this program. The quality of a traverse program can be maintained only if the program reflects the interests and background of people familiar with the local region. Therefore, the grass-roots aspects of the program are very important. The Panel believes that the U.S. Geodynamics Committee could provide the much-needed neutral ground for such an undertaking. However, its membership would need to be expanded to include more stratigraphers and sedimentologists.

Communication and Coordination

We recommend that a committee be established to set up four regional working groups to draft the details and costs for geotraverses and related work in the East Coast, Gulf of Mexico, West Coast, and Alaska. Those who direct the project should enlist scientists from government, academe, and industry, selected for their competence in the field. The COCORP (Consortium for Continental Reflection Profiling) Program and the Deep Sea Drilling Project are excellent administrative models.

We again emphasize that the geological–geophysical part of the program is an independent activity that can provide valuable scientific insights. Drilling may be needed in some cases to solve key problems, but, in our view, drilling without adequate geological and geophysical preparation is wasteful.

We see the following as problems deserving a concentrated effort:

- Rifting on passive margins;
- Stratigraphy, diagenesis, rate of subsidence, paleooceanography during rifting of passive margins;
- Evolution of continental fragments;
- Continent–ocean boundary on all margins (see detailed recommendation below);
- Thermal history of passive margins;
- Formation of accretionary wedge and forearc basins on active margins;
- Geochemical evolution of active margins (see detailed recommendation below);
- Formation of marginal basins;
- Effective monitoring of seismicity on all domestic margins.

Atlantic Transects

Over the next decade, some carefully designed transects should include the Florida–Blake Plateau, North Carolina to Cape Hatteras offshore, New Jersey and offshore, and New England and offshore. Information from these transects would be most useful for the study of mature passive margins.

Gulf of Mexico Transects

In the near future, numerous multichannel reconnaissance lines in the Gulf of Mexico will be completed. Following their publication and interpretation, a detailed plan can be worked out to integrate these lines with in-

formation from the landward side of the Gulf. Deep crustal refraction work is needed to complement the reflection lines.

West Coast Transects

West Coast transects should be planned as a follow-up of the cross-section project of the Plate Boundaries Group of the U.S. Geodynamics Committee. Many of these sections were based on surface geology. Now, multichannel seismic data and crustal and mantle refraction work are needed to complement the surface data. Traverses should contrast two important areas: (1) Southern California and the California Borderland—a classical transform-fault plate boundary; and (2) Oregon and Washington onshore and offshore—a typical subduction boundary.

Alaska Transects

An intensified earth-science research effort onshore and offshore Alaska is needed. More than 50 percent of the marine continental margins of the United States are off the shores of Alaska. Alaska should be an integral and important part of any national program. An Alaskan multisensor transect should extend all the way from the Aleutian Trench across the mountains to the Beaufort Sea. On land, a transect should include detailed geological and geochemical studies, multichannel reflection surveys to map from the surface to the upper mantle, refraction surveys, and so on. An all-water transect across the Bering Sea is also needed. The Alaskan transects may be viewed as part of a joint Arctic Ocean program with Canada, or part of the U.S.–U.S.S.R. project for plate-dynamics traverses.

Primary Components of the Transect Program: Continent–Ocean Boundary Surveys

We recommend that systematic geological and geophysical studies be made to study the transition between oceans and continents on passive margins. Such studies would search for adequate descriptions of the boundary zone, based on multichannel seismic-reflection and -refraction data. Bedrock mapping, using submersibles, would help to calibrate some of the seismic data. Surveys of simple young margins, young margins drowned with sediments, old starved margins, and related microcontinents and old mature margins need to be made.

Costs for a typical survey of the continent–ocean boundary are on the order of the following:

2000 miles for multichannel reflection lines @ $500/mile	$1.0 million
1000 miles refraction (20 profiles)	$0.8 million
Submersible operations	$0.5 million
TOTAL	$2.3 million

Primary Components of the Transect Program: Seismic Stratigraphy

We recommend that great care be taken to provide sufficient coverage to map transgressive and regressive cycles in detail. We recommend that future deep-sea drilling on continental margins be aimed at coring and testing well-described seismic–stratigraphic sequences. In our judgment, this will provide valuable and reliable control for paleooceanographic studies.

Secondary Components of the Transect Program: Drilling

Drilling should be considered only after comparing and contrasting the results of different transects. The work on any given transect may well conclude with no drilling proposal. On the other hand, should drilling be justified for better understanding of the geology, then estimated costs would be in the following ranges:

Drilling 3–4 deep holes on land and on the continental shelf	$45 million to $60 million
Drilling 2–3 holes in deeper water	$60 million to $90 million
TOTAL	$105 million to $150 million

OUTFITTING MODERN GEOPHYSICAL RESEARCH VESSELS—HIGH PRIORITY

Many vessels in the U.S. oceanographic fleet are inadequately instrumented for the resolution and deep penetration needed to solve significant geological problems on continental margins. Cost studies for marine surveys show that the major expenditures are for the vessel, precise navigation, and scientific and technical personnel. Once committed to these expenditures, the incremental cost for the best geophysical and geochemical equipment is comparatively small and is desirable for cost-effective acquisition of geological–geophysical information. To work with obsolete equipment is a waste of money, scientific talent, and time. The significant geological and geophysical research on the continental margins requires the best multisensor state-of-the-art geophysical and geochemical instrumentation, including multichannel seismic reflection and refraction, and onboard data processing. (See also p. 11.)

Therefore, we recommend that funds be made available to outfit two such vessels to operate principally (1) on the East Coast and Gulf of Mexico margins and (2) on the West Coast and Alaska margins. The two ships should have reinforced hulls to permit summer work in Arctic waters. The *Hollis Hedberg* (described in Appendix C) is an example of such a vessel.

The vessels should be equipped with onboard digital processing capable of real-time monitoring of all sensors, including seismic deconvolution and stacking of sufficient channels (at least half of those recorded) to ascertain that the data recorded are qualitatively and quantitatively adequate to solve the objectives of any mission. To contemplate outfitting vessels now with less than 48 to 96 seismic-reflection channels would fail to take advantage of available technology.

While a substantial amount of processing would take place onboard, it would be necessary to plan facilities and/or contracts for postcruise playback processing to support detailed interpretations.

From industry experience, it is estimated that to build (or purchase) and equip vessels as described above would cost $12 million to $14 million each (in 1977 dollars). Total annual costs, including shipboard operations, data acquisition, onboard and postcruise processing, and interpretation are estimated at $9 million to $11 million for each vessel. The availability of two modern geophysical vessels would substantially reduce the demand on ships currently used for geophysical research and permit them to be used more heavily in other fields of oceanographic research or permit a reduction in the size of the existing fleet.

The geophysical vessels should be highly available to and shared by qualified scientists, primarily members of the academic community. Industry and government scientists have access to their own modern geophysical vessels. The ships should be national facilities based at two selected oceanographic institutions. Disposition of the vessels and their long-term programming would be a major responsibility that should respond to national scientific goals, while component projects should be selected on the merit of the science proposed. Planning and advice could be structured in analogy to other "big-science" efforts, such as IPOD-DSDP, IDOE, or COCORP.

SECOND-PRIORITY RESEARCH FOR THE 1980'S

GEOTRAVERSES AND OCEAN–CONTINENT BOUNDARY STUDIES ON FOREIGN CONTINENTAL MARGINS—SECOND PRIORITY

We recommend that foreign geological and geophysical traverses be carried out if the specific problem cannot be studied on U.S. margins or if the problem is complementary to the domestic work. If the same class of problems can be studied on domestic margins, they should be studied there. In general, most problems outlined for the domestic geotraverse program can also be studied on other continental margins. The sequence of activities and the research methods would be similar, and the relevance is consonant. In order of priority, the following areas are of particular importance.

Arctic Ocean Margins

We recommend that an international effort be undertaken to better understand the continental margins of the Arctic Ocean.

• We recommend that aeromagnetic surveys of the type currently being pursued by Canadian and American scientists be extended to cover the entire Amerasian Basin and surrounding continental margins. When the technology for airborne gravity surveys is realized, we recommend that these measurements be included in the surveys.

• We recommend crustal seismic-refraction experiments with multichannel reflection lines and gravity measurements to determine the nature of the crust in Arctic basins (particularly the Amerasian Basin).

• We recommend that borehole information from Arctic wells be related to stratigraphy as it is shown on seismic-reflection lines.

• We recommend that efforts be made to develop instruments that will make geophysical measurements both under the ice with tethered or remote vehicles and through the ice.

• We recommend more paleomagnetic studies from any portion of the Amerasian Basin, the surrounding landmasses, margins, and continental fragments.

• We strongly recommend that the effort in the Arctic be a joint project between Canadian and American scientists. Furthermore, Russian and Scandinavian participation is highly desirable and should be actively sought.

• Part of this program may overlap with the traverse program in Alaska. The cost for a typical Arctic Ocean transect is about $5 million. To make some five transects would require about five years and a total cost of about $25 million. (See Chapter 7 for further discussion.)

Caribbean Margins

We recommend that the Caribbean, where the structural and stratigraphic evolution is not understood and is the subject of intensive scientific speculation, be studied in many transects.

Conjugate Passive Margins

We support studies of conjugate passive margins. To compare and contrast passive margins that are more or less symmetrical about midocean ridges can be most useful. Insights gained on one margin may be applied to its symmetrical counterpart. On the other hand, serious mismatches between conjugate passive margins could lead to substantial revisions or modifications of currently popular plate-tectonics concepts.

The most obvious candidates for such studies are the Northwest African Atlantic margins, which could be compared with North America margins. Couples of other passive margins could be studied on the margins of the South Atlantic and the Indian Ocean. Such studies would be fruitful if the adjacent coastal nations would be willing to share in the project and make existing subsurface information available.

West Pacific Margins and Their Marginal Seas

This type of margin is not well represented in the United States. Study should be continued because the formation of marginal seas is a principal, still unsolved plate-tectonics puzzle.

We encourage continuation of the work on the East and Southeast Asia transects (SEATAR, or Studies in East Asia Tectonics and Resources) that are now carried out under joint CCOP (Committee for Coordination of Joint Prospecting for Mineral Resources)–IDOE auspices. In the same vein, we encourage the plans for joint U.S.–U.S.S.R. transects in the North and Northeast Pacific.

DRILLING ON CONTINENTAL MARGINS—SECOND PRIORITY

The following summary of where we stand today is based on a memo to the National Science Board written in spring 1977.*

IPOD is the international phase of the Deep Sea Drilling Project (DSDP). There are five contributing foreign countries (the Soviet Union, Germany, the United Kingdom, France, and Japan). IPOD drilling with the *Glomar Challenger* is scheduled to end in 1979. This will be followed by a phasing out of DSDP in 1980. By that time, DSDP costs will have totaled $137 million, including $19 million in foreign contributions.

It has been determined that *Glomar Challenger* capabilities are not sufficient to achieve the desired scientific targets on continental margins and oceans in southern latitudes. To reach many (although not all) of these targets, a dynamically positioned drill ship is needed that can deploy a 12,000-ft riser and a well-control system for safe drilling on continental margins. Commercial drilling ships are not expected to have riser capabilities much in excess of 6000 ft by the mid-1980's (see Figure 13.3). Ships such as the *Discoverer Seven Seas* could drill in water depths up to about 6000 ft. Such ships are contracted for years ahead. However, if they are available, they could help substantially to solve some continental-margin problems (i.e., the calibration of seismostratigraphic sequences on continental shelves, drilling into the top of the marginal highs of passive margins, and drilling deep into subduction complexes and into the ridges in front of forearc basins). The ships' operating costs are on the order of $80,000 to $100,000 per day.

---

*Memorandum to Members of the National Science Board (April 1977). Subject: "The Future Drilling in the Deep Ocean for Scientific Purposes in the 1980's." NSB-77-169, signed by R. C. Atkinson, Acting Director, National Science Foundation, Washington, D.C.

An international planning conference held in spring 1977 at Woods Hole, Massachusetts, weighed many alternatives, including drilling by commercial ships. That group developed a preferred program. It calls for continued use of the *Glomar Challenger* for seven years and *Glomar Explorer* (equipped with riser and well-control system) for six years. A three-year overlap in the operation of the two ships is anticipated. The overall program is estimated to cost $400 million, excluding the $50 million necessary to convert the *Glomar Explorer* for riser drilling. It should be kept in mind that the proposed program also foresees drilling in the abyssal portions of the oceans and in high southern latitudes. Therefore, the merits of the program should not be judged only in terms of what it can do for continental margins.

The budget of the proposed preferred plan breaks down into $113 million for seven years of *Glomar Challenger* operations; $162 million for six years of *Explorer* drilling; $83 million for geophysical work; $38 million for analysis, interpretation, and synthesis; and the already mentioned $50 million to convert the *Glomar Explorer*.

The members of our Panel were particularly gratified to see the emphasis on geophysical work preceding the drilling project and the stipulation that logging should be carried out on all drill holes. Substantial portions of the plan proposed by our Panel are aimed at providing the background that would enable scientists to choose the best drill sites for solving key problems. In particular, the recommendations for domestic transects will help future continental-margin drilling focus on targets that are on our own nation's continental margins. A number of problems will remain that are best solved on continental margins outside North America.

We would like to stress once more that the proposed geological and geophysical programs stand as important research items independent of the drilling. Drilling, of course, would immensely enhance the geological and geophysical programs by providing the much-needed ground truth, but drilling cannot stand alone.

Therefore, we fully and emphatically concur with the recommendation of the JOIDES *Subcommittee on the Future of Scientific Ocean Drilling (*FUSOD*) that such an ambitious drilling program be undertaken, but* **only if adequate funding is assured for scientific studies** *that include (1) broad-scale problem definition, (2) small-scale site examination and preparation, (3) sample analysis and well logging, and (4) interpretation and synthesis.*

This Panel also agrees with FUSOD that the high total cost of the proposed plan for continued *Glomar Explorer* drilling (about $450 million) dictates that the best possible geological and geophysical reconnaissance be undertaken to provide a selection from which to choose the best drilling sites.

Because continental margins extend onto land, geophysics and land drilling should be part of the same program. Costs for that part of the program can be estimated only after the committees concerned with land aspects of continental-margin drilling have been consti-

tuted and have reviewed the widely scattered geological and geophysical information.

By doing both the science and the drilling on the marine and landward portion of U.S. continental margins, the United States could, for the first time, have a comprehensive approach to solid-earth research on domestic continental margins.

The recommendations of the Panel on Continental Drilling of the FCCSET *Committee on Solid Earth Sciences do not address the problem of continental-margin drilling on land. Such drilling could be complementary to and provide continuity with the program envisaged in the* JOIDES/FUSOD *proposal. We recommend that a committee be appointed and funded to work within the framework of the U.S. Geodynamics Committee and to develop specific plans for geological and geophysical traverses on the landward extensions of continental margins (see pp. 185–186) with a view toward continental-margin drilling on land. The approximate costs for such a program need to be established.*

In principle, this Panel agrees with the FUSOD recommendations. However, we believe that our recommendations concerning sediment dynamics, domestic traverses, and outfitting modern geophysical vessels have a higher priority and are more important for the healthy growth of our continental-margin research efforts than drilling plans, which we rank as only second priority.

It is difficult to assess the importance of drilling for scientific purposes. There is no doubt in our minds that drilling key research wells on continental margins will be the fitting ultimate test of some of our ideas. The prospect of having the necessary tools available in time to test new concepts is exciting. We also recognize the important role of such a drilling project for international scientific cooperation. Nevertheless, the Panel gave the following reasons for giving a second priority to the drilling program:

1. The fact that we are so insistent on adequate and intensive preparatory geological and geophysical work preceding any drilling suggests that, at this time, the ideas for a drilling program are not very specific.

2. In our judgment, the high-priority efforts that we have recommended stand as valuable research targets quite independent from any drilling plans. Some of us are particularly concerned that a traverse program as well as other oceanographic research projects might be refashioned into a simple surveying process focused mainly on finding a sufficient number of drill sites in time, while other scientific objectives are downgraded.

3. We believe that, while some drilling targets could be reached only by the proposed 12,000-ft riser technology, many objectives could probably be reached with the current commercial drilling technology. We judge that, with proper planning, technology could be made available for scientific drilling in the 1980's. Note also that the FUSOD report (*The Future of Scientific Ocean Drilling*, a report by an *ad hoc* subcommittee of the JOIDES Executive Committee, 1977) identifies some sites with 13 km of

sediment over basement and deep crustal holes in 18,000 ft of water with 9000 ft of penetration. The proposed 12,000-ft riser technology will not be able to reach these deep targets.

4. The overall logic inherent in a research program leading from a geological–geophysical reconnaissance to detailed surveys and then drilling suggests that drilling should have a second priority. In other words, first the high-quality geophysical research, then the drilling.

To sum up, we believe that it is far preferable that the basic science be healthy and adequately funded before more expensive drilling is planned.

ENDORSEMENT OF RECOMMENDATIONS OF THE PANEL ON SEISMOGRAPH NETWORKS

Effective monitoring of continental-margin seismicity, studies of earthquake dynamics, and investigations of the deep structure of continental margins depend critically on the collection of high-quality, broadband seismic data from both global and local networks of instruments. *We strongly support the recommendations contained in the recent National Research Council report by the Panel on Seismographic Networks of the Committee on Seismology—specifically their recommendations regarding the global networks:*

Recommendation 1. Stable funding . . . should be established to assure . . . continuing operation of the WWSSN as a basic research facility for U.S. investigators.
Recommendation 2. Stable funding should be established . . . to continue operation, maintenance, and improvement of the [digitally recording] HGLP; SRO; ASRO; and IDA seismograph stations. Further, [a number of] WWSSN stations should be upgraded to include digital recording capability. . . . Facilities for the organization, storage, retrieval, and distribution of digital data from the above observatories [should be provided].

The study of the structure and seismicity of the continental margins on both local and regional scales requires the use of ocean-bottom seismometers. *Therefore, we recommend that a strong national program in ocean-bottom seismology be established and maintained. Elements of this program should include sufficient funds for instrument development.*

CONCERNING PROPOSAL WRITING AND THE DISTRIBUTION OF GRANTS

Writing proposals for research, attending committee meetings, and particularly the ever-increasing budget monitoring and administrative work required to verify the time (in minutes!) spent on each project are fast becoming dominant and time-consuming activities for many of the nation's best researchers. The time many scientists spend on various planning committees with overlapping scopes is time they cannot devote to their research.

We recommend that the National Science Foundation (NSF) take the lead to review and streamline current research funding procedures with a view to

(a) Designing a more standardized format for research funding procedures that NSF and most other federal agencies could use;

(b) Minimizing the length of research proposals;

(c) Limiting the required length of vital statistics and scientific pedigree of the requesting researcher(s) to one page (and for the bibliography, only six key references of the researcher's own choice);

(d) Standardizing budget forms; reducing budget details to the minimum acceptable to federal auditors; and

(e) Streamlining committees and their procedures to avoid duplications and to limit the scope and size of such advisory groups.

We further recommend that NSF study the feasibility of "progressive grant status" for institutions engaged in well-circumscribed basic research fields, e.g., continental-margin research, as follows:

1. Project support—for individual projects;

2. Coherent—for groups of projects at an individual institution; and

3. Institutional—for large programs at institutions that are generally acknowledged as having a broad base of competent activity in a given field.

A move in this direction could help to streamline procedures and, at the same time, provide more meaningful relations between grantor and grantees.

APPENDIXES

Organization and Funding of Marine Continental-Margin Research

THE ROLES OF THE PLAYERS

We sketch what we perceive to be changes in the roles played by the three main constituencies involved in continental-margin research: industry, academe, and government.

INDUSTRY

"Industry," for all practical purposes and as used in this report, stands for the petroleum industry. Other extractive industries do have some interest in continental margins, but the petroleum industry is predominant on the continental-margin scene. The chapters (11, 13, and 14) on drilling and geophysics indicate that, today, the petroleum industry has by far the most advanced technology relevant to continental-margin exploration and research.

Industry—if allowed to explore competitively—gradually will develop the technology to move into increasingly deeper waters. However, the exploration pace today is not dictated by technical limitations. Instead, political and economic realities hold the key to progress. To illustrate: Industry clearly will not move into deepwater continental-margin exploration offshore of the United States until shallower portions of the U.S. continental margins have been explored adequately. The results of shallow-water exploration will influence the advance into deeper water because, as described earlier, these areas are geological extensions of continental shelves.

The exploration pace is almost entirely in the hands of the government, and present practice in the United States suggests a slow exploration tempo on the shallow continental margins, with deepwater areas to become prospects for exploration sometime in the mid- or late-1980's. Thus, although drilling and geophysical technology is ready and available to do much of the work needed to understand non-Arctic continental margins of the United States, exploration proceeds at a slow pace.

The technological prowess of today's industry is not due to research concentrated in a few laboratories run by the major petroleum companies. It was developed in a competitive mode by the industry as a whole, with contractors doing much of the technological development. Much of the technology is available to anybody willing to pay the price. Over the years, research in industrial laboratories moved away from a more broadly based basic approach to clearly applied research. This coincided with the increased government support of research at federal mission-oriented agencies and at academic institutions.

Industry supports some academic oceanographic research through associate contributions to most major oceanographic institutions. These range—depending on the size and scope of the institution—from $5000 to

$50,000 annually per institution. In fact, it appears that today these are often the only discretionary funds available to the directors of these institutions.

In addition to this, some companies selectively support specific projects of special interest to them. These vary quite widely, from building new instruments to collecting regional background information. A fine example of an industry-supported academic project is the Gulf of Mexico program of the University of Texas Marine Science Institute (UT/MSI) at Galveston. UT/MSI's research vessel (the R/V *Ida Green*) was obtained and equipped with industry support. Most of the regional reconnaissance of the Gulf of Mexico program was shot with support from industry.

There are some who would like to see industry increase its financial support of academic research on continental margins. Industry is actively pursuing its own research and has developed techniques that are available to the community. While the academic world invented plate tectonics, industry developed multichannel seismic-reflection technology, and the products of these activities are mutually beneficial.

Industry is not likely to provide long-term financial commitments for academic continental-margin research; it has to maintain flexibility in the disposition of its research dollars.

We would like to see industry–government relations change consonant with the following quote from the Stratton report*:

Because the commercial exploitation of the sea's resources is the task of profit-oriented industry, the national plan should create a climate in which industry can operate effectively with assistance from the Federal Government in those areas of scientific research and technological development where private investment cannot be expected to assume the full burden.

To sum up, industry would like federal research efforts to shift to areas that are complementary to its own activities. Industry does support some academic efforts on continental margins and is likely to continue this support, but increased and more long-range support is not likely to be forthcoming from industry.

ACADEME

Geological and geophysical research in the marine portions of continental margins was very limited prior to World War II. The need for information during the war and the techniques developed in response to this need led to a greatly increased effort in the postwar years. This effort was expended in the United States primarily by the academic institutions because the U.S. Geological Survey (USGS) and the Bureau of Land Management (BLM) did not yet have mandates offshore; the Coast and Geodetic Survey, now incorporated in the National Oceanic and Atmospheric Administration (NOAA), was concerned principally with mapping and charting the topography of the shelf; and the Navy obtained some of its data through grants and contracts with the academic institutions. The depth of water in which oil wells could be completed was quite limited, so industry concentrated on the nearshore regions. The academic institutions thus had the dual role of being effectively the main source of geological and geophysical data from the offshore parts of continental margins as well as the source of trained investigators to respond to future needs.

In subsequent years, government agencies developed new missions in the offshore United States. The interest of the Navy in supporting external oceanographic research waned as a result of the Mansfield amendment* and was never adequately replaced by support from the National Science Foundation (NSF). The capacity of industry to operate in greater water depths increased, and the interest of the oceanographic institutions was drawn to the deep ocean by the exciting opportunities offered by the newly developed plate-tectonics model.

When new techniques and intellectual challenges drew ocean scientists in academe back to the continental margins, they found industry and government agencies there in force, and, in the U.S. margins, many administrative and regulatory inhibitions had been developed that had not previously existed. Furthermore, changes in the economic structure and in federal hiring practices enacted by the Congress altered the interchange of professional personnel between the academic institutions and government agencies to a dominant flow from academe to government.

Most academic oceanographic institutions have derived most of their support from government grants and contracts (see Figures A.1, A.2, and A.3). Of late, some significant shifts in the pattern of funding have occurred. In the past, oceanographic institutions obtained only a small number of large contracts; today, this number has increased dramatically, but the grants are smaller and short-lived.

Oceanographic research cannot be separated from teaching and training. Funding agencies that claim that their responsibility is limited to research support fail to understand the vital role of teaching and training in the successful accomplishment of research. (1) Universities do not have the money to support young researchers. (2) Continuing research is not possible without a steady supply of young and imaginative researchers. In our modern world, the best of these researchers need at least a minimum of job stability, or they will move to industry or government.

---

*Our Nation and the Sea: A Plan for National Action (1969). A report of the Commission on Marine Science, Engineering and Resources (J. A. Stratton, Chairman), U.S. Government Printing Office, Washington, D.C. (p. 10).

*Public Law 91–121, Section 203 (Military Authorization Act for 1970), and Public Law 91–441, Section 204 (Military Authorization Act for 1971).

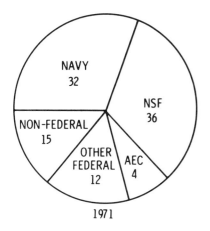

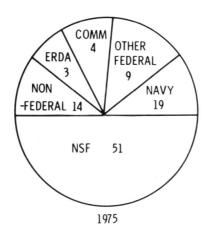

FIGURE A.1 Percentage of support for oceanographic institutions by source of support. (Courtesy of the Ocean Sciences Division, National Science Foundation.)

In the recent past, total funding of academic institutions has leveled out and is now barely keeping up with inflation. As one consequence, equipment maintenance is the first to suffer. Furthermore, it can be fairly anticipated that if the trend continues and funds are reduced, fewer scientists will enter the field, and many of these scientists will stay for shorter periods, only to move to government or industry. This will happen precisely at the time when talents in the academic institutions are needed the most to solve problems related to environment and resources.

Let us ask whether academic research is really necessary. After all, industry has been acting independently for years and has developed a technology vastly exceeding expectations. Major petroleum companies recruit people from academic institutions only to put them into their own schools, to train them for their own needs. Eventually, many leave to work for other companies and contractors or for the government. Government institutions concerned with continental margins are also doing much research, and they have effectively increased their staff by successfully hiring from industry and academe. So one could easily conclude that academe has been overtaken by its own success. True, they did lead the way, but bigger and better efforts are now undertaken elsewhere, and continental margins are well taken care of.

We disagree with such as assessment, specifically because in continental-margin research, academic institutions have amply demonstrated the pioneering quality of their work. They were the first to use refraction techniques to demonstrate the existence of large sedimentary basins on the ocean. They led in the early development of nonexplosive sources for reflection seismic surveys. Multichannel processing techniques developed by industry are solidly based on computer research in academe. Academic oceanographers and seismologists led the way to plate tectonics. The Deep Sea Drilling Project (DSDP) is a spectacular success. While we have already mentioned some of the exciting scientific results, a negative conclusion of great importance is often overlooked: DSDP has shown that vast areas of the deep oceans cannot be reasonably viewed as potential hydrocarbon provinces.

Industry and government institutions have also contributed, but the plurality of approaches that academic institutions offer is responsible for the variety and quality of the earth sciences today. In many cases, work done as "basic" research has led to unexpected practical conclusions. In fact, the unexpected benefits are leading to the difficulty that we have in separating basic from goal-directed research.

We realize that academic continental-margin research is but a minor concern in the face of a variety of much larger problems that academic institutions face today.* In our view, however, academic research related to continental margins has demonstrated its creativity and has delivered one of the finest scientific products of the recent past. To stunt its growth by reducing financial support and fragmenting that support into smaller and smaller grants would indeed be a tragedy.

In our judgment, multi-institutional and multidisciplinary projects on continental margins are here to stay. On March 29, 1976, the following educational institutions incorporated in the State of New York to promote, encourage, develop, and support efforts to advance knowledge and learning in the science of oceanography:

University of California
Columbia University
University of Hawaii
University of Miami
Oregon State University
University of Rhode Island
Texas A & M University
University of Washington
Woods Hole Oceanographic Institution

The nonprofit corporation is known as Joint Oceanographic Institutions, Incorporated (JOI, Inc.), and the di-

*For a detailed study see B. L. R. Smith and J. J. Karlesky, 1977. *The State of Academic Science: The Universities in the Nation's Research Efforts*, Change Magazine Press, New Rochelle, N.Y., 2 vols.

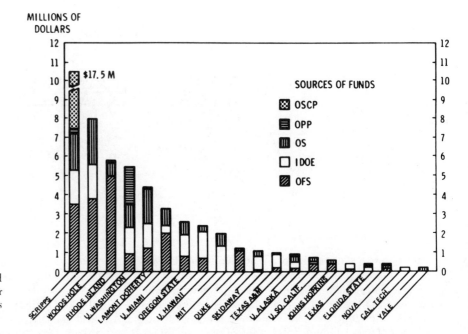

FIGURE A.2 Top 20 NSF-funded oceanographic institutions, fiscal year 1975. (Courtesy of the Ocean Sciences Division, National Science Foundation.)

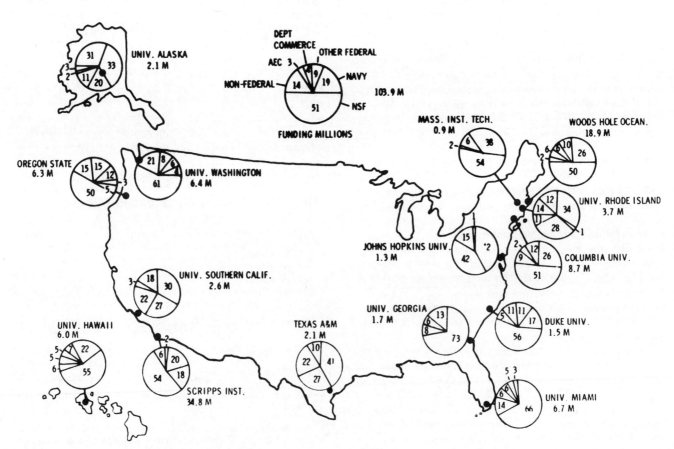

FIGURE A.3 Sources of funding for major private/academic oceanographic institutions, fiscal year 1975. (Courtesy of the Ocean Sciences Division, National Science Foundation.)

rectors, deans, or department heads of the affiliated oceanographic organizations constitute the Board of Governors of JOI. The Board has had considerable experience in planning and coordinating multi-institutional experiments through Joint Oceanographic Institutions for Deep Earth Sampling (JOIDES). JOIDES is now an executive committee of JOI, Inc., and continues to provide basic scientific guidance to the DSDP. JOI, Inc., also provides a mechanism whereby other cooperative studies of the ocean can be undertaken, using the corporation as a vehicle through which funding is sought and management provided, while actual programs are carried out by scientists at affiliated and other institutions.

GOVERNMENT

Much of the remainder of this Appendix is devoted to a discussion and review of the federal oceanographic program. We will limit ourselves to relations between government and academe and between government and industry as they affect continental-margin research. To list government simply as part of the cast is not quite correct. In fact, the government is the producer and director of the whole show, the taxpayers being the backers who support the production.

More and more people are concerned with the future of our energy and mineral and water resources, and they worry about the environment. As scientists, we share these concerns. To take care of anticipated problems, new laws and institutions are designed. Some of these designs are good, others are flawed, but they have the common quality that they invariably signal a significant government involvement in the activities concerned. Research and exploration on continental margins was done by academicians and industry. Now, the government has, in part, taken over many of these activities.

A first problem relates to government organization. In 1975, a report of the Comptroller General of the United States to the Congress noted that some 11 departments and agencies are involved in one form or another of oceanographic work. In regard to the study of the geological structure and composition of the ocean floor, the same report states that some 15 programs are administrated by 7 agencies and noted: "In requesting the agencies to comment on this situation, GAO suggested that their many areas of common interest must, axiomatically, lead to ineffectiveness and inefficiencies. A number of agencies argued persuasively that this did not necessarily follow." The report eventually joins in the popular call for a comprehensive national ocean program, pointing out that such a need was recognized by Congress in 1966 and bewailing that little progress had been made since. The spectacular successes of the earth sciences and particularly of oceanography in the past two decades provide convincing proof that government organization did not prevent scientists from doing good science.

Many government agencies interested in continental-margin work have active cooperative programs with academe. Examples are the cooperative geophysical effort between the USGS and the University of Delaware on the Atlantic margin and the survey of the Blake Plateau done jointly by the USGS and the University of Texas at Galveston. A substantial increase in joint programs between government and academe would be most helpful, particularly if such programs are on the "pure" side of the mission-oriented activity. Long-term arrangements could help to provide stability for oceanographic institutions. The appeal of these arrangements is (1) that it gives smaller institutions a chance to participate in research, which otherwise is dominated by a small number of large academic oceanographic institutions, and (2) that it gives the agency the flexibility to select the best researcher to work on a given project.

A third point has to be made before we describe some of the details of the federal involvement in oceanography. Some of us view the increased involvement of government in continental-margin research with some apprehension. However, in the case of the USGS we can only praise the quality of their product. Members of the USGS have obtained and interpreted much information in a short time. The teams of the USGS have much cohesion, and they have made some very substantial contributions to our understanding of continental margins. So here, as with academe and industry, we can clearly recognize that past performance and the product amply justify the continuation of the effort.

HISTORY OF FEDERAL OCEANOGRAPHIC PROGRAMS

THE BEGINNINGS

The increasing interest in the continental margins stems partly from their potential as a source for certain raw materials, but also they are important in evaluating, refining, and extending the plate-tectonics model. Plate tectonics has provided geologists, for the first time, with a credible working model for the earth as a whole. Such diverse specialties as seismology, paleontology, rock magnetism, stratigraphy, gravity, and igneous petrology have been brought together into a single focus, thanks to this concept. The plate-tectonics revolution exploded almost without warning; yet, in retrospect, its roots can be traced back several decades.

Prior to the nineteenth century, science was concerned largely with the present. It was not until geologists began to consider the earth in light of its history that it became possible to develop a systemized knowledge of the more remote past. With these new perspectives open to them, geologists spent the next three quarters of a century studying the continents. Geological pioneers compiled a tremendous amount of descriptive information about the earth, its history, the evolution of life, and the composition and structure of the rocks that appear on the earth's surface.

Then, starting in the 1930's with the high-pressure apparatus of Perry Bridgeman at Harvard, physicists and chemists developed new instruments that were destined to drastically alter the course of geological research. The timing was fortunate. Geologists had gathered enough field data to ask the critical questions. Some of the new instruments enabled them to reproduce earth processes in the laboratory. Other tools made it possible for geologists to measure earth properties in the field with meaningful and increasing precision. The traditional tools of the geologist—hammer, Brunton compass, and hand lens—were gradually supplemented with increasingly sophisticated instruments—x-rays, mass spectrometers, infrared spectrometers, spinner magnetometers, electron probes—just about any type of gear that you would see in a physics or chemistry laboratory, plus sensitive field gear such as seismometers and gravimeters.

Gradually, geochemical experiments became progressively more earth-oriented. From them, geologists began to gain insight as to how rocks crystallized from molten magma. Armed with better instruments, geologists began a fresh study of modern processes in order to better interpret ancient rocks.

Finally, geologists went to sea to explore the 72 percent of the earth's crust that is covered with water. Earlier, they had been baffled by the logistics of exploring or mapping beneath all that water. It was more profitable to stick to the exposed continents. (Besides, were not most of the ocean floors just vast, flat, featureless plains?)

Virtually all of our knowledge of the seafloor and oceanic crust postdates World War II. It was not, for example, until the International Geophysical Year (IGY) in 1957–1958 that we learned that the Mid-Atlantic Ridge was a continuous physiographic feature, let alone a part of a worldwide rift zone. Scripps Institution of Oceanography had been established in 1912, but it was a small laboratory attached to the University of California that dealt mostly with marine biology and physical oceanography (which was also in its infancy). Woods Hole Oceanographic Institution was not even chartered until 1930. Within the government, the U.S. Geological Survey (established in 1879) was limited by its Organic Act to the shoreline, and the Coast and Geodetic Survey was concerned largely with bathymetric charts and precise geodetic controls.

The first real ventures into the submarine realm came in the 1930's. Although singling out any one or two pioneers is always risky, it is probably no exaggeration to say that modern marine geology in the United States began in 1935 with the grant from the Geological Society of America to Maurice Ewing to do seismic profiling on the continental shelf. This was followed by Harry Hess' work with Vening Meinesz on submarine gravity measurements. At about the same time, Francis Shepard started his long career studying marine sediments and submarine canyons.

Yet, in spite of these auspicious beginnings, there was no concerted effort—and certainly no mad rush—to the new frontier.

THE POSTWAR BOOM

World War II brought about, among other things, both an awareness of the oceans and an increased ability to study them logistically. Activities at Scripps and Woods Hole were expanded, and new laboratories were established.

For nearly a decade, the major support for the expansion came from the U.S. Navy in general and the Office of Naval Research (ONR) in particular. The National Science Foundation was chartered in 1950 but did not have the resources to play a significant role until it became the funding agent for IGY in 1957. The Atomic Energy Commission (AEC) had definite but restricted interest. Oceanic research actually lagged after its postwar spurt until the IGY opened the door to study oceans on a worldwide basis.

The first emphasis in the expanded postwar effort was on physical oceanography. But people like Ewing, Revelle, Hess, and Von Arx were also involved and would have their say, so an effort in submarine geology and geophysics gradually evolved. The work was paid for by the federal government and carried out by scientists at the major oceanographic institutions.

Within the government, several separate events of the 1950's were to have long-range significance. First, in 1953, the Geological Survey's purview was extended to include the continental shelf. This allowed the Survey to move "offshore" and eventually led to its present program of marine geology.

Second, G. F. Jordan, of the Coast and Geodetic Survey, demonstrated the potential use of bathymetric charts for geologic interpretations and thereby stimulated a new program in that agency.

Third, in March of 1957, the Advisory Panel for Earth Sciences of the NSF—especially Walter Munk and Harry Hess—hatched the idea that became Project Mohole. It is interesting to note that when Mohole was first suggested, no one, including Hess, foresaw the coming geological revolution of plate tectonics. Indeed, Mohole was conceived partly out of frustration. Many hoped that Project Mohole would lead geology out of the doldrums and into new thinking. Only half a decade later, Hess* and another scientist, Robert Dietz,† published the seafloor-spreading papers that touched off the explosion.

Mohole itself never came to fruition, but much of its pioneer thinking carried over into other geological and geophysical projects, especially the spectacularly successful DSDP. DSDP's list of accomplishments and discoveries is long and needs no repetition here. Suffice to say that we believe that no other single geological project has ever contributed so much new and exciting information. It came at a time when the whole picture of continents and ocean basins was being repainted. It is interest-

*H. H. Hess, 1962. History of the ocean basins, in *Petrologic Studies: A Volume to Honor A. F. Buddington*, A. E. J. Engel, ed., Geol. Soc. Am., New York.
†R. S. Dietz, 1961. Continent and ocean basin evolution by spreading of the sea floor, *Nature* 190:854–857.

ing to note that when Mohole was planned, and DSDP was first proposed as a companion project (1961), one of the main objectives of the drilling was to "sample a continuous sedimentary record back through geologic time, into the early pre-Cambrian and perhaps even to the time of the origin of life itself."

Finally, an unpretentious event that attracted little attention (then or since) greatly affected the marine program. In 1957, Gordon Lill (then with ONR) stimulated the formation of the Coordinating Committee on Oceanography (CCO). This was an informal interagency group composed of people with operational responsibilities who met once a month to exchange information. CCO reported to no one, had no status, no charter. But it was effective. It provided the vehicle for getting interagency backing of the proposed National Geographic Data Center. Working quietly behind the scenes, CCO helped NSF to get approval to build the research vessel *Atlantis II* for Woods Hole Oceanographic Institution. Most importantly, CCO fostered the formation of and furnished support for the National Academy of Sciences–National Research Council's Committee on Oceanography (NASCO).

Few reports of advisory committees have had the impact of NASCO's *Oceanography 1960 to 1970* (published in 1959). For NASCO, the time was ripe and the audience ready. Realizing this, the committee took great care that the report could be understood by nonoceanographers. Also, NASCO made specific recommendations in this report, outlining a ten-year plan for increasing the U.S. fleet of oceanographic ships, stipulating what kind of education for new scientists was needed and how much that would cost, and drawing up a year-by-year budget for the proposed new oceanographic activities. Committee reports are seldom hallmarks of progress for a science, but *Oceanography 1960 to 1970* may be the exception. Some members of Congress—in those days, people from Capitol Hill were invited to attend NASCO meetings—took a keen interest in this report. It was certainly more than a jest when NASCO Chairman, Harrison Brown, received the American Miscellaneous Society's humorous and prestigious Albatross Award for "Outstanding Accomplishments in Political Oceanography."

THE TRANSITIONAL 1960'S

By the mid 1960's, today's pattern of marine programs had pretty well emerged. The popular and political support for oceanography made it possible for both NSF and ONR to expand and then to consolidate the expanded position of academic and nonprofit laboratories.

The International Indian Ocean Expedition provided some additional monies. Later, the International Decade of Ocean Exploration (IDOE) helped to solidify this level of support at a time when science budgets, in general, were in trouble. Finally, Sea Grant, started in NSF and later moved to the Department of Commerce, added another dimension by supporting applied research and training.

The troubled Mohole Project was cancelled in 1967, but its companion, NSF's Ocean Sediment Coring Program, flourished after a slow start. Its first project was a small coring effort in 1963 off Jamaica by the University of Miami. Then, in 1965, JOIDES was formed and undertook the Blake Plateau Project. Finally, in 1967, the larger and continuing DSDP was launched.

Other than NSF and ONR, significant federal support for the academic laboratories came from the waste-disposal programs of the AEC and the Geological Survey's Atlantic Shelf program cooperative with Woods Hole.

In-house activities had also increased. In the Navy, the Hydrographic Office became the Naval Oceanographic Office and started an expanded program of seabed surveying. In Commerce, the Coast and Geodetic Survey and the Weather Bureau were melded into the Environmental Sciences Services Administration (ESSA). Marine geology was expanded under ESSA's Institute for Oceanography (later to become NOAA's Atlantic and Pacific Laboratories).

In the Department of the Interior, the Geological Survey's authority was extended to cover the continental shelves. The Survey started, in 1962, a study of the topography and surface sediments of the Atlantic Shelf in cooperation with Woods Hole. Since then, the Survey's activities, under its Office of Marine Geology, have expanded to cover work off the Atlantic and Pacific coasts, the Gulf of Mexico, and the Arctic Sea.

Also within Interior, the Bureau of Land Management was given responsibilities for evaluating offshore lease areas, and the Ocean Mining Administration was formed to help foster exploitation of submarine minerals (to date, mostly manganese nodules).

Along with all these activities, oceanography began to have money problems. By the late 1960's, the post-World War II enthusiasm for science had waned. The nation had other pressing problems. Funds for new projects were increasingly hard to come by, so the federal ocean program barely kept up with inflation, and there is some evidence that the fraction of the federal program devoted to research has been decreasing. This problem has contributed to a reluctance to initiate new research on the continental margins and may continue to do so in the future.

THE SCENE TODAY

Our knowledge of continental margins lags behind our understanding of both the deep-ocean floors and the continents. Some reasons for this lag are obvious from the history of the science just presented. To summarize:

1. Geologists did their early work on the exposed rocks of the continents.
2. When oceanography (including marine geology) expanded, the work was concentrated in oceanographic institutions the chief interests of which were the deep oceans. Since the nature of oceanic sediments and crust

was virtually unknown, finding out about it absorbed most of the early marine efforts.

3. Although the oceanic crust is much more complex than once was postulated, it still may be far simpler than continental crust and simpler still than the transitional zones of the continental margins (where even the Moho commonly disappears). Moreover, many of the available instruments are not so effective for work on margins as for work in the deeper oceans. Therefore, it has been "convenient" to defer work on the margins until their importance can no longer be ignored.

A summary of the current activities of federal agencies having significant programs in marine geology follows. Most of the basic research is still deep-sea oriented—and most of the work on the continental shelf and slope involves only the upper parts of the sedimentary column.

CURRENT FEDERAL PROGRAMS

INTRODUCTION

It would be useful to have some statistics that illustrate the support of research done on continental margins. However, such figures are hard to come by because (1) it is not quite clear what, in people's minds, constitutes research, and (2) we are including in our purview the landward continuation of the subsea margins. But even if we looked only at submarine margins, it would be well-nigh impossible to separate marine continental-margin expenditures from overall oceanographic expenditures.

To get at least some perspective, we have leaned on some internal NSF reports, a study made by the Ocean Sciences Board (OSB), and annual reports of the President to the Congress on the nation's efforts to comprehend, conserve, and use the sea—*The Federal Ocean Program*, issued annually by the U.S. Government Printing Office. Tables A.1 and A.2 show

1. A very substantial increase in the category of the nonliving resources, reflecting Interior's $90 million increase from 1974 to 1976 directed toward appraisal for leasing and environmental studies on the OCS;

2. Another increase of $54 million, in the Department of Commerce, going to Outer Continental Shelf baseline studies, the Coastal Zone Management Program, fish and marine mammal assessment and protection, and evaluation of satellite data;

3. An increase in the NASA program caused by the SEASAT Program;

4. Overall research dollars in oceanography just barely stay ahead of inflation, although there has been a shift of research money to ERDA (which has recently become the Department of Energy).

The general impression is that, since 1974, there has been a substantial increase in federal ocean program budgets (say, a 43 percent increase from 1974 to 1976). There has been a shift of money to mission-oriented ef-

TABLE A.1 Federal Ocean Program—Budget by Department and Independent Agency

Department or Agency	Estimated by Fiscal Year				
	1975	1976	TQ	1977	1978
1. Department of Defense—Military	212.4	195.5	52.3	242.6	243.7
2. Department of Defense—Civil Works	31.9	29.2	7.1	32.1	63.8
3. Department of Commerce	213.2	235.2	66.5	274.2	315.9
4. National Science Foundation	69.2	68.8	15.8	74.8	79.9
5. Department of Transportation	75.6	36.4	9.3	46.6	49.1
6. Department of the Interior	106.8	142.0	37.5	153.8	161.7
7. Environmental Protection Agency	22.4	24.3	6.1	25.1	22.7
8. Department of State	11.6	12.7	5.9	14.9	18.2
9. Department of Health, Education and Welfare	11.2	16.4	2.8	17.3	18.0
10. Department of Energy[a]	12.8	18.9	6.2	28.9	40.6
11. National Aeronautics and Space Administration	12.5	20.9	8.0	30.2	18.5
12. Smithsonian Institution	2.9	2.6	0.7	2.8	2.9
TOTAL	782.5	802.9	218.2	943.3	1035.0

[a]Formerly the Energy Research and Development Administration.

forts, while money allocated to oceanographic basic research barely keeps abreast of inflation. An informal OSB study looked at federal oceanographic research over a longer time-span, and, in terms of constant 1967 dollars, noted the following:

	1966	1971	1976
Total Research Funds	$118,000	$137,000	$136,000
Academic Funding	$ 43,000	$ 53,000	$ 52,000

The same OSB study also suggests that the proportion of funds directed to basic ocean research within the total federal program is shrinking. To compensate for this, there appears to be an increasing role played by state and industry funds. In 1975, some 15 percent of the support of six major academic oceanographic centers in the United States came from states and private sources. These same institutions increased their own budgets from a total of $31.5 million in 1965 to $59.8 million in 1971 and $79.3 million in 1975.

The total picture of distribution of funds to oceanographic institutions is shown on Figures A.1, A.2, and A.3. Figure A.1 shows the shift from Navy funding to NSF funding. The actual distribution of the funds is illustrated in Figures A.2 and A.3 and reveals the dominant position of some of the large institutions. The picture on Figure A.2 is a little distorted, because it involves sediment-coring program money allocated to Scripps; and the University of Rhode Island got $4 million in fiscal year 1975 from "Facilities" to construct a research ship—thus, the expenditure does not reflect the size of the activity at that university.

Simple conclusions based on the statistics are:

1. We cannot really get meaningful figures for work done on continental margins. It is, however, quite obvious that mission-oriented efforts on continental margins have dramatically increased in response to resource and environmental needs. These mission-oriented efforts are carried out mostly by federal agencies.

2. Research, as a part of the total program, has been decreasing. Activities that, in the past, would have been considered to be research, today are no longer classified as research, because they form part of a mission-oriented effort. In other words, the category, "oceanographic research," lacks a convincing definition.

3. Funding of academic oceanographic institutions is barely staying ahead of inflation.

In short, it does not look as if there is a lack of funds for work on continental margins; but there *is* a problem with the funding of academic oceanographic research.

The following pages discuss some of the marine geological–geophysical activities that the federal government supports.

DEPARTMENT OF THE INTERIOR

U.S. Geological Survey (USGS)

The authority of the USGS to work in the marine environment derives from the Survey's Organic Act (1897), the Outer Continental Shelf (OCS) Lands Act (1953), the National Environmental Protection Act (1969), the Mining Policy Act (1970), and the Secretarial Initiatives on OCS lease sales. Both the Geologic and Conservation Divisions are involved in marine work.

TABLE A.2 Federal Ocean Program—Budget by Major Purpose Category

Major Purpose	Estimated by Fiscal Year				
	1975	1976	TQ	1977	1978
1. International cooperation and collaboration	11.7	12.7	5.9	14.9	18.2
2. National security	93.0	81.8	21.6	99.6	94.4
3. Living resources	113.7	87.6	20.4	106.4	114.4
4. Transportation	37.9	31.9	6.2	38.1	47.0
5. Development and conservation of the coastal zone	116.5	124.8	31.1	133.4	188.3
6. Nonliving resources	82.9	121.2	31.9	131.4	140.5
7. Oceanographic research	124.1	129.3	37.7	146.4	158.4
8. Education	11.3	10.5	4.7	12.6	14.2
9. Environmental observation and prediction	35.2	40.0	11.4	47.8	46.6
10. Ocean exploration, mapping, charting, and geodesy	99.4	87.7	22.4	111.4	111.9
11. General-purpose ocean engineering	40.9	62.6	20.0	86.5	87.1
12. National centers and facilities	15.9	12.8	4.9	14.8	14.0
TOTAL	782.5	802.9	218.2	943.3	1035.0

The *Geologic Division* provides geological, geochemical, and geophysical information for other government agencies and the general public on land resources, energy and mineral resources, and geological hazards within the nation and its territories.

The Office of Marine Geology (OMG), which is in the Geologic Division, has the prime responsibility for offshore programs. Its activities are coordinated with other Survey offices and branches. The two branches of the OMG are the Atlantic and Gulf of Mexico Branch (Atlantic/Gulf), located at Woods Hole, Massachusetts; and the Pacific and Arctic Branch (Pacific/Arctic), located at Menlo Park, California.

What grew to be OMG started in 1962 as a survey of the upper sediments of the Atlantic shelf in cooperation with Woods Hole Oceanographic Institution. In 1969, similar work began in the Gulf of Mexico. This early work produced more than 200 papers on the sediments, structures, biology, and bathymetry of the Atlantic and Gulf margins. The Pacific/Arctic Branch was organized in 1971, formalizing work that began in the 1960's. Starting with a budget of less than $0.5 million, this branch grew rapidly after the 1974 decision to lease OCS lands for petroleum exploration.

The current program of OMG includes three major activities:

	FY77 $	Permanent Employees
(1) Offshore oil and gas resource assessment	$9,155,000	98
(2) Offshore, energy-related, environmental investigations	$4,040,000	34
(3) Marine geological investigations	$1,206,000	25

(1) The offshore oil and gas resource assessment program gathers data to identify and study the geology of potential OCS lease sale areas in support of the Department of the Interior's OCS oil and gas leasing program. The geophysical techniques used include multichannel seismic reflection, seismic refraction, and magnetic (air and sea) and gravity data-collecting tools.

The Atlantic/Gulf Branch buys most of its seismic data from commercial geophysical companies. It maintains a six-channel reflection system on a long-term chartered ship. Data processing is done on contract or by the Survey's Oil and Gas Branch on its Phoenix System in Denver.

The Pacific/Arctic Branch operates a 24-channel reflection system on the R/V *Lee*, a Naval AGOR on long-term loan to the Survey. Data are processed, again, either by commercial contract or in Denver, but on-board processing is planned for the future.

The information collected by OMG's two branches is prepared and submitted to various agencies as precall for nomination reports, as open-file geological or technical reports, as research reports, and as maps and circulars for tract selection prior to a lease sale.

(2) The offshore, energy-related, environmental investigations are designed to maximize the safety of exploring and developing oil and gas resources on the OCS. These investigations seek to identify features that may pose such hazards as faults, slumps, and sand waves. This and other information on sediment dynamics are used by the Bureau of Land Management (BLM) to prepare environmental impact statements prior to lease sales, for lease-tract decisions, and as baseline data to assess and monitor the effects of offshore development.

The Atlantic/Gulf Branch charters a variety of vessels on a short-term basis and one large vessel on a long-term basis to make the offshore investigations. Work continues in the Georges Bank area, the Baltimore Canyon Trough, the Southeast Georgia Embayment, and in the Gulf of Mexico off Texas and Louisiana. Funding is partly from BLM.

The Pacific/Arctic Branch charters the *Sea Sounder* to gather environmental data. This ship is equipped with high-resolution geophysical equipment. However, *Sea Sounder* is used primarily for sampling (piston-core, grab, and box samples). The Alaskan program is funded largely by NOAA, which acts as a manager for BLM. The West Coast program is partly funded by BLM.

(3) Programs funded under "Marine Geology Investigations" support the other two programs. The mission here is to develop better techniques with which to collect, analyze, and interpret data. Studies of engineering properties in the Gulf of Mexico are funded under Marine Geology. Coastal processes along the Texas shoreline and interdisciplinary geological/hydrologic studies in San Francisco and Monterey bays in California and Willapa Bay in Oregon are also supported by OMG.

The Office of Earthquake Studies is primarily concerned with onshore earthquake hazards, as these affect populated areas. Recently, however, the Office has begun to analyze OCS earthquake risks. This Office is funded by the Survey and by BLM to prepare seismic-risk maps for the OCS. Earthquake data are essential in order to determine design criteria for drilling platforms and pipelines in areas susceptible to earthquake activity. The Office of Earthquake Studies program in southern California is monitoring earthquake activity around active drilling platforms. This study is using a net of ocean-bottom seismometers.

The *Conservation Division* (CD) of the Survey is responsible for classifying and evaluating OCS resources; for supervising and regulating operations associated with exploration, development, and production of OCS lands under lease; and for licensing and projection permits. The CD concentrates on identifying areas for future lease sales, advising the BLM on selecting tracts for sale and on estimating the value of oil and gas on each tract offered.

In carrying out its responsibilities, the CD gets its data from (primarily) industry and from the OMG. OCS regulations stipulate that all industrial geophysical and geological data gathered on the shelf are subject to acquisition by the Survey on a proprietary basis. Publically available regional resource and environmental data gathered by OMG are used to identify future lease-sale areas and to pinpoint regions where there may be geological hazards. If the data are not good enough for final tract selection, CD may contract directly with industry to get more information.

Under the supervisory and regulatory responsibilities, CD reviews and approves operational plans, such as drilling new wells or abandoning sites. CD also inspects operations and computes rents and royalties.

The total budget and personnel for CD's activities for fiscal year 1977 were $37,400,000 and 930 permanent employees.

The Future

Present resource assessment efforts on the continental shelf should continue at about the present level for at least the next three to five years. Program emphasis will gradually shift to the continental slope and rise and to the small ocean basins adjacent to U.S. territory. Environmental studies will probably extend into the developmental phase of the OCS and eventually follow or parallel resource investigations in deepwater parts of the margin.

(1) *Ocean-Bottom Stability.* Unstable ocean-floor conditions exist in many OCS areas already leased or being considered for leasing in the near future. Consequently, the Survey plans a major effort to gather and interpret data that bear on engineering-design criteria in all the proposed OCS lease areas. These include engineering properties of sediments, cyclic accelerations from storm waves, wind and wave stress, and the nature of slump and creep movement.

(2) *Ocean Minerals Program.* Because manganese-nodule mining may soon be under way, the Survey has proposed a program of resource analysis to define potential sites for mining and to set up realistic regulations. There is, however, still some uncertainty as to which federal agency should do this job.

(3) *Deep-Ocean Resource Assessment.* As development of the OCS progresses, interest in the continental slope, rise, and adjacent small ocean basins will increase. The Survey plans to shift its emphasis on resource analysis and environmental studies from the continental shelves to deeper waters as soon as activity on the Outer Continental Shelf permits.

Bureau of Land Management (BLM)

Operating primarily under the Outer Continental Shelf Lands Act and associated laws, the BLM is charged with assuring that the government receives fair value for federal lands leased, promotes orderly development of those lands, and protects the environment. In the early 1970's,

the obligations of both the National Environmental Resource Assessment (NERA) and a massive program of frontier leasing were assigned to the Interior Department and BLM. This combination and a variety of other pressures resulted in the initiation of an environmental studies program (not research, in BLM's definition). Initially, the program was related to one lease area at a level of about $1 million for the first year. The program has now grown to a current annual budget of about $50 million and relates to approximately 15 lease areas. The funds are used for work in all the subfields of oceanography. About half of the funds cover the Alaskan lease areas and are passed directly to NOAA for environmental assessment work.

BLM's work consists of literature surveys, reconnaissance work, baseline or benchmark studies, monitoring, and other special studies, such as fates and effects of pollutants. All the work is contracted out, most of it to other federal agencies and private consulting firms.

Annual reports are anticipated for each of the frontier lease areas. The reports will synthesize environmental information for Department of the Interior responsibilities.

Ocean Mining Administration (OMA)

OMA was established in 1975 and has the responsibility to plan policy for the development of mineral resources of the deep sea (beyond the continental shelf), plus Antarctica. OMA is also concerned with jurisdictional issues in international negotiations.

OMA is charged to maintain coordination with other Interior Bureaus and with other agencies in the federal government. To date, OMA's activities have been concentrated on manganese nodules; its data acquisition and research are contractive. Its future plans are uncertain.

DEPARTMENT OF DEFENSE

United States Navy

Within the U.S. Navy, two categories of research are conducted at continental margins: in-house research [at the Naval Research Laboratory (NRL) in Washington, D.C., and at the Naval Oceanographic Research and Development Activity (NORDA) in Mississippi] and contract research [at academic institutions through the Ocean Research Office of the Office of Naval Research (ONR)]. The scope of work is worldwide and extends from the continental margins into the deeper ocean basins.

In-house research concentrates on the acoustic behavior of marine sediments deposited along the margins, the nature and source of gravity anomalies, and the relation between magnetic anomalies and the fabric and tectonics of the ocean floor. Out of a total fiscal year 1978 Navy oceanography basic research budget of $45 million, approximately $8 million is allocated for marine geology and geophysics. Of the $8 million, $1 million is dedicated to in-house research. Of the $1 million, approximately

$0.3 million is planned for aeromagnetic research projects, the bulk of which will be done in the Arctic basin.

The aeromagnetic studies are unique in that long-range, Navy P-3 aircraft, re-instrumented with research magnetometers, are used as platforms. This technique allows a thorough survey of an oceanic area, usually with a line spacing of 5–8 km and a flight elevation of 200 m above sea level. Airborne research has been particularly useful in identifying magnetic anomaly patterns over relatively inaccessible regions such as the continental margins of the Arctic basin and over magnetically complex regions such as the Pacific margin island arcs.

Coupled with the aeromagnetic studies are geophysical traverses made aboard vessels such as the USNS *Hayes* (NRL). These vessels are equipped to do acoustic propagation studies, through both the water layer and the oceanic crust. The acoustic work concentrates on propagation frequencies below 100 Hz and is implemented in the field by towed and ocean-bottom-mounted sensors. Navy researchers do very little multichannel subbottom profiling on continental shelves. The Navy's continental margins work is mostly done along the deeper parts of the continental slope and upper rise. In-house marine geology and geophysics work at NRL and NORDA is all unclassified, but directed, research. It ties in closely to Navy missions. Although the budget for in-house research work is relatively small, much progress has been made, especially in the fields of aeromagnetic studies and acoustic propagation, because expensive field platforms (e.g., ships and airplanes) are funded largely out of Navy operations funds.

The *contract research* program of ONR in marine geology and geophysics has an annual budget of approximately $7 million. The Navy contracts with oceanographic research groups in the U.S. academic community. Contract research is administrated through proposals, proposal reviews, and annual site visits of the contract institutions. Research supported by contracts is largely shipboard open-ocean, and is carried out through team efforts rather than through individuals. Research cruises tend to be devoted to integrated marine geological and geophysical investigations that are worldwide in scope.

The object of the program is to provide an oceanwide ability to predict or assess the ocean-floor environment for future naval systems or operational requirements and to provide quantitative data on the physical properties of the oceanic crust for the Navy's antisubmarine warfare programs. The program views the ocean-floor environment as the product of an oceanwide geological process. The morphology of seamounts, the variations in acoustical and physical properties of marine sediments, the variability in magnetic anomalies observed above the oceanic crust, and the variations in the marine gravity field are viewed not as random phenomena but as components of discrete and predictable natural processes.

Particular emphasis within the program is directed toward studies of ocean-floor acoustic propagation, that is, the physics of high-frequency reflectivity of the ocean floor, the physics of low-frequency acoustic propagation through the ocean floor, the formation of ocean-floor acoustic reflectors, and the areal variations in these acoustic phenomena. A secondary, but significant, part of the program is directed toward studies of the origin and location of major bathymetric, gravity, and magnetic anomalies.

Another part of the program examines the dynamics of sediment transport. This includes the effect of bottom currents on microtopography and rates of sedimentation and erosion, the effects of sediment composition and chemical changes on the mechanical properties of the ocean floor and on the ocean-floor mechanical stability, and the relationship of sediment parameters to acoustic parameters.

Principal contract institutions include Lamont-Doherty Geological Observatory (Columbia University), Woods Hole Oceanographic Institution, Scripps Institution of Oceanography (University of California), Massachusetts Institute of Technology, the University of Hawaii, Oregon State University, the University of Rhode Island, the University of Washington, the University of Texas, and the University of Connecticut.

In funding the research, approximately 15 percent of the total budget is devoted to supporting ship-time on academic research vessels.

The ONR contract research program is closely tied to similar basic research programs of the NSF. The tie, however, is a complementary one. For example, the NORDA contract research program supports very little work in biostratigraphy, but it supports a major effort in ocean-bottom seismology. Under this program, a concentrated effort has been directed toward sediment thickness studies of the continental margins of the Atlantic and Pacific through single-channel seismic-reflection techniques, sonobuoy refraction (using Navy sonobuoys), ocean-bottom seismology, and gravimetry. This work resulted in published sediment isopach maps of the Pacific, the Atlantic, and the North American continental margin. The work was done by the Lamont-Doherty Geological Observatory.

Other studies that ONR supported included ocean-bottom seismograph investigations of the West Coast margin and the Juan de Fuca Ridge by the Oregon State University; geophysical studies of the west Pacific island arcs by the Scripps Institution of Oceanography (SIO); a sedimentological investigation of the Bay of Bengal by SIO; and gravimetric studies of the Aleutian, Kuril, Japan, Bonin, and Mariana trench systems of the Pacific Ocean by the Lamont-Doherty Geological Observatory. Sediment dynamics studies of the slope and upper rise areas of the East Coast offshore have been conducted on Navy submersibles by the Woods Hole Oceanographic Institution and the Lamont-Doherty Geological Observatory.

The Naval Arctic Research Laboratory (NARL) is a Navy-owned, contractor-operated, research facility, at Barrow, Alaska, on the Arctic Ocean. It consists of 116 semipermanent and permanent buildings located on land withdrawn from the Navy Petroleum Reserve Number Four.

The Laboratory supports both Navy/Department of Defense and other agency research on a reimbursable basis. The logistic capability of NARL includes a Twin Otter aircraft, two C-117s, and smaller single-engine aircraft. A light-weight warping tug modified for oceanographic research and the 76-foot *Natick* comprise the fleet.

The NARL role includes both the support of studies conducted at Point Barrow and the logistics and support services for field parties that conduct research at field stations that may range from the Bering Strait to the coast of Greenland and out on the Arctic pack ice.

In the *future*, most of the projects on continental margins that ONR will fund will concentrate on the behavior of sound in the marine sediments and in the oceanic lithosphere. The projects will be carried out by using both ocean-bottom seismometers and long-line seismic-refraction techniques. These studies will be supplemented by marine gravity measurements in locations of special interest. Continental-margin sedimentation studies will be carried out through the increased use of submersibles.

The fiscal year 1978 budget breakdown is as follows:

Navy oceanography (basic research) total	$35 million
Navy marine geology and geophysics	8 million
Navy marine geology and geophysics in-house research	1 million
Navy marine geology and geophysics contract research	7 million
Navy marine geology and geophysics contract research (continental margins)	2 million

DEPARTMENT OF COMMERCE

National Oceanic and Atmospheric Administration (NOAA)

NOAA's involvement in marine geology and geophysics started with its ancestors, the U.S. Coast and Geodetic Survey (USC&GS) and the U.S. Weather Bureau. The Survey's efforts were confined largely to geodesy and some geophysics until the mid-1950's, when G. F. Jordan demonstrated that the detailed bathymetric contouring of the continental shelf, produced primarily for nautical charts, could be used for inferring some shelf geology. Then, in the early 1960's, following the recommendations of NASCO and the Interagency Committee on Oceanography (ICO), the USC&GS undertook systematic deep-sea surveys of the North Pacific, using echo-sounders, magnetometers, and gravity meters.

The first descriptions of seafloor magnetic striping—the now-familiar zebra patterns of seafloor magnetic anomalies—indicated that systematic magnetometer surveys provided a new and exciting tool for marine geophysical research. Magnetic surveys by Scripps scientists aboard the USC&GS ship, *Pioneer*, resulted in papers by Raff, Mason, Vacquier, and others that helped to develop the hypotheses of plate tectonics and seafloor spreading. The systematic survey (by those who possessed the tolerance for tedium to carry it out) became an accepted research tool.

In 1965, the Survey and the Weather Bureau became the major components of the new Environmental Science Services Administration (ESSA). The marine geological and geophysical research activities became part of the ESSA Institute for Oceanography (IO). With the formation of NOAA in 1970, this research went to IO's successors—mostly to NOAA's Atlantic Oceanographic and Meteorological Laboratories (AOML), which moved from Washington, D.C., to Miami, Florida, in 1967. A much smaller part of the research remained with NOAA's Pacific Marine Environmental Laboratory (PMEL) in Seattle, Washington.

The National Ocean Survey (NOS) is the successor to the USC&GS. The present program of NOS in the shelf–slope–upper rise region is mission-oriented. It focuses on making detailed bathymetric and oceanographic surveys and mapping of estuarine regions and the continental shelf and upper rise out to a distance of 200 miles from U.S. territorial landmasses.

The surveys use 12-kHz and 3.5-kHz echo-sounding and water-mass measurement devices installed on 25 NOAA vessels. Ten of these vessels are devoted entirely to hydrographic surveys in U.S. territorial waters. The standard survey ship is the *Mt. Mitchell*, 240 feet in length, 3000 tons in displacement. The NOS program provides detailed bathymetric data from ship tracks spaced ½ to 1 mile apart over the continental shelves and margins of the United States. The southeastern section of the continental shelf of North America has been resurveyed. Accompanying these surveys are tidal measurements and shelf–ocean water-circulation studies. Combined, these studies are aimed at predicting tidal and current fluxes on the shelves and the effect of those fluxes on the development of bathymetry.

The BLM, the USGS, and NOS use these survey data to assess hydrocarbon lease regions of the OCS. NOS contracts with academic institutions to do research aimed at assessing the impact of bathymetry and water circulation on ocean dumping off the East Coast and in the Gulf of Mexico. Detailed bathymetry surveys with track lines of spacing closer than ½ mile are being conducted along the approaches to ports and jumbo-tanker mooring facilities.

At present, five NOAA ships are assigned to research [as carried out by NOAA's Environmental Research Laboratories (ERL)]. Nautical charting and hydrography costs NOS about $10 million annually. Three-hundred and twenty employees are assigned to charting and hydrography. Ocean and coastal mapping costs approximately $5.7 million/year, and about 175 people work on it.

The nation's increased need for bathymetric and oceanographic data over oceanic regions under U.S. jurisdiction has led NOS to develop innovative survey techniques, such as the shelf swath-mapping, heave-compensated, sonar (BOSN) that is being fitted to the continental-shelf hydrographic vessels. This technique will be used extensively for detailed mapping and study of the extent and origin of complex bathymetric features. A program of deepwater, swath-mapping, microtectonics is being developed through retrofitting of a 3-km-wide swath, deep-sea mapping facility on the *Discoverer* and *Surveyor*. This program will work closely with academic institutions and will use the *Alvin* in developing an understanding of deepwater microbathymetric and tectonic trends. Also, NOS is digitizing all its bathymetric data so that they can instantly be retrieved.

Laboratory analyses and process-oriented studies that are associated with the NOS mission programs are done largely by contract and through cooperative studies with NOAA's ERL and by contracting with academic institutions. The current contract research budget of NOS with academic institutions is at a level of $4 million per year.

NOS also has a comprehensive, satellite-based, ship-based gravity-and-geodetic program. The present problem being investigated is the origin of the continental-margin geoidal fluctuation. This program will be expanded through the use of SEASAT satellite altimeter data for the study of the relationship between continental-margin geoidal fluctuations, the deep geological structure of the margins, and the water masses circulating over the site of the geoidal anomalies. Fifteen people are involved in these studies. The program has an annual budget of $1 million.

Environmental Research Laboratories

Pacific Marine Environmental Laboratory (PMEL), Seattle, Washington

Partly with in-house capability, partly on contract with others, and in cooperation with the Geological Survey, PMEL is working on the Alaska shelf as part of NOAA's Outer Continental Shelf Environmental Assessment Program (OCSEAP) for BLM. PMEL's major marine geological effort is assessing the potential environmental impact of deep-sea mining for manganese nodules in the Pacific. Called Project DOMES (for Deep-Ocean Mining Environmental Studies), the in-house work has been carried out primarily from the NOAA ship *Oceanographer*.

Atlantic Oceanographic and Meteorological Laboratories (AOML), Miami, Florida

AOML's Marine Geology and Geophysics (MG&G) Laboratory has two major programs on continental margins: Continental Margin Sedimentation (COMSED) and the Trans-Atlantic Geotraverse (TAG).

COMSED is composed of four projects:

1. Continental Shelf Sedimentation (CONSED) is an attempt to understand sedimentary processes and sediment dynamics of the Atlantic continental shelf. The project looks at interactions between shelf waters and the underlying sediment and studies offshore waste disposal. A goal is to provide a portion of the environmental understanding that is needed for operations on the seafloor, e.g., siting for offshore breakwaters for floating nuclear power plants and offshore drilling rigs.

2. The Marine Ecosystems Analysis (MESA) Program is a multidisciplinary study of the effects of man's activities on the marine ecosystem. The MESA New York Bight Project studies an area where man's depredations of the marine environment have been particularly severe. This project includes a marine geological component designed to determine the character and distribution of surficial sediments in the area seaward of New York Harbor and to determine the sediment flux, estimate the sediment budget, and determine the sediment-transport mechanisms and routes of anthropogenic and naturally occurring seafloor materials.

3. Rational Use of the Sea Floor (RUSEF) is a project aimed at determining the sedimentary framework of the Outer Continental Margin, with particular emphasis on processes currently affecting the erosion, transport, and deposition of sediment along the continental margin. Particular emphasis has been on the role of submarine canyons as potential conduits for sediments from the shelf to the deep sea.

4. Marine Geotechnique investigates the geotechnical, or mass physical properties, of selected seafloor areas in order to assess the geological conditions and processes related to marine sediment stability. Studies have been made off the Mississippi Delta jointly with the USGS, several universities, and industry (Project SEASWAB). This project tries to understand the processes within the sediments that affect sediment stability—again, a particular concern of those involved in emplanting offshore rigs or laying subsurface pipelines. Additional studies have been made in recently discovered major slump areas on the continental slope just north of Wilmington Canyon.

The second program of AOML's MG&G Laboratory is only indirectly related to continental margins. The TAG is the seaward extension across the Atlantic of the Transcontinental Geotraverse undertaken as part of the Upper Mantle Program. TAG is one of the few instances in which a project has been based on the assumption that geology does *not* stop at the coastline. Because the Navy has covered most of the western Atlantic with detailed bathymetric and geophysical surveys, the TAG program has concentrated more on the eastern Atlantic and on the Mid-Atlantic Ridge. Where the TAG corridor crosses the Ridge at 26° N, the most recent TAG work focuses on hydrothermal processes that concentrate metals in the oceanic crust at the Mid-Atlantic Ridge. In 1974, the TAG program discovered a hydrothermal manganese field within the midoceanic rift valley. This program also plans to produce a joint U.S.A.–U.S.S.R. atlas of the central North Atlantic ocean basins and continental margins.

With the present overflow of proposals to reorganize the oceanic agencies within the Executive Branch, any *future plans* must be considered as no more than the roughest of speculations. NOS plans to continue its nautical charting activities as long as they are needed. Those who say "Haven't you got the shelf charted yet?" simply do not realize that maintaining and updating charts must be done not only because bottom topography changes with time but also because the increasing draft of large tankers demands new charts. NOS plans to work even closer with the NOAA/ERL system and the academic community, especially on such new NOS responsibilities as the ocean-dumping program.

AOML and PMEL plan to accelerate their efforts to understand the dynamics of sedimentary processes on the continental margin. AOML anticipates that this work will contribute to a much better understanding of the processes affecting sediment stability or, more importantly, instability.

In the Pacific, PMEL probably will become increasingly involved in the sedimentological aspects of a Puget Sound study within the MESA Program. PMEL will continue work on the DOMES Project and—at least for the near future—it will make additional studies (cooperatively with the USGS) that relate to the OCS Environmental Assessment Program for BLM.

With the exception of DOMES and the TAG program, most of NOAA's marine geological and geophysical work for the next several years most likely will be concentrated on processes that act on the surficial and uppermost portions of the U.S. continental margins.

The Sea Grant Program

The Sea Grant Program is based on the Sea Grant College and Program Act of 1966 and is designed to support activities that "benefit the United States, and ultimately the people of the world, by providing greater economic opportunities, including expanded employment and commerce." The Sea Grant is modeled after the Land Grant Program.

The national Sea Grant Program has been described as a unique federal–state–university partnership. It is, in fact, the only federal grant program that mandates and supports in American universities the integration of research, education, and public service in a practical approach to developing marine resources and technology. Sea Grant is oriented toward applied research. It encourages the formation of interdisciplinary teams to define problems and to respond to local, state, and regional needs in research. The emphasis on application to specific problems gives Sea Grant a strong grass roots base. That base is further reinforced by Sea Grant's Marine Advisory Services, which are aimed at transfer of knowledge and the search for and definition of problems.

In its relation to universities, Sea Grant introduced the concept of the progressive status of its grantees, encouraging growth from several projects through several stages to the status of a Sea Grant college. Sea Grant support comes in three forms:

1. Project—for individual projects;
2. Coherent—for groups of projects at an individual institution; and
3. Institutional—for large programs at institutions generally acknowledged as having a broad base of activity in marine affairs and that have committed themselves to positive, long-term objectives in the field.

To strengthen the relation between investigators and the organizations for which they work, Sea Grant has created program directorates within universities. The Sea Grant director is a university employee who controls the program. This procedure restores authority and guidance to the university and fosters local support of the program by being sensitive to local and regional programs.

Finally, one third of the cost of all programs must be provided from nonfederal funds, thus providing an incentive for support from other sources. This has been so successful that in some programs nonfederal sources exceed 50 percent, making the federal government the junior partner in some Sea Grant programs. Today, there are 12 Sea Grant colleges and universities.

The Sea Grant Improvement Act of 1976 directs the Secretary of Commerce to issue guidelines for national projects, calls for the development of international cooperative assistance programs, and includes mandates for greater emphasis on regional consortia and educational fellowships. Of particular interest is the proposed development of programs by which universities under Sea Grant sponsorship may work with sister universities in other countries. This is to be done with experts in international oceanographic affairs.

The Sea Grant role in geological and geophysical research on continental margins has been relatively minor because this work is more within the scope of other federal agencies.

NOAA's Environmental Data Service operates the National Geophysical and Solar-Terrestrial Data Center (NGSDC) located in Boulder, Colorado. This Center serves as the national focus for marine geological and geophysical data management and dissemination to users. NGSDC has a constantly expanding data base that currently consists of some 6,000,000 nautical miles of underway geophysical data (gravity, magnetics, and seismic) and some 75,000 geological data entries. The Center is prepared to archive new continental-margin data as well as to provide historical data-support and data-management services. The IDOE continental-margin data are included in the Center's data files, as well as USGS data, and the NOS is cooperating with the NGSDC in digitizing that vast accumulation of sounding data and surficial-sediment characteristics data that NOS has obtained over the years on the U.S. continental margin.

DEPARTMENT OF ENERGY

Introduction

The shape and scope of the new Department of Energy (DOE) is not yet clear to this Panel. There is little doubt, however, that the programs of this department will have considerable impact on research of continental margins.

Because the Energy Research and Development Administration has been absorbed by the new Department of Energy, we include our summary under this heading.

Energy Research and Development Administration (ERDA)

ERDA and its predecessor, the Atomic Energy Commission (AEC), have a history of supporting research that focuses on migration of energy-related wastes or by-products, particularly radionuclides, in the marine environment. Most of this research has been concerned with transport in the ocean itself, but quite a bit deals with the movement and fixation of wastes in marine sediments.

ERDA's current interests in the marine environment of the continental margins include potential resources of petroleum, uranium, thorium, and geothermal energy; utilization of ocean thermal gradients to generate electricity; and environmental consequences of energy-related activities.

The nonrenewable fossil-fuel and nuclear-energy resources are of particular importance with respect to the nation's immediate (to about year 2000) energy requirements. The sediments and sedimentary rocks of the U.S. continental margins have considerable promise regarding fossil-fuel energy resources. To realistically evaluate the potential of such resources will require a major research effort involving both fundamental and applied research activities. ERDA is cooperating with the USGS to provide the basic information required to assess the petroleum potential in rocks of the continental slope, rise, and adjacent deep ocean.

The DOE is expected to support research directed toward making effective and efficient use of the nation's energy resources. Basic research will be an essential component of this effort. It is necessary to provide the scientific base for future energy technologies. In view of the mandates of other federal agencies, DOE no doubt will coordinate its research activities with and support those of its federal siblings.

Exploration, extraction, ultimate consumption, and final disposition of waste energy and materials are all encompassed in using energy resources. If energy resources are used, and waste products disposed of, effectively, man's physical and biological environment need not be significantly degraded.

DEPARTMENT OF STATE

The Department of State's interest in geological and geophysical research on continental margins relates primarily to the delimitation of maritime boundaries. Negotiations about the Law of the Sea deal with national claims to jurisdiction over the submerged areas of the continents. For example, if the continental margin comprises the seabed and subjacent sediments of the shelf, slope, and rise, then what provisions must be made for a nation whose margin extends beyond 200 nautical miles? Another important issue in the Law of the Sea negotiations is the question of national jurisdiction over scientific research (and the concomitant distinction between basic and applied research) within maritime zones (see the section on "The Law of the Sea and Scientific Research" at the end of this Appendix).

The Office of the Geographer (Department of State) is given special funds to conduct a resource assessment of U.S. maritime boundary areas. The Geographer must compile, for each area, all known information on the distribution of resources, both living and nonliving, both present and potential.

The Office of the Geographer has funded the processing costs of previously derived geophysical data, and, via the USGS, has purchased proprietary seismic data (which remain proprietary). The Office has also obtained data from private industrial firms and from university-sponsored research. The monies available to this Office are too small to support new field research.

In the assessment of maritime boundary areas, the Office of the Geographer reports directly to the Office of the Legal Adviser. Other offices or bureaus within the Department of State with interests in the maritime boundary project include the Executive Secretariats, the National Security Council Interagency Task Force on the Law of the Sea, the Bureau of Oceans and International Environmental and Scientific Affairs, and the Office of Oceans and Fisheries Affairs. In addition, the various regional bureaus are concerned with pertinent maritime boundary zones affecting their geographic or political areas of interest.

No significant support for continental-margin research comes from the Department of State.

NATIONAL AERONAUTICS AND SPACE ADMINISTRATION (NASA)

In the chapter on remote sensing (Chapter 12), some of NASA's projects that are particularly relevant to continental-margin studies are discussed. A number of specific projects related to earth science are carried out in NASA Space Flight Centers (e.g., the Geodynamics Branch of the Goddard Space Flight Center). Overall coordination occurs under the Earth and Ocean Dynamics Program (EODAP) in the Office of Application. EODAP originated with Geodesy and the National Geodetic Satellite Program.

Today, EODAP is attempting to define user needs in the marine and continental earth sciences; to develop tools to meet those needs; and to demonstrate tools, techniques, overall aerospace technology programs, analysis, and modeling. Finally, EODAP is concerned with the transfer of that technology.

EODAP elements—on applications research and technology—include the following: measurement systems, forecasting techniques, modeling, data analysis, tectonic-plate motion/validation, EODAP surface truth, advanced applications flight experiments, Shuttle experiment definition, application analysis and studies, and application systems verification tests.

In addition to this, EODAP is concerned with the following flight missions: Skylab, Geodynamics Experimental Ocean Satellite (GEOS), Apollo/Soyuz Test Project, Laser Geodynamics Satellite (LAGEOS), Ocean Dynamics Satellite (SEASAT), Magnetic Field Satellite (MAGSAT), Space Shuttle/Spacelab, Earth Surface Surveyor (ESS), and Gravity Field Satellite (GRAVSAT).

EODAP had an Ocean Dynamics Advisory Subcommittee working under the Application Steering Committee. That committee was an outgrowth of a SEASAT User Working Group. An Earth Dynamics Advisory Subcommittee was appointed with the assistance of the Committee on Earth Sciences of the National Research Council's Space Science Board. The Earth Dynamics Advisory Subcommittee had about 50 members, approximately half of whom were university scientists and another half, federal scientists; two scientists were from industry. As a result of the Executive Office directive to reduce the number of government advisory committees, both the ocean and the earth dynamics advisory groups have been disbanded. At this writing, discussions about what advisory mechanisms should replace these are in progress.

NATIONAL SCIENCE FOUNDATION

Introduction

The National Science Foundation (NSF) was established in 1950 to promote the progress of science in order to advance the national health, prosperity, and welfare. The Foundation fulfills this responsibility primarily by sponsoring research in all the sciences, encouraging and supporting improvement in science education, and fostering scientific information exchange. NSF does not, itself, conduct research. Recipients of NSF support have full responsibility, in accordance with the terms of their grants or other agreements, for the conduct of their projects and for the results produced. Proposals for support are assigned to the appropriate directorate, division, section, and program for review and evaluation. The basic organization of the NSF is shown in Figure A.4.

Research projects that pertain to geological and geophysical studies of continental margins reside within the Astronomical, Atmospheric, Earth and Ocean Sciences Directorate (AAEO). This Directorate is subdivided into five Divisions. The three Divisions with major projects in continental-margin studies are the Division of Ocean Sciences (OCE), the Division of Earth Sciences

(EAR), and the Division of Polar Programs (DPP). The Divisions are composed of Sections, which cover broad disciplinary areas. An exception is EAR, which has only programs. The final organization unit of the Foundation is the Program. All research proposals ultimately go to a Program for review and evaluation. The primary Program offices that support continental-margin geological–geophysical research are the following:

1. *Division of Ocean Sciences:* Submarine Geology and Geophysics Program (Oceanography Section) and Seabed Assessment Program (Office for the International Decade of Ocean Exploration).
2. *Division of Earth Sciences:* Geochemistry Program, Geology Program, and Geophysics Program.
3. *Division of Polar Programs:* Polar Earth Sciences Program and Polar Ocean Sciences Program (Polar Science Section).

In addition to these, three "offices" provide logistics and specialized facility support: the Office of Oceanographic Facilities and Support (OFS), the Ocean Sediment Coring Program (OSCP), and the Polar Operations Section (part of the Division of Polar Programs). More detailed descriptions of the varying divisions, sections, and programs, with emphasis on the geological and geophysical research programs, follow. It is important to note that the research programs all have responsibilities for research areas broader than continental-margin research alone. Proposals for support of continental-margin studies are evaluated in the context of overall program responsibilities and priorities.

The approximate level of NSF support for geological and geophysical research on continental margins in fiscal year 1977 was $10.7 million, exclusive of Antarctic logistics support. The general distribution of funds by programs was as follows:

	$ Thousands
Division of Ocean Sciences	
Submarine Geology and Geophysics Program	$1,800
Seabed Assessment Program (IDOE)	1,000
Office of Facilities and Support (ships)	2,000
	$4,800
Division of Earth Sciences	
Geophysics Program	—
Geochemistry Program	1,200 (est.)
Geology Program	—
Ocean Sediment Coring Program (DSDP)	4,000
	$5,200
Division of Polar Programs	
Polar Earth Sciences Program	400
Polar Ocean Sciences Program	400 (est.)
Polar Operations Section (logistics)	—
	$ 800

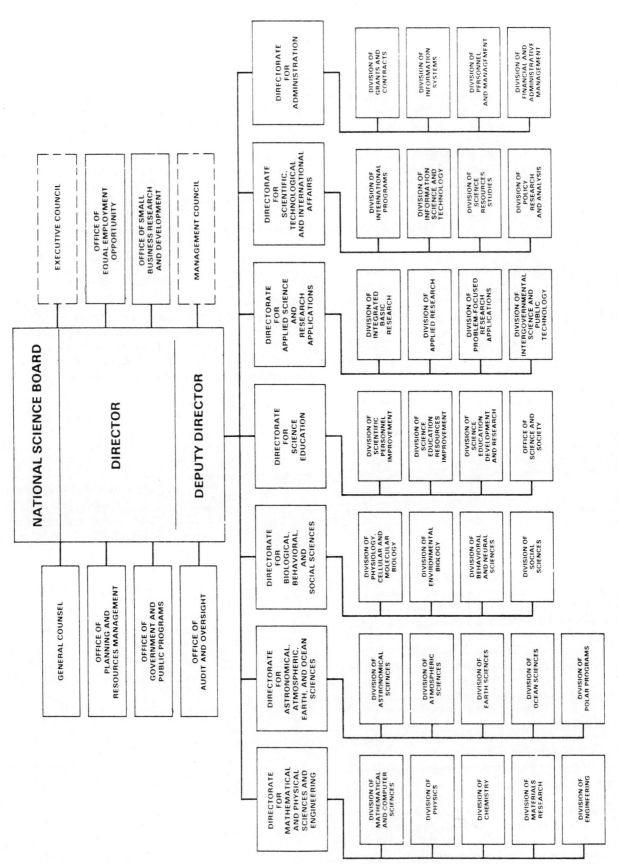

FIGURE A.4 Organization chart of the National Science Foundation.

Division of Ocean Sciences

The overall objective of the Ocean Sciences Program is to improve understanding of the nature of the ocean, its influence on human activities, and human impact on the marine environment. This is accomplished (1) by supporting individual scientists' basic research projects (Oceanography Section) and (2) by funding large projects within the Office for the International Decade of Ocean Exploration (IDOE). The Oceanographic Facilities and Support Program provides acquisition and operating funds for research vessels to carry out these research programs.

Submarine Geology and Geophysics (SG&G) Program

The SG&G program is very broad. It supports all areas of geological, geochemical, and geophysical research that pertain to the nature of the oceanic sediments and rocks and to the genesis and evolutionary history of the ocean basins. Research supported by this program includes the following areas:

• Geologic structure; tectonics—geophysical and geological investigations of the structure of the oceanic crust and overlying sedimentary layers. This includes studies of continental margins, oceanic rise systems, ocean basins, deep-sea trenches, fracture zones, seamount chains, and aseismic ridges. Projects range from local detailed structural studies to the development of global geodynamics models. Primary data sources include seismic, gravity, magnetic, bathymetric, thermal, and electromagnetic measurements, often coupled with physical property determinations.

• Igneous and metamorphic processes—geological, geophysical, and geochemical studies of the rocks and sediments that form the ocean floor. Basic emphasis is on describing and understanding the origin, development, and subsequent alteration of the ocean crust. Includes studies of IPOD and dredge samples from oceanic rises, fracture zones, and sediments; basalt–seawater interaction; mineral chemistry; age-dating techniques; and petrogenetic modeling. Primary data types include major, minor, and trace-element determinations; bulk chemistry; and petrologic and petrographic studies, coupled with experimental laboratory studies and geochemical modeling.

• Sediment deposition; diagenesis; composition—geological, geophysical, and geochemical studies of oceanic sedimentary sections. Basic emphasis is on understanding the physical processes of sediment deposition and erosion, sediment distribution patterns, chemical changes and alteration of sediments with time, sediment–seawater and sediment–igneous rock interactions, and dominant geological and oceanographic processes controlling sedimentary systems. Individual projects range from studies of the interaction of bottom water currents and sediment bedforms to examination of the changes in organic constituents with temperature and pressure from DSDP cores. Data sources include bottom-current measurements, physical property measurements, organic geochemistry, physical chemistry, and stratigraphic and lithologic measurements.

• Geologic history; paleoenvironments; biostratigraphy—geological, geochemical, and geophysical projects directed toward understanding the evolution of the ocean basins and the interaction of the geologic systems to other environmental factors. Includes studies of past oceanic circulation patterns, paleoclimates, evolution of microfossil groups, water chemistry and temperature controls on fossil assemblages and sediment types, and interaction of land and oceanic geologic processes. Primary data sources are micropaleontologic and geochemical measurements coupled to oceanographic, biological, and physical modeling efforts.

Seabed Assessment Program

The International Decade of Ocean Exploration (IDOE) is a long-term, international, cooperative program to improve the use of the ocean and its resources for the benefit of mankind. Late in 1969, the Vice-President of the United States, in his capacity as Chairman of the National Council on Marine Resources and Engineering Development, assigned the responsibility for planning, managing, and funding the U.S. program to NSF. With regard to geology and geophysics on continental margins, the Seabed Assessment Program was relevant, and the following goals were stated: "Expanded seabed assessment activities to permit better management—domestically and internationally—of marine mineral exploration and exploitation by acquiring needed knowledge of seabed topography, structure, physical and dynamic properties, and resource potential, and to assist industry in planning more detailed investigations."

Seabed Assessment Programs are divided into three broad areas: continental-margin studies, metallogenesis and plate tectonics, and manganese-nodule studies. Specific projects tend to cut across the first two themes, e.g., Nazca Plate Study and Studies in East Asia Tectonics and Resources (SEATAR). Seabed Assessment also supports use of the submersible *Alvin*, e.g., Project FAMOUS and the Galapagos Rift.

In general, IDOE supports a few large-scale projects that focus on major scientific problems. The problems must be amenable to solution. In other words, the study must be based on a hypothesis that can be tested. The CLIMAP program, although part of the Environmental Forecasting, was essentially a geological project. It is considered one of the more successful "big science" efforts. The preparation of a temperature map 18,000 years B.P. served as the goal for a large and varied number of measurements.

IDOE was to focus on large-scale projects, i.e., on "big science." As the following tabulation compiled by the OSB shows, this was at one time true, but today the average grant size is similar to that of typical NSF grants.

Grant size	1972	1976
Oceanography Section	$ 42,000	$48,000
	$274,000	$79,000
Grant size in 1970 dollars		
Oceanography Section	$ 41,000	$33,000
IDOE	$263,000	$54,000
Man-months per grant		
Oceanography Section	8.8	8.6
IDOE	85.2	18.5
Equipment per grant		
Oceanography Section	$ 2,800	$ 2,900
IDOE	$110,000	$ 4,200
Equipment per grant in 1970 dollars		
Oceanography Section	$ 2,700	$ 2,000
IDOE	$106,000	$ 2,900

Major research projects supported during fiscal year 1977 are as follows:

• Studies in East Asia Tectonics and Resources (SEATAR)—SEATAR investigates the tectonic and geological processes that produce subduction zones, island arcs, and marginal seas and attempts to relate these processes to metallogenesis and hydrocarbon genesis. Studies of selective areas along the trench from Sumatra to the Banda Arc are concerned primarily with hydrocarbon genesis; those along the Banda Arc–Philippine Sea trench, with metallogenesis. Regional studies include gathering and synthesis of seismicity, heat-flow, gravity, and magnetic data. Maps of total sediment thickness and tectonic synthesis are being prepared.

• Nazca Lithospheric Plate: Plate Tectonics and Metallogenesis—This study uses a wide range of marine geophysical, geological, and geochemical experimental techniques directed at determining the relation between the tectonic cycle of an oceanic plate and metallogenesis and eventual ore emplacement. This tectonic cycle is being considered in relation to (a) the primary oceanic source of metal at a divergent crustal plate boundary; (b) the transport and metallogenesis enrichment phase on the moving oceanic plate; and (c) the zone of subduction, where the oceanic plate is assimilated beneath the continent, with consequent mountain-building, seismicity, volcanism, and intrusion of material into the overlying crust to create mineralization zones. All fieldwork has been completed on this project. A final synthesis and analysis phase is being supported.

• Galapagos Rift: Hydrothermal Processes and Metallogenesis—This project studies seawater hydrothermal systems that extract and transport heat and metals from newly intruded or erupted oceanic crustal rocks. Hydrothermal processes influence the composition of seawater, form extensive geothermal reservoirs, and lead to the formation of a variety of economically important ore bodies now preserved on the continents. Specific objectives are (a) to locate active hydrothermal vents on the seafloor in the Galapagos Rift region, relating their distribution to the local and regional crustal tectonic fabric; (b) to determine the heat and seawater budgets of oceanic hydrothermal systems; (c) to understand the interaction between circulating fluids and seawater, and the processes of precipitation, dispersal, and sedimentation of resultant solid phases; and (d) to relate the observations to models of the kinetics and thermodynamics of water–rock interaction.

• Manganese Nodule Program (MANOP)—This project endeavors to understand the processes controlling the distribution and composition of deep-sea ferromanganese nodules. Specific goals for 1977–1978 are (a) to relate the composition of nodules to the flux of transition metals at the nodule sites; (b) to relate the nodule composition to the chemistry of the solution in which they grow—both pore and bottom waters; and (c) to determine constraints for theories of nodule accretion and growth, based on chemical and nodule distribution observations. The research centers on an integrated series of bottom ocean-monitor experiments to be carried out on the seafloor in the eastern equatorial Pacific.

Past projects included partial support for Project FAMOUS and support for regional geological and geophysical studies of the South Atlantic (including the African Atlantic margin and the South American Atlantic margin).

Office of Oceanographic Facilities and Support (OFS)

OFS provides a major share of the total support for 30 research ships and a number of specialized facilities operated by 15 oceanographic laboratories (UNOLS). The continuing objectives of this program are (a) to provide operating costs for the research vessels needed to carry out NSF-supported research, (b) to maintain and improve the material condition of the ships, and (c) to coordinate Foundation ship support with other agency programs.

Division of Earth Sciences (EAR)

EAR endeavors to provide a basic knowledge of the structure and composition of rocks and sediments that comprise the earth's crust and the processes that form and modify these rocks. EAR supports basic research programs of individual scientists, and research consortiums by the Geology Program, the Geochemistry Program, and the Geophysics Program. In addition, the Ocean Sediment Coring Program sponsors the acquisition of geological samples from the floor of deep-ocean basins by means of rotary drilling and coring in sediments and underlying crystalline rocks.

Geology Program—supports field-oriented studies for testing and modifying geological theories along with studies of natural geological processes and the near-surface environment.

Geochemistry Program—directed toward determining the chemical nature of the earth, the ages of rocks and minerals, distribution of past temperatures and pressures in the earth, and the formation of ore deposits.

Geophysics Program—supports research in earthquake seismology; gravity variations; heat-flow; internal electric currents; and past and present magnetic fields. Specific projects range from theoretical studies of elastic-wave propagation to an extensive field study of deep continental structure.

Ocean Sediment Coring Program—The major activity under OSCP is the Deep Sea Drilling Project (DSDP), which is currently in its International Phase of Ocean Drilling (IPOD). The major portion of OSCP monies goes to operational support for the drilling ship, *Glomar Challenger,* to obtain oceanic sediment and rock samples. OSCP also supports advisory panels for site selection and advance planning, sample-curating activities, and site surveys for specific drilling locations. The program is largely ship-operations support. Although support for the direct operation of the drilling program proper comes from EAR, support for research on samples obtained by DSDP/IPOD primarily comes from the SG&G Program and is a significant component of that program.

Division of Polar Programs (DPP)

The DPP supports projects in all relevant sciences for the Antarctic and Arctic regions, including studies in human behavior, biology, cartography, geology, glaciology, meteorology, oceanography, solid-earth geophysics, upper-atmosphere physics, and magnetospheric physics. Because there are extreme environmental conditions, a considerable effort is devoted to logistics support.

Polar Earth Sciences—Objectives in the Antarctic are

• To understand the geology and history of the Antarctic;
• To assess the mineral resources potential in the Antarctic; and
• To determine and understand the relationship of Antarctica to global geodynamics.

Areas of special interest are the McMurdo Sound vicinity, the central Transantarctic Mountains, the Scotia Arc, the Lassiter Coast, the Pensacola Mountains, Marie Byrd Land, the coast of Ellsworth Land, the Ellsworth Mountains, and the continental shelf. Emphasis is being put on the completion of medium- and large-scale maps of areas where an analysis and a display of geological data are needed.

Objectives in the Arctic are

• To define the tectonic and geologic history of the Arctic Basin and the continental shelves;
• To synthesize existing data on the geological development of the Arctic in relation to the distribution of non-renewable resources and to reducing environmental impacts; and
• To determine the origin, state, and dynamics of subsea permafrost.

The Arctic Research Program provides support for geological and geophysical investigations in paleontology, paleomagnetism, permafrost, seismology, tectonics, and crustal structure. Research is conducted in Svalbard with scientists from Norway, Poland, West Germany, and the United Kingdom. Svalbard is near a junction of the Mid-Atlantic Ridge, the Eurasian and the American tectonic plates, and the Arctic Basin.

The NSF is interested in defining the governing factors that regulate the Arctic environment to minimize any detrimental effects of human activities in the Arctic. There is a need to understand and predict offshore permafrost so that the least environmental damage will be done when natural resources are developed. Little is known of the temperatures, the hydrological characteristics, and the sediment types that control the existence of permafrost. Investigations on the origin, the cause, and the dynamics of offshore permafrost are encouraged as a part of the overall need to learn more of the fragile Arctic environment.

Polar Ocean Sciences—Objectives in the Antarctic are

• To complete a physical and geological–geophysical survey of the ocean waters surrounding the Antarctic continent;
• To determine the dynamics of formation and distribution of Antarctic bottom-water, currents, sea-ice, atmospheric circulation systems, and biological productivity of the Weddell Sea;
• To understand the physical, geological, and chemical oceanography of the Ross Sea region; and
• To investigate air–sea interactions by examining the combined heat mass and momentum transport in the coastal regions.

The Southern Ocean is a large source of the intermediate-depth and deepwater masses that circulate throughout the world. Nutrient-rich waters support high biological activity. The large, annual variation in temperature and sea-ice extent exert profound effects on energy transfer processes.

The seafloor around Antarctica presents fundamental unsolved problems in marine geophysics. A circum-Antarctic survey, begun in 1962, is nearing completion, but data analysis and some additional ship-based research will be continued. Weddell Sea oceanography, Ross Sea sediments, analysis of coastal structures, and marine chemistry are topics of special current interest. In addition to ship-based programs, surveillance by air and the deployment of buoys for year-round data collection are new techniques being explored.

Objectives in the Arctic are

• To increase knowledge of Arctic water-mass interactions, of oceanfloor characteristics, and of the Arctic Ocean's role in global climate;

• To study the Nansen Cordillera and its relation to seafloor spreading; and

• To perform energy-budget studies of Arctic sea ice and observe water transport to isolate factors that could determine the structure of the Arctic sea-ice cover.

Areas of special interest are investigations of the formation, movement, heat and salt exchange, and mixing processes of the Arctic water masses and sea ice; analyzing magnetic anomalies, heat-flow, sedimentary history, and gravitational data of the ocean floor; and assessing the role of the Arctic Ocean in global climate. Seafloor-spreading studies on the Nansen Cordillera, a continuation of the Mid-Atlantic Ridge into the Arctic Ocean, are planned, as is sediment coring.

Polar Operations—This Section funds logistics and operational support at four permanent Antarctic stations and at field sites, aboard the research ships *Hero* and *Islas Orcadas*, aboard the icebreakers, and at several sub-Antarctic field sites.

Operation support elements include U.S. Navy, U.S. Coast Guard, Military Airlift Command, Military Sealift Command, contractors, and other specialized units that may be assigned by the Department of Defense.

The Navy, with funds provided by NSF, provides the largest part of the operational support. This support includes five LC-130 aircraft and six UH-1N helicopters during the austral summer field season. The Military Sealift Command and the Military Airlift Command transport personnel and cargo between the United States and Antarctica.

The U.S. Coast Guard provides icebreaker support services, including logistics and marine sciences support in ice-covered areas that research ships are unable to reach.

A contractor operates and maintains South Pole, Siple, and Palmer Stations and the Foundation's research ship *Hero*. The contractor is responsible for designing and constructing replacement science laboratories and support facilities and for renovating and modifying existing facilities.

Islas Orcadas (formerly, USNS *Eltanin*), a 266-foot research ship, operates under a cooperative arrangement with Argentina. Fiscal year 1978 will be the final year of this ship's operation under the current arrangement.

The Operations Section provides no logistics support in the Arctic, although it can assist grantees in arranging their transportation in Greenland.

ADVISORY GROUPS

The National Academy of Sciences (NAS) was established over 100 years ago to provide advice to the government "upon any subject of science or [technological] art."* The operating agency of the NAS, the National Research Council (NRC), is currently structured into Assemblies and Commissions. Quoting from Article III, Section 1, of the NRC's amended (1974) Articles of Organization, "Assemblies shall be concerned with furthering the advancement of their component disciplines, and with the general furtherance of science and engineering as components of the intellectual resources of the nation, and with their contribution to national life. Commissions, as multidisciplinary bodies, shall be concerned principally with those problems of the nation and society that require the application of a broad range of interdisciplinary knowledge in scientific, engineering and other fields."

Advisory mechanisms have been established nationally and internationally to provide a sense of direction for future programs of research on continental margins. There are several groups within the Commission on Natural Resources offering recommendations and advice relevant to research on the continental margins. This Panel is within the Assembly of Mathematical and Physical Sciences (AMPS), which has a number of boards and committees that make recommendations for research on continental margins. Among these are the Geophysics Research Board and one of its subcommittees, the U.S. Geodynamics Committee; the Office of Earth Sciences; the Committee on Seismology; the Committee on Geodesy; the U.S. National Committee on Geology; the U.S. National Committee on Geochemistry; the Space Science Board through some of its subcommittees; and, of course, the Ocean Sciences Board (through this Panel).

The scientific community directly involved in continental-margin studies is not large, and scientists from this community serve on these advisory groups, so there is considerable overlap in committee membership. Studies carried out by these committees are usually confined to limits prescribed in their charters. Broad coordinating functions in this area are carried out by the Geophysics Research Board, whose members are the chairmen of those Assembly boards and committees having interests in the earth sciences. Recently, the Office of Earth Sciences has included members on its Advisory Committee who are concerned with the hydrosphere and atmosphere as well as with the solid earth. The U.S. National Committee on Geology concerns itself mostly with international geology, including International Geological Congresses and the International Union of Geological Sciences rather than domestic geological affairs.

These committees and boards have issued a number of reports that include aspects of geophysical and geological research on the continental margins, but few look at the total picture. This situation is not surprising, as the sponsoring agencies, themselves, have limited jurisdictions. This situation is cause for concern, because if sources for

---

*From: An Act to Incorporate the National Academy of Sciences, 1863.

funding research follow similar jurisdictional patterns, opportunities for multidisciplinary research that crosses the ocean–continent boundary tend to be lost for lack of suitable avenues for support.

A similar problem exists with international advisory bodies. The Scientific Committee on Oceanic Research (SCOR) advises the Intergovernmental Oceanographic Commission (IOC) on oceanographic matters—but stops at the shoreline. Within the International Union of Geodesy and Geophysics (IUGG), there are associations that deal with various aspects of research on continental margins in terms of individual disciplines. In the International Union of Geological Sciences (IUGS), there are committees and commissions concerned, again, with aspects of continental margins—but not with their full scope. The Inter-Union Commission on Geodynamics (the coordinating body for International Geodynamics Project) takes a problem approach, thus *does* concern itself with integrated research across continental margins. It transmits its concerns to member countries through national committees.

The National Advisory Committee on Oceans and Atmosphere (NACOA) was legislatively established to advise the President and the Congress on these matters; the President being advised through the Secretary of Commerce. NACOA does not concern itself to any great extent with scientific research, *per se*, but it does consider research in terms of national policy and the effects of the Law of the Sea negotiations, treaty, and other jurisdictional uncertainties on marine research. By definition, its activities do not include the land portions of continental margins.

Two principal bodies established to advise the Congress are the Office of Technology Assessment (OTA) and the Congressional Research Service (CRS). OTA was created to examine, largely through extramural contracts, significant technological and governmental programs and to assess their problems and opportunities. OTA must select its projects with the approval of a board of 12 members of Congress (the Technology Assessment Board); therefore, the political process is intimately involved in its work. The purpose of CRS is to respond to specific Congressional requests for information and policy analysis. In doing this, it sometimes anticipates potentially pertinent issues. Of the approximately 500 professionals in CRS, there are five geologists and one oceanographer holding advanced degrees. Other staff members who handle issues of concern to geological scientists—including energy, mining and minerals, Outer Continental Shelf, and deep seabed assessment—are lawyers, economists, physicists, biologists, chemists, and geographers. One might expect that reports produced by the NAS/NRC committees would automatically be sent to pertinent CRS divisions so that the information and conclusions would be readily available, but this does not appear to be the case.

This is not intended to be a comprehensive review of the numerous advisory mechanisms that are available but rather a sampler that suggests that existing mechanisms may be less than adequate for those who are concerned with scientific research on continental margins.

INTERDISCIPLINARY PROGRAMS

The International Geophysical Year (IGY), 1957–1958, gave scientists an opportunity to study the floor of the world ocean. These were primarily exploratory studies, since little was known about the sediments, the crust, or the mantle beneath ocean waters. At the same time, the first standardized worldwide seismograph network was established to look at the deeper earth. The IGY program established mechanisms for international communication among scientists and for data exchanges. While not emphasized in the IGY programs, geological and geophysical investigations on margins were made in some parts of the world.

During the early 1960's, the International Upper Mantle Project (UMP) was organized, originally along specific disciplinary lines. As this project proceeded, it became apparent that the world rift system was associated with ocean ridges, and that, consequently, continental margins and island arcs deserved special attention. UMP established subcommissions to concentrate on the research problems and opportunities in these areas. Crustal seismic refraction and other geophysical studies were made on the continents and their margins, and the World-Wide Standard Seismograph Network was extended to include these areas.

During the UMP decade, the plate-tectonics model developed. This model provided a conceptual framework into which could be fit the various types of continental margins—and, indeed, global geological structure. It was a major step that led naturally to new research directions promising to increase understanding of the nature and history of continental margins. The Deep Sea Drilling Project (DSDP), subsequently to become the International Program of Ocean Drilling (IPOD), was begun during this decade. The UMP made major contributions to our knowledge of the ocean basins. However, safety considerations involving open-hole drilling precluded drilling deep holes in the thick sediments of continental margins.

The UMP was followed by the International Geodynamics Project (IGP). This provided continuity for the communication flow (and data exchanges) begun during IGY. Unlike IGY, which was exploratory, and UMP, which was initially discipline-oriented, IGP was oriented toward problems and processes. Its prime objective was to develop the plate-tectonics concept. This required a better understanding of the kinematics of the model and of the properties of and the processes acting in the earth's interior. Regional working groups were organized to examine the properties of active and inactive continental margins and, on a problem basis, to examine the driving mechanism of plate movement. Also in this period, the International Decade of Ocean Exploration (IDOE) took place. This program was designed to support large-scale oceanography in areas related to societal needs. One IDOE program—Seabed Assessment—led to investigations of continental margins. For administrative, political, logistic, or intellectual reasons, these investigations have been almost totally confined to the oceanic portions of continental margins and to margins *other* than those of the United States.

Anticipating the scheduled termination of IGP, IDOE, and the International Program for Ocean Drilling (IPOD) in the late 1970's, planning is under way to determine how, in the 1980's, to take best advantage of the potential challenges and opportunities. Since the societal and scientific interest in continental margins has significantly increased in recent years, study of margins will figure strongly in planning for future work in oceanography.

The International Council of Scientific Unions (ICSU)—through two of its member unions, the International Union of Geodesy and Geophysics (IUGG) and the International Union of Geological Sciences (IUGS)—is currently considering a program of earth dynamics, but it focuses upon the crust rather than the deep interior. At the same time, ICSU's Scientific Committee on Oceanic Research (SCOR), together with the Committee on Marine Geology (CMG) of the IUGS, is looking at future opportunities in marine geology and geophysics research.

Meanwhile, back home, the U.S. Geodynamics Committee (USGC) is developing a program on crustal dynamics that includes studies of the evolution of continental margins. Complementing USGC's study is this Panel's study, sponsored by the Ocean Sciences Board of the National Research Council (OSB-NRC). Although this report is primarily oriented toward oceanographic problems, it contains recommendations for research on both the land and the ocean portions of continental margins.

The future of drilling for ocean science was considered in early 1977 by representatives from an international group involved in planning international programs. This group's document makes a strong case for using ocean drilling to solve some of the problems of continental margins. Meanwhile, the Committee on Solid Earth Sciences of the Federal Coordinating Council on Science, Engineering, and Technology (FCCSET) has produced a report (April 1977) containing ways to implement a program of land drilling based on the scientific objectives set forth in the report, *Continental Drilling* (E. M. Shoemaker, editor, published by the Carnegie Institution of Washington in June 1975).

So much for current and proposed scientific programs that deal with the geology and geophysics of continental margins. We did not include studies related to resource evaluation or utilization carried out by the NAS/NAE/NRC, government agencies, or industries. We described the framework in which geological and geophysical investigations of continental margins have been carried out or coordinated in the past and some of the mechanisms through which plans for future studies are being developed. It was significant in the trend of thinking that the

Committee on Solid Earth Sciences, created under FCCSET, explicitly considered that the scope of its activities included all solid-earth investigations whether under the continents or under the oceans. Following further reorganization in 1978, the status of this committee has become informal.

MARINE RESEARCH IN CANADA

The formal and informal organization of the earth sciences in Canada is rather different from that in the United States because the dominant activities are different. The differences arise from a variety of obvious causes. For example, Canada has 0.2 square mile of landmass per person; the United States has only 0.015. Whatever the causes, there are several effects:

1. Canadians emphasize land geology, because they earn money from land-based resources. Consequently, enthusiasm for direct participation in a project like IPOD is limited to a relatively small number of individuals.
2. There are effectively no private universities in Canada. With a few significant exceptions, such as the Department of Oceanography at Dalhousie University, university geology or geophysical departments have substantial undergraduate programs. These programs allow relatively little time for research with graduate students and by the professors themselves. Thus, probably more attention is paid to the needs of the students than in many of the U.S. institutions. The price paid for this is a relatively smaller research output.
3. There are no "private" oceanographic institutions like Woods Hole. All marine research is done by government agencies (which usually have rather specific tasks) or by universities.
4. Much industry is foreign branch-plant industry, and this industry does, effectively, no or very little research as such in Canada. However, industry has, in contrast with the United States, drilled a large number of wells off eastern Canada, and these wells have become a very valuable research resource.
5. The community of scientists is relatively small, which has obvious advantages, but it also makes it easy for science to become inbred. Recently, this inbreeding tendency has been aggravated by the restriction of travel funds, which further limits the opportunity for Canadian scientists to keep in contact with their colleagues abroad.

Canadian government departments concerned with marine research are the Departments of Energy, Mines and Resources; Environment; Transport; National Defence; and Fisheries and the Canadian National Research Council. In addition, most of the provinces have their own Departments of Mines (or their equivalents) and Provincial Research Councils.

Federal earth science activities are under the administration of the Minister or an Assistant Deputy Minister. In the Department of Energy, Mines and Resources, there are several departments concerned with the earth sciences: the Geological Survey of Canada, the Earth Physics Branch, the Polar Continental Shelf, the Canada Centre for Remote Sensing, the surveys and mapping Branch, the Mines Branch, the Explosives Branch, and various advisers.

Federal research activities in the earth sciences in Canada are largely mission-oriented and are focused on specific tasks. This has the effect that some important surveying tasks are done with dispatch. The land geology in the Canadian Arctic has been mapped much more systematically than in the United States. Canada, some years ago, produced an aeromagnetic map of the country, a project about which the United States is still debating. Subsurface studies trying to incorporate the results of industry drilling are quite common in the Canadian Survey. The charter of the Canadian Geological Survey differs somewhat from the charter of the USGS in that the Canadian Survey has a mission actively to lead the way in the resource development of the country, a task they performed splendidly with their early reconnaissance mapping in the Arctic. It paved the way for industry exploration in that area.

There are several nongovernmental (or quasi-governmental institutions) in Canada. These include (1) the Science Council, which advises the federal government and the scientific community; (2) the Association of Scientific and Technologic Societies (SCITEC), which tries to get scientists to speak with one voice; (3) the Royal Society of Canada, to be viewed as an Academy; and (4) the Canadian Geoscience Council, an umbrella of earth societies, which tries to get earth scientists to speak with one voice. These societies include the Geological Association of Canada, the Canadian Geophysical Union, the Canadian Society for Petroleum Geologists, the Well Drillers Association, and the Canadian Society for Exploration Geophysics.

Finally, there is the National Research Council, with its laboratories in Ottawa and outside Ottawa (i.e., the Atlantic Regional Laboratory). The NRC has been the main funding agency for research in Canadian universities, but recently this funding role has been taken away from NRC and given to a new Natural Science and Engineering Research Council.

There are two main federally funded oceanographic complexes. (1) Bedford Institute of Oceanography, which is formed from several laboratories: Atlantic Geoscience Centre (Geological Survey of Canada); Marine Ecology Laboratory; Atlantic Oceanography Laboratory; Canadian Hydrographic Service Regional Office; Institute Facilities. These laboratories cover rocks, biological matters (especially fish), physical and chemical oceanography, charting, and ships. (2) Patricia Bay Institute of Ocean Sciences, Vancouver Island; this is developing in a similar way. The principal universities engaged in marine research are Dalhousie University, Halifax, Nova Scotia; McGill University and a number of other Quebec universities such as the University of Quebec at Rimouski; University of British Columbia.

From all this, there emerges a picture of a strongly mission-oriented effort in oceanography. Canada's government, university, and industry activities on the continental margins commenced in the early 1960's and expanded to the present level in the late 1960's and early 1970's. Two main government laboratories, on the east and west coasts, provide ships not only for their own work but also for university programs. These laboratories together have about 60 scientists (at the MSc and PhD levels) plus a large number of support staff. These are near and cooperate closely with major universities that have active programs in marine geology and geophysics. Because Canadian oceanographic ships concentrate their efforts in Canadian waters, most of the programs are at the moment directly connected with the continental margin. Canada has also contributed to drilling in young oceanic crust (DSDP and other projects).

Industrial activity has been focused on the eastern continental margin, in the McKenzie Delta, and in the Canadian Arctic Islands. The total activity has decreased in the past two years because of the failure to discover significant petroleum deposits off eastern Canada, with the possible exception of the Labrador Shelf.

An interesting sidelight is the activity of the Ad Hoc Advisory Committee (AHAC) to the Geological Survey of Canada. Following the suggestion of the Geological Survey of Canada, the Canadian Geoscience Council agreed to form an *ad hoc* committee to review the activities of the Geological Survey. The AHAC will report directly to the appropriate Assistant Deputy Minister (ADM). The composition of the committee reveals the representative nature of the review, consisting as it does of representatives from university, petroleum industry, mineral industry, geological engineering industry, and provincial government. In its first year of operation (1976–1977) the committee visited all but two of the divisions of the Geological Survey and met with nearly all its management and half of its research scientists. Reports were produced for each Division, and a consolidated report was sent to the Assistant Deputy Minister responsible for the Geological Survey.

The committee may prove valuable in two ways. First, it will provide specific and well-informed comments on what is right—or what is wrong. Second, the committee will become knowledgeable about the GSC and, understanding its overall problems, be able to aid in the general process of getting earth scientists in Canada to work together.

In comparing Canadian efforts with U.S. efforts, several aspects are noteworthy:

1. Because of favorable offshore regulations, industry aggressively pursued the exploration for hydrocarbons during the last 15 years. Unfortunately—with the possible exception of the Mackenzie Delta and the Labrador Shelf—no major reserves were found. Nevertheless, it is

fair to say that, as a consequence, Canadian continental margins are far better known than U.S. continental margins. Sensible regulations provided for release of the geological results of exploration wells, some 2 years after they were drilled; however, seismic work is not released. Probably the most outstanding fact is that on the Canadian Atlantic shelf there have been some 200 wells drilled for petroleum exploration, while so far none have been drilled on the U.S. Atlantic shelf. (The three COST tests being essentially stratigraphic tests.)*

2. The problem of coordination and communications between various oceanographic missions was resolved with the creation of oceanographic complexes at the east coast and the west coast. Presumably, the physical concentration of differing mission-oriented government efforts in a single complex has the advantage of preserving the integrity of the missions, while, at the same time, facilitating coordination through communication on the working level.

3. Because the main sources of funds have been with mission-oriented government departments, most of the work has been done in a "surveying" mode. In other words, the accent was more on coverage than on problem-solving. It would appear that some of the creative excitement inherent in the multiplicity of U.S. academic approaches is missing in Canada. Thus, the efficient accumulation of knowledge in Canada is not accompanied with corresponding debates about that knowledge. And so, Canada offers an example of continental-margin studies with little creative flavoring from academe.

The future of Canadian geological–geophysical research on continental margins can be seen only through a rather cloudy crystal ball, but the following seems a reasonable set of predictions for the next five years:

1. There will be little or no expansion in the number of people or funding of projects for continental-margin research either in government or in universities.

2. Industry will not provide any major input to continental-margin research, mainly because of the nature of Canada's "branch-plant" economy. However, the Canadian-owned oil company, PETROCAN, is likely to expand its operations in petroleum geology.

3. Individuals or groups of scientists will make more efforts to participate in international research projects.

4. There will be more cooperative programs between government, universities, and small Canadian-owned industries to develop technology that can be manufactured and sold by Canadian companies (e.g., the HUNTEC Project's development of a high-resolution, deep-tow seismic system for engineering studies).

5. There will be more attention to the nearshore environ-

*This picture has changed during 1978 by the drilling of a limited number of essentially unsuccessful wells in the Baltimore Canyon area.

ment, with a view to developing tidal power and to tackling environmental problems that may arise if the pipelines are built.

6. There will be an extensive attack on problems on Arctic continental margins.

THE LAW OF THE SEA AND SCIENTIFIC RESEARCH

In 1968, Arvid Pardo, the Ambassador from Malta to the United Nations, proposed to the UN General Assembly that the seabed and its resources "beyond the limits of present national jurisdiction" should be declared the common heritage of humanity. In response to that call, the UN began to work out a set of principles to govern seabed-related activities, procedures for negotiating a treaty for a new international regime, and mechanisms for managing and administering the resources of the deep seabeds for the common heritage of humanity.

On March 8, 1968, the President of the United States proposed "an historic and unprecedented adventure—an International Decade of Ocean Exploration for the 1970's." In December 1968, the UN General Assembly endorsed "the concept of an international decade of ocean exploration to be undertaken within the framework of a long-term programme of research and exploration. . . ." In 1971, a treaty prohibiting the emplacement of nuclear weapons and other weapons of mass destruction in the seabed was signed. In 1973, the present Conference on the Law of the Sea commenced. One of the important issues selected for inclusion in that charter was marine scientific research.

To date, the Conference has held seven sessions (New York, December 1973; Caracas, summer 1974; Geneva, spring 1975; New York, spring 1976, summer 1976, and summer 1977; the seventh session was held in Geneva, spring 1978, and resumed its work in New York, summer 1978). The eighth session will be held in Geneva in spring 1979. As this Panel's concern is the question of scientific research on the continental margins, we will discuss only those aspects that are relevant to such questions.

Before dealing with specifics, however, we believe that it is important that the reader gain some insight into the political and economic pressures under which decisions are being made. These are in part reflected by a summary by E. Mann Borgese (1977):

It is evident. . . that progress has not been linear. The law of the sea is being shaped by a host of contradictions, by the dialectical interaction of nationalist and internationalist trends, clashes of personalities, perceived interests and technological imperatives and, of course, by the law (and lawlessness) of the land.

Borgese continues:

The dialectic between nationalism and internationalism operates in curious ways. We have seen how the attempt to protect the common heritage of mankind through a moratorium may have accelerated nationalistic expansiveness. According to many observers, the unilateral assertion of national claims to a two hundred-mile fishery of "exclusive economic zone"—an escalating process characterizing the period between the fifth and sixth session of the United Nations Law of the Sea Conference—may undermine the authority of the Conference by confirming the thesis that the primary source of international laws is state practice.

Seen from the outside, it is indeed a curious spectacle to watch developing nations first enthusiastically embrace the concept of a common heritage and then take an active role in appropriating that very heritage to themselves through declarations of exclusive economic zones. As of September 1977, the situation was as follows: Approximately 51 states claimed jurisdiction over 200-nautical-mile maritime zones. Approximately 36 nations claimed (expressly stated or implied) direct or indirect jurisdiction over scientific research within these zones, including 14 territorial seas (Argentina, Benin, Brazil, Ecuador, El Salvador, Ghana, Guinea, Liberia, Panama, Peru, Sierra Leone, Somalia, Togo, and Uruguay); 20 economic zones (Bangladesh, Burma, Comoros, Cuba, Dominican Republic, Guatemala, Guyana, Haiti, India, Maldives, Mauritius, Mexico, Mozambique, New Zealand, Pakistan, Portugal, Senegal, Seychelles, Sri Lanka, and Vietnam); and two fishing zones (Norway and an unofficial statement of France, but not specifically mentioned in French law).

As of June 15, 1978, approximately 66 states claimed jurisdiction over a 200-nautical-mile zone. Twenty-seven states clearly claim jurisdiction over marine scientific research; 9 more states could exert control over marine scientific research under the conventional rights of exclusive jurisdiction within the territorial sea; and presumably, 15 more states can claim direct or indirect jurisdiction over marine scientific research within their 200-mile zones. The rough breakdown of those states that may or actually do claim jurisdiction over marine scientific research is as follows:

(a) The actual law or decrees specifically cover marine scientific research (27 states): Argentina, the Bahamas, Barbados, Brazil, Burma, Cape Verde, Cuba, Dominican Republic, Equador, Guatemala, Guyana, India, Ivory Coast, Japan, Maldives, Mauritius, Mexico, New Zealand, Norway, Pakistan, Peru, Portugal, Seychelles, Sri Lanka, Uruguay, Vietnam, and Yemen, Aden.

(b) Although not specifically stated, it can be assumed that jurisdiction applies to marine scientific research under the conventional rights of exclusive jurisdiction within the territorial sea (9 states): Benin, Congo, El Salvador, Ghana, Guinea, Liberia, Panama, Sierra Leone, and Somalia.

(c) Where jurisdiction applies to "activities" related to fisheries or living and nonliving natural resources within the maritime zone, then it can be assumed that the jurisdiction covers marine scientific research where it impacts on the fisheries or natural resources (15 states): Angola, Bangladesh, Canada, Chile, Comoros, France, German Democratic Republic, Haiti, Iceland, North Korea, Mozambique, Nicaragua, Poland, Senegal, and Togo.

Figure 3 shows that the 200-mile belt, for all practical purposes, includes most continental margins of the world It appears likely that as a result of the Law of the Sea negotiations, 12 miles will be established as the breadth of the territorial sea. The resource zone in which coastal states will exercise jurisdiction over living and nonliving resources, as well as for other specific purposes, will extend to 200 miles beyond the baseline from which the breadth of the territorial sea is measured.

It is now clear that some form of consent regime for research will be included in a Law of the Sea treaty for the area extending seaward to 200 nautical miles and including the continental shelves, although the details are still under discussion. We concur with the NAS position that the Law of the Sea be modified "to establish the right to conduct all research beyond the territorial sea (except for carefully specific and limited types)." We also concur with the original U.S. position that marine scientific research within the exclusive economic zone should fulfill the following obligations for scientific research: (1) advance notification, (2) participation, (3) data sharing, (4) assistance in data interpretation, (5) open publication, (6) compliance with environmental standards, and (7) flag-state certification.

At the Caracas session of the Law of the Sea negotiations, four alternative texts emerged. One, endorsed largely by developing countries, required consent for the conduct of all research within the economic zone and would require marine scientific research conducted in the international seabed area to be either conducted directly by the International Seabed Authority or through service contracts with the Authority. This was the most restrictive approach. The second, a U.S. position, proposed that consent would be required in the territorial sea, while in areas beyond the territorial sea where the coastal state would have resource rights, a series of obligations would be placed on all individuals and agencies conducting marine scientific research. A third text recognized the right to freely conduct marine scientific research within the economic zone and beyond, with the exception of research concerned with the exploration and exploitation of living and nonliving resources. The fourth text would require coastal state consent before scientific research could be conducted in the economic zone, but that the coastal state shall not normally withhold consent when the requesting state fulfills a series of obligations.

During the Geneva session of the Law of the Sea negotiations, another position emerged as a possible area of compromise: that consent would be needed to conduct research related to resources, but that only fulfillment of the eight obligations was necessary for research not so related. While that approach has some commendable points, the obvious drawback is to substantiate the distinction between resource-related research and research unrelated to resources.

The negotiations, however, did not stop at that point. They have moved the text much closer to a total consent regime. Under the now-existing form, the text (which is not final) states that a coastal state, in the exercise of its jurisdiction, has the right to regulate, authorize, and conduct marine scientific research in the exclusive economic zone and on the continental shelf. Research undertaken there is subject to the coastal state's consent; however, this consent shall be granted "in normal circumstances." Coastal states may, at their discretion, withhold this consent when the research (a) is of direct significance for the exploration and expoitation of natural resources; (b) involves drilling into the continental shelf, the use of explosives, or the introduction of harmful substances; or (c) involves the construction and use of artificial installations. Consent may also be withheld if the proposal for the project is believed to be inaccurate or if the researching state has outstanding objections (e.g., assessment of data or data-sharing) to the coastal state from a prior research project. There is also an implied (tacit) consent provision whereby, if the coastal state does not respond to a request within four months, the research may begin six months from the date upon which the requisite information was provided to that state. There is, in addition, a new Article making consent unnecessary for research undertaken by a regional or global organization if the coastal state in whose zone, or upon whose shelf, the research is to be done approves the proposed project or is willing to participate in it. In such a case, notification alone is required.

The problem for the researcher has become magnified in the light of U.S. funding practices in recent years. In the past five years or so, most of the substantial research on U.S. continental margins was justified on the basis of its practical importance, i.e., resource evaluation and concerns for environmental conditions. Such justification gave mission-oriented government agencies a dominant role to play in continental-margin activities. Increasing pressure to demonstrate the social relevance of scientific research caused university scientists to submit proposals containing rhetoric extolling the potential social impact (i.e., the application of, or societal need for, the project). Adding another layer to the problem, the Seabed assessment Program of the International Decade of Ocean Exploration was a program designed to aim directly at resource evaluation. The distinction between research for science and research conducted to evaluate resources became blurred.

Because of the emphasis placed on justifying research as having significance for resource evaluation, great strain may be placed on the credibility of the researcher if he proposes to do similar work on foreign margins, claiming it to be "pure" scientific research—that is, research that does *not* have direct significance for resource evaluation. Although the difference between "prospecting" and "scientific research" is fairly well known, both types of research are often engaged in the same generic activities. The tactics of execution vary widely between the two because of the differing problem definition. In prospecting, one systematically narrows down on a prospect; in scientific research, one focuses on geology that is not adequately understood. The problem for the foreign bureaucrat remains the same: how to judge the project before it is carried out.

One possible answer may be to convince foreign skeptics that U.S. academic institutions are not implicitly interested in prospecting, that their interest is to contribute to the understanding of principles related to resources and the environment. Such research activities could be termed, "conceptual resource evaluation." Here again, performance would be the test. Much credibility could be gained by having a strong academic domestic scientific research program on U.S. continental margins.

There is another aspect of the Law of the Sea that involves the use of scientific research and that is of direct interest to users of continental margins. According to the present version of the negotiating text, the continental shelf is defined as the area of the seabed and subsoil that extends beyond the territorial sea throughout the natural prolongation of the land to the outer edge of the margin, or to a distance of 200 nautical miles if the margin does not naturally extend that far. However, the precise debate centers upon the so-called "Irish formula," which proposes to a two-part test. The coastal state would be able to delimit the outer edge of the margin by a series of short, straight lines placed at a distance of 60 miles from the foot of the margin, or, if it prefers, it can place those lines at a point where the depth of sediments does not exceed one percent of the distance from the foot of the margin. Thus, under the optional sediment test, if a coastal state wishes to draw its lines at 100 kilometers from the foot of the margin, it must demonstrate that the sediment is at least 1 kilometer in thickness at that point.

This dual formula has been roundly criticized from a scientific point of view, but it is a compromise between those broad-margin states that seek to enclose the "last grain of sand" and others that wish to place some discreet limit on the extension of national jurisdiction over the shelf. This item will receive further attention in the next session.

At this stage, however, it is apparent that most developing countries, as well as the Soviet Union, are not inclined toward compromises. The ultimate result may well be a form of consent regime. In that case, scientists will have to consider if they want to support ratification of a treaty containing provisions such as those in the present text.

In conclusion, it is our judgment that exclusive economic zones are here to stay, whether or not a new Law of the Sea treaty is concluded. State practice may have already preempted that negotiation. Consequently, work on foreign continental margins will become the subject of bilateral, regional, or global negotiations over the precise manner in which marine scientific research will relate to the coastal state and how it will be conducted. Many of the nations insisting on consent have been working constructively with industry and contractors from the United States and other nations to obtain information regarding their own continental margins.

The challenge for U.S. oceanographic institutions is to demonstrate that they are capable of adding that extra dimension that differs from prospecting for resources but that will lead to understanding principles governing the formation and evolution of continental margins. Judging from past contributions from academe, we have full confidence in the resilience of our oceanographic institutions and the international oceanographic community to develop scientifically creative transnational programs on foreign continental margins.

REFERENCES AND BIBLIOGRAPHY

Agnew, A. F. (1975). The U.S. Geological Survey, Congressional Research Service, Library of Congress, Environmental Policy Division, U.S. Government Printing Office, Washington, D.C.

Atkinson, R. C. (1977). Memorandum to the Members of the National Science Board on the Future of Drilling in the Deep Oceans for Scientific Purposes in the 1980s.

Borgese, E. M. (1974). The Law of the Sea, *The Center Magazine*, 7(6):25–34.

Borgese, E. M. (1976). Law of the Sea, *The Center Magazine*, 9(5):60–70.

Borgese, E. M. (1977). A Ten-Year Struggle for Law of the Sea, *The Center Magazine*, 10(3):52–62.

Brown, H. (1960). Chapter 1, Introduction and Summary of Recommendations, in *Oceanography 1960 to 1970*, National Academy of Sciences–National Research Council, Washington, D.C.

Browne, M. A. (1977). Law of the Sea Conference, Issue Brief Number IB74104, The Library of Congress Congressional Research Service, April 18.

Clingan, T. A. (1977). The Plight of Marine Scientific Research in the Law of the Sea. Contributed (not for publication or attribution).

Commission on Marine Science, Engineering and Resources (1969). *Our Nation and The Sea*, U.S. Government Printing Office, Washington, D.C., January.

Committee on Mineral Resources and the Environment (COMRATE), Commission on Natural Resources, National Research Council (1975). *Mineral Resources and the Environment*, National Academy of Sciences, Washington, D.C.

Committee on Seafloor Engineering, Marine Board, Assembly of Engineering, National Research Council (1976). *Seafloor Engineering: National Needs and Research Requirements*, National Academy of Sciences, Washington, D.C.

Galey, M. E. (1977). From Caracas to Geneva to New York: The International Seabed Authority as a Creator of Grants, *Ocean Development and International Law J.* 4(2):171–211.

Intergovernmental Oceanographic Commission (1976). Report of the Second International Workshop on Marine Geoscience (Workshop Report No. 9), Mauritius, August 9–13, 1976, UNESCO, Paris, France.

International Decade of Ocean Exploration (1976). *Deep Sea Searches: The Story of the Seabed Assessment Program*, National Science Foundation, Washington, D.C.

JOIDES Subcommittee on the Future of Scientific Ocean Drilling (1977). Based on a meeting held March 7–11, 1977, at the Swope Center, Marine Biological Laboratory, Woods Hole, Massachusetts.

LeBlanc, L. (1977). Nations scramble for unclaimed seabed, *Offshore Magazine* March:41–46.

Library of Congress, Congressional Research Service (1976). Effects of Offshore Oil and Natural Gas Development on the Coastal Zone, U.S. Government Printing Office, Washington, D.C., March.

Miles, E. (1977). Developments in the Law of the Sea, *Nature* 267:760–769.

National Advisory Committee on Oceans and Atmosphere (1974). A Report to the President and the Congress, Third Annual Report, Washington, D.C., June 28.

National Science Board (1976). *Science at the Bicentennial: A Report from the Research Community*, National Science Foundation, 154 pp.

National Science Foundation (1975). Geodynamics, Director's Program Review, June 13.

Ortega y Gasset, J. (1966). *Mission of the University*, W. W. Norton & Company, New York, N.Y., 94 pp.

Smith, B. L. R., and J. J. Karlesky (1977). *The State of Academic Science: The Universities in the Nation's Research Effort*, (two vols.), Change Magazine Press, New Rochelle, N.Y. Based on a study sponsored by the National Science Foundation.

Talwani, M. (1977). Memorandum to Attendees and Invitees at the Ocean Crustal Dynamics Workshop held at Lamont-Doherty Geological Observatory, January 17–19.

B.1 Advances in Geophysical Methods for the Detection of Hydrocarbons

CARL H. SAVIT

INTRODUCTION

For many years I have been privileged to meet and confer with the highly skilled earth scientists of your progressive and well-staffed oil and gas organizations. From the nature and contents of our discussions, I must conclude that you are all well acquainted with the past, present, and probable future advances in the profession, which we share. It would, therefore, be somewhat presumptuous on my part to attempt to deliver an address based on the concept that I am offering knowledge and information that is not already known to you.

Instead, this paper will be devoted to my personal views of the ways in which progress is made in geophysical exploration. Within this framework I will attempt to describe the present situation and, from that point, to project our future course for the next decade or two. Because negative predictions, even from the greatest scientists, usually turn out to be spectacularly wrong, I will make no statements about what we cannot do. I will also refrain from estimating the time it will take to achieve any

particular level of advance because such time estimates have also generally been quite wrong. Times to realize technical advances depend only to a minor extent on technical considerations but instead are largely dependent on political, social, or economic factors.

HOW ADVANCES HAVE BEEN MADE

Now that I have finished telling you what I do not intend to say, I can begin with what I am going to talk about— how advances have been realized in geophysical exploration for hydrocarbons.

At present, more than 90 percent of the expenditures for petroleum geophysics are devoted to the reflection seismic method. Because there does not at present seem to be any advance in other methods that might tend to decrease the share of reflection seismic activities in the worldwide exploration budget, we will devote this discussion entirely to advances in reflection seismic exploration.

Late in the year 1917, a proposal was published suggesting that acoustic waves should be introduced into the earth in order that information could be obtained about what was in the earth from reflections of those acoustic waves. That proposal was the text of a United States of America patent issued to Reginald Fessenden,

This paper was originally presented in Spanish to the 25th meeting of ARPEL (Association for Reciprocal Assistance among Latin American State Oil Companies), Santa Cruz, Bolivia, July 27, 1976.

who had applied for it in January of that same year.

It was, however, not until 1926 that J. C. Karcher took the first reflection seismograph crew into the field and actually recorded reflections from underground strata. Twelve more years passed before the number of reflection seismograph crews operating throughout the world reached 200.

The point I want to make is that the advance in hydrocarbon exploration represented by the reflection seismograph did not comprise a single, easily identified, particular event. One man conceived the basic idea. It took almost a year for that idea to be published. Eight more years passed before another man had made the idea into a concrete reality and proved that the idea worked. Finally, it took another 14 years before enough people were using the idea to have a major influence on an entire industry.

As a matter of fact, 20 to 25 years for a completely new idea to mature into widespread use is quite a short period as these things go. Only in time of war could we expect a much shorter interval.

Let us return to reflection seismic prospecting and look at the introduction of new developments and improvements. The time interval between the conception of an idea and its widespread use may be much shorter than 20 years, but it is, nevertheless, longer than we would like to believe.

Digital recording and processing of reflection seismic data was proposed in the early 1950's. The first report of the actual deconvolution of a seismic wavelet in the files of the Western Geophysical research department is dated 1952. Henry Salvatori, then president of Western, in his report on advances in seismograph prospecting at the World Petroleum Congress in Rome in 1955, suggested that the best method for recording seismic data in the field would be digital. He also pointed out that such a recording could be used as direct input to a digital computer. The report had been written in 1954 in order to be available for publication in time for the Congress. It was about 8 years later that digital field recording was introduced on a few field crews, and by 1970 there were enough digital units in the field to represent widespread use. It seems that any improvement requiring the building of a large number of complex pieces of machinery still takes much time.

At this point I must interject the observation that the speeds of the introduction of new developments that I am discussing apply only to the nations outside the Eastern Bloc. In the Eastern Bloc new developments are usually scheduled only in accordance with a master plan based on social and political as well as economic considerations. Their rates of development do not, therefore, result from the interplay of natural economic and technical forces. A single example will illustrate the point. In the Soviet Union, at present, nearly all seismic-reflection crews record analog data only. Only a very few crews are equipped with digital recorders, in contrast to the western world in which conversion to digital is now essentially complete.

Even where the production and distribution of large quantities of new machinery are not involved, a substantial amount of time is usually required before a new idea has a sufficiently wide distribution and sufficiently wide acceptance to have a substantial effect on an industry. A recent case in point is the introduction of the so-called "bright-spot" technique as a direct indicator of the presence of natural gas, either alone or associated with petroleum. The idea that seismic reflections from sands saturated with hydrocarbons might be stronger than reflections from the adjacent nonpetroliferous formation first appeared in a paper that I published in *Geophysics* in 1960. Results of experimental work that had been done in the previous two years demonstrated that at least in two cases reflections from a given zone were stronger where the zone supported commercial production than where the same zone was dry. The results presented were empirical. No theory was offered to support them.

Within the next two years, however, publications appeared in the Soviet Union giving the results of theoretical analyses of reflection coefficients under a variety of conditions. The results clearly demonstrated that a gas-saturated, poorly cemented sandstone should produce an unusually strong reflection. A whole succession of papers appeared in different publications reinforcing the general idea that the amplitudes of seismic reflections could be diagnostic of the presence of hydrocarbons.

What happened next is not contained in the public record. We do know, however, that in the late 1960's one major international oil company began using the bright-spot technique. That company was sufficiently convinced of the technique's efficacy to acquire drilling rights and to drill wells based on the results. By the 1970's, another company had also discovered and begun to use the technique. Finally, in 1972, an independent consultant group, founded by a man who had left the second oil company, began advising the use of the bright-spot technique to its clients.

By January of 1973, seismic-data-processing companies were being asked to carry out processing in such a way as to preserve amplitude relationships among reflections. Many of these processing companies were unable to do so because their computer programs operated in such a way as to remove indications of reflection signal strength at the very first stage of the processing sequence. They used programs that had the same effect as the old analog automatic volume controls. The remainder of their processing programs were of the fixed-point type and, hence, could not effectively deal with and preserve the amplitude information necessary for the operation of the bright-spot method.

A few of us had faith in the theory and in the earlier results, so we had arranged our programs to preserve relative amplitude values. Those organizations that had the appropriate programs were able immediately to process data for bright-spot analysis, but a full year or more passed before the ability to process for bright spots became generally available in the industry.

Here again, a long period of time passed before an idea went through the necessary stages to have a major influence on operations in general. Fourteen or fifteen years from idea to general practice does not seem to be unusual.

WHY WE HAVE DELAYS

Many of us, especially when we were young, have been impatient with the delays that we experience in the real world. It is easy to assume that new ideas are not adopted because of human resistance to new ideas, stupidity, a desire to protect vested rights or painfully acquired knowledge and skills, or other evil factors.

In reality, a large number of reasons can be found, all perfectly sensible, why delays are inherent in the process of introduction of innovation into industry. One of the prime reasons has already been mentioned in passing. That reason is the substantial time required for the design, testing, construction, and distribution of complicated and expensive machinery. It is also quite reasonable to delay the replacement of expensive machinery by newer machinery until one is quite certain that the results will be sufficiently improved to compensate for the throwing away of the still useful old equipment. Perhaps the old equipment can still produce results that are good enough for the purposes of the user.

A second and somewhat analogous delay is the time required first to discover how to operate the new equipment or technique and to use the results produced and then to teach a large number of people to be builders, operators, and, above all, users of the new methods.

Victor Hugo said: "Greater than the tread of mighty armies is an idea whose time has come." This saying expresses in a simple way the concept that an idea must have the proper environment to grow, flourish, and be accepted. In geophysical exploration, all the circumstances must be right for the success of a new development. First, of course, the general state of man's knowledge must be such that the idea itself can be built out of existing parts. Not only do the separate concepts of which the idea is built have to exist, but there must also exist a need for the idea. That is, there must be a problem to be solved or a difficulty to be overcome.

The great Sir Isaac Newton expressed his appreciation for those who had come before him and established the foundations upon which his ideas were based when he said: "If I have seen a little farther than others, it is because I have stood on the shoulders of giants." Indeed, the calculus could not have been invented before Descartes invented analytic geometry, nor could the three laws of motion have been formulated in the absence of Kepler's empirically discovered laws of planetary motion.

In reflection seismology, the primary factors that determine whether an idea can be made to work are the general levels of instrument technology and the relative difficulty of finding oil and gas. The basic ideas for improvement are far ahead of the instruments to carry them

out. Fessenden's 1917 idea could not have been turned into a useful tool before the invention of the vacuum-tube amplifier. No one would have needed a powerful oil-finding tool before the automobile became popular and created a great demand for fuel.

Once created, the worldwide demand for fuel has increased continuously year after year. Today, the demand for hydrocarbon fuel and for ever more powerful geophysical techniques to find that fuel is greater than ever before. Every successful new technique results in a spurt of discoveries, but soon afterwards, the number of barrels found per seismograph crew month begins to decline. The demand for new developments appears insatiable.

The dominating reason that implementation of digital recording of reflection seismic data was delayed until the early 1960's is that the necessary instruments and components did not become available until then. The bright-spot technique could not easily be carried out until digital electronic circuits became fast enough to follow rapid changes in signal amplitudes in each of the dozens of channels of seismic data customarily recorded. Another governing factor was the availability of digital computers able to process floating-point data rapidly and inexpensively. To have carried out the programs needed to produce seismic record sections in which relative amplitudes are preserved with computers available in 1965 would have been prohibitively expensive. With computers generally available today, such processing is hardly more expensive than the older, gain-limited processing.

WHERE WE ARE TODAY

Now that we have reviewed reflection seismology from its beginnings through its development to the present time, we have a basis upon which to analyze where the profession is today and to make some informed guesses as to where we will be tomorrow.

At present, there are just over 500 field crews operating on land outside the Eastern Bloc. Between 50 and 70 crews operate in the sea. The fewer marine crews, because of their greater productivity, survey nearly two and one-half times as many kilometers of line as do the more numerous land crews. Slightly less than one half of all land crews and more than nine tenths of the marine crews use nonexplosive sources to produce seismic energy. The remainder use explosives in one form or another. There does not seem to be any great tendency to change these proportions rapidly, although the proportion of Vibroseis crews on land has been increasing steadily, but slowly, from year to year.

A typical crew today operates with spreads of 1 to 3 km or more in length and with 48 groups of geophones or hydrophones. There are still in existence a sizable number of 24-trace crews, but there are also an increasing number of crews with 96 or more active groups. The trend toward increasing numbers of groups in a spread has been continuous and inexorable. There is no sign that this tendency will not continue. On the other hand, spread lengths after increasing from about 100 m to the present 3 or 4 km usual maximum show no sign of further increase largely because of geometrical constraints growing out of the fact that we are not interested in exploring the earth to depths of more than about 8 to 10 km.

Group lengths, or more generally group dimensions, of 20 to 200 m have been used for at least 25 years. The specific dimensions are selected on the basis of the local conditions of surface noise and subsurface dip.

Our seismic data are typically recorded on standard computer tape, one-half inch wide in one of several formats. The recording instruments are of a type that can accept and faithfully digitize signals within a dynamic range more than adequate to encompass anything we are likely to encounter in reflection seismology. Most of the recorded data are of the instantaneous floating-point type, although many of the older binary gain instruments are still in use. Instruments with limited dynamic range (that is, instruments depending on programmed or automatic gain controls) have been virtually eliminated from the mainstream of exploration technology. The precision with which these instruments convert seismic signals into binary numbers is generally more than necessary on the basis of information theory analysis of signal and noise levels.

I shall not speak at all about such important problems as positioning at sea or transportation on land nor of all the other essential logistic and ancillary activities essential to the operation of field crews. To discuss those matters fully would require more space than we can devote to this entire paper.

Seismic data are today reduced to a form intelligible to a human being by means of digital computers. Data reduction is performed either on a so-called "small" computer in a field or a central office or on a "large" computer in a major computer center. The terms "large" and "small," when applied to computers, do not have a meaning that is constant with time. Today's small computer would have been yesterday's large computer. It has been said, for example, that the most powerful computer in the world at any given date has power equal to the combined power of all computers in the world 15 years earlier. Studies indicate that computers tend to increase in power about one order of magnitude every 2.7 years. That trend shows no signs of diminishing.

WHERE WE ARE GOING

To analyze what are the current trends today we can examine the ideas that have been published but have not yet been carried to a point that they have a significant effect on our profession. Some of these ideas have been carried out in part, others are in the stage of design and experiment, and still others must await a major advance in the general level of technology before they can be applied to actual operations.

Before we examine some of the specific ideas that will be the subject of the remainder of this paper, let us try to decide on what our ultimate goals as geophysicists should be. As I have stated in many previous speeches and published papers, I believe that our ultimate goal is to be able, anywhere in the world, to determine from physical measurements on the surface exactly what the drill would find at any specified location and depth.

Please do not misunderstand me. I do not say that we can attain this goal in the next decade or two. I merely say that this is the direction in which we must go, and that when we reach that goal, there will be no more need for research into geophysical exploration methods except possibly to make the process cheaper.

At this moment the most widely pursued short-term goal in exploration seismology is to improve resolution, that is, to be able to detect and map finer details. This short-term goal is a step toward our ultimate goal because at its present stage of development the reflection seismograph has a resolving power adequate to map layers a few tens of meters thick at depths of a few thousands of meters. In the horizontal direction the resolution is of the order of 100 m. Our resolution of lithologic properties is coarser by at least a factor of 10, that is, we can, at best, measure the velocity and the attenuation of sound with useful accuracy in a layer not less than several hundred meters thick. To achieve our ultimate goal we would have to be able to resolve properties of rock to resolutions of better than 1 m in the vertical dimension. At present, however, we are only striving to improve the resolution by a factor of 2 or 3.

To this end, we are in the process of completely redesigning the entire reflection system.

We are quite confident that when the entire new system is in operation, we will indeed improve our resolution by at least a factor of 2.

The ideas on which the redesign of our system is being based were, at least in part, enunciated in publications in 1958 and 1960. In one of those publications I suggested that to achieve the higher resolution required to find stratigraphic traps it would be necessary to abandon filtering as a means of improving signal-to-noise ratios, but instead to use compositing. In the other, I suggested that the outputs of individual phones in a group be separately recorded for later combination in data-processing apparatus.

Recent experiments and theoretical studies have shown that every step of the seismic-reflection exploration method we currently use acts as a high-cut filter. Here then is the heart of our problem, because to improve resolution we must pass the higher frequencies instead of rejecting them. It is a well-known principle of mathematics and physics that the resolution of any measurement is determined by the frequency of signals used in making that measurement.

Because the seismic signal must pass through the various parts of the system in succession, the overall frequency passband of the entire reflection seismic system is the product of all the individual passbands of the parts. If each part has a passband that is a high-cut filter, the composite passband of the system must be highly destructive of exactly the high frequencies we seek.

The succession of individual steps that combine to yield the seismic-reflection data for us to interpret begins with the generation of the initial signal. Most sources in wide use today are band-limited on the high side. Only explosives yield more energy above 100 Hz than below it. Some of the nonexplosive sources have relatively more high-frequency energy than others. In our experiments and special survey work, we have used small explosive charges of 225 g in both marine and land work. Also, at sea we have successfully used our Aquapulse source in high-resolution work. We have recently been using high-pressure air guns in marine situations. We have found that the high-frequency output of air guns can be improved by increasing the operating pressure from the conventional 140 atm (or bars) to about 350 atm.

After the seismic signal is produced, it must pass through the earth. The earth is a low-pass filter. We cannot (as yet) modify the earth. Our only recourse is to modify our surveying system so that its response is biased heavily in favor of the higher frequencies to compensate for the earth's lack of compliance with our desires.

After traversing the earth, the seismic signal is received by a geophone or hydrophone. Most commercially available phones have a fairly flat response above a certain minimum frequency and, hence, pose no problem to us *per se*. On land, however, the coupling of a geophone to the earth usually acts as a high-cut filter with erratic characteristics. It is highly destructive of high frequencies. In the very-high-resolution work, which we have been doing during the last year, we have been putting specially designed pressure phones into water-filled holes a few meters deep. Using such a detection system, John Farr was able to demonstrate, in a paper presented in June 1976 at the meeting of the European Association of Exploration Geophysicists, continuous reflections with a peak frequency greater than 500 Hz.

One of the steps most destructive of high frequencies in present-day seismic practice is the combination of signals from many phones in a group of considerable size into a single output. We are accustomed to using groups of 20 to 60 phones over group lengths of 30 to 200 m for the purpose primarily of cancelling horizontally traveling energy. Such groups have the unfortunate property of discriminating against high-frequency information. When a group is deployed over a large distance, any signal wave front is likely to reach the different phones that make up the group at different times. On land, weathering and elevation differences produce a random scatter of time differentials. On both land and sea, normal moveout produces systematic differences while dip moveout produces differences that depend both on the geological structure and the direction in which the survey is conducted.

Differences in the times of arrivals of signals mean that combinations of those signals will be out of phase. Dif-

ferences of 5 msec across a group are common. At 25 Hz a 5-msec difference is a phase difference of 45 deg. Two signals added 45 deg out of phase will reinforce each other. But at 100 Hz, 5 msec is 180 deg, and two signals out of phase by that amount will cancel. It is not hard to see why large groups are extemely destructive of high frequencies—they are powerful, high-cut filters and we remember that one of the ideas we want to carry out is to eliminate filtering from our system.

In Dr. Farr's experiments and in specialized marine operations conducted during 1976 in Far Eastern waters, we have used very short groups spaced near to each other and have successfully recorded unusually high frequencies at depths of interest in petroleum exploration.

In the next generation of field equipment that is currently under development we will be using phone groups whose dimensions and spacing along the line are of the order of a very few meters. If we are not to sacrifice the ability to cancel multiples, the immediate consequence is that we must increase the number of groups on the cable to several hundred. In my 1958 publications, I suggested that such large numbers of groups were impractical within the bounds of the then existing recording systems. Today technology has advanced to the point where recording systems of adequate capacity can be provided.

This remark leads us to examine the next phase of the seismic process, namely, the recording system. Present-day electronic techniques are adequate to amplify and digitize hundreds of signals to the required specifications. But, what are those specifications? We know from the Nyquist theorem that in digitizing we must sample our data more than twice for each cycle of the highest frequency to be expected in the data and that we must filter the data to prevent the phenomenon known as aliasing. From a practical standpoint, these considerations limit us in frequency to about 60 Hz for 4-msec sampling, 120 Hz for 2-msec sampling, 240 Hz for 1 msec, etc.

Our work in the experiments and special projects that I have mentioned confirms these values. We have concluded that for the next stage of development our systems must at the least be capable of digitizing at 1-msec intervals.

All of this analysis points to a major problem. Our tape-recording system will require a capacity of between 10 million and 20 million bits per second.

The highest-capacity magnetic-tape system commonly available today has a packing density of 1600 bytes per inch. There is no conceivable way of running such a tape sufficiently rapidly to record 10 million bits per second. Even if there were a way to transfer such data rates to the common 1600-bytes-per-inch tape, one reel of tape could only accommodate three 6-sec records at 20 million bits per second or six such records at 10 million bits per second. Even the highest-capacity instrumentation tapes available today would be completely overwhelmed by the data produced by the system that we have been describing.

In order, therefore, to make possible the use of our new

system, recourse will have to be made to entirely new methods of tape recording. The method that has been adopted is described in a paper by Paul E. Madeley, which appeared in the *Oil and Gas Journal* on June 21, 1976. The particular type of tape-recording system is one that has been adapted from high-fidelity color-television recording. It is known as the quadruplex, transverse recording system. It permits recording data at rates of 10 million to 20 million bits per second, while, nevertheless, driving the 2-inch-wide tape forward at a rate of 10 to 20 cm per second. Since the standard video tape reel contains something more than a kilometer of tape, one reel can record several hundred 6-sec records.

Once we have recorded data at these unprecedented rates, we will have the problem of finding a computer or computers that can accept such rates. The capacity of present-day, large-scale computers such as, for example, the IBM 370/168 is adequate to handle such data flows, but one marine crew would occupy the major part of the capacity of one computer. In the initial stages it will, therefore, be necessary to design and build special-purpose computers that will do the first-stage computations to reduce the volume of incoming data to a manageable level.

Our research workers have been carefully examining the various steps of data processing that have been used in the past and are currently still being used. They have come to the conclusion that nearly all of the approximations and algorithms that are currently in use lack sufficient accuracy to be valid for signals in the 100- to 200-Hz range. Many of the present algorithms have a filtering effect that produces a rather severe high cut. For example, a common-depth-point stack has the same need for the alignment of signals in phase as does the seismometer group in the field. Inaccuracies amounting to 5 msec in computing and applying normal moveout or static corrections will be totally destructive of signals at 100 Hz. The effect is negligible at 25 Hz.

As a matter of fact, the inadequacies of the common-depth-point method become painfully apparent when we try to improve our computational accuracy. It will not be many years before we will have abandoned geometric optics as the mathematical tool for reducing seismic observations. Instead, we will be using algorithms based on acoustic theory, that is, the scalar wave equation. We can even predict that ultimately we must be using the equations relating to real solids and work with tensor equations if we are to achieve our ultimate goal.

The data rates that I have been describing will require computers of vastly greater capacity than are currently available merely to accept the data, not to speak of the need to carry on the much more complex computations involving many more steps. We will need a new generation of computers, at least 1000 times more powerful than the present ones. We at Western are working in close association with a major computer manufacturer to realize as soon as possible the necessary computer power. If we assume that present computer speeds are based on 1974

technology, and if the power of computers continues to increase at the rate of one order of magnitude every 2.7 years, we should expect the necessary 1000-fold improvement by 1982.

In keeping with the promise I made at the beginning of this paper to refrain from predicting times for accomplishment, I am merely saying what could happen if past trends continue into the future.

My remarks to this point have been a description of what is, what is about to be, and what will be sometime in the future. As time goes on, more and more of the things I have been describing will become available. At first, the availability will be on an experimental or test basis. Later, the techniques will be transferred to a few regular operational crews in areas in which those techniques are especially desirable. Still later, the methods will become generally available on operating crews throughout the world. Without saying when any given technique will have reached a particular step of development, I will assert that all that has been described is already beyond the stage of being a mere idea.

IDEAS FOR THE FUTURE

I would now like to share with you a few of my speculations as to developments for the more distant future, ideas upon which no serious work is currently being done.

These speculations follow naturally from what has gone before. We are indeed going to resolve subsurface information into much finer detail than is now available, and more information is to be extracted from those data. Consequently, the final presentation to the user will have overwhelmingly more information than at present. The question immediately arises, "Where will we find the increased numbers of skilled people to read, analyze, and interpret the much greater amounts of data that we will be creating?"

The increased complexity of our civilization is making more demands on our population for skilled, trained, intelligent workers. Shortages of the highly skilled are increasingly apparent. I predict that there will be no way to fulfill the demand for human analyzers and interpreters of the data that will be produced when the present high-resolution trends have become mature. Only the computer will be able to step into the breach.

For many years, mathematical techniques have been studied that will enable a computer program to make some sort of judgment as to whether a given body of data describes some object or not. Twenty-five years ago the burning topic in programming these techniques was whether a program could be designed to distinguish the printed letter A from the letter O. Papers on that subject received great attention at international conferences. Today computer programs can, in fact, distinguish all the letters of the alphabet under a wide variety of conditions.

It is but a simple step forward to imagine a computer program that can examine a vast body of geophysical data and determine, within specified limits of probability, whether these data describe an oil field. The probable reserves of such an oil field could be a simple extension of the determination of its probable existence. The calculations of probable payout would then be mere child's play to the computer program, which could then simply indicate where one should or should not drill. Such a program, or a computer programmed with such a program, could very well be called an "analyst." Today an analyst is a human being. Tomorrow an analyst may be a machine.

For those who are inclined to doubt, I would point out that in 1900 a "typewriter" was a human being who operated a machine for writing in type. In 1948, I went to work in the Western Geophysical Company offices and was classified as a "computer," along with many of my fellow employees. Today a computer is a machine. Can we not believe that in the not too distant future, probably within the lifetimes of many of the people in this room, an analyst, a geophysicist, and a geologist may all be machines?

$B.2$ | Multichannel Seismic-Reflection Systems with Some Applications to Continental Margins

EDGAR S. DRIVER and N. WAYNE LAURITZEN

DEFINITION

The multichannel seismic-reflection technique calls for the summation of multiple reflection recordings at different shot-to-detector distances from the same subsurface reflection point. An essential feature of the technique is the application of static and normal moveout corrections prior to summation. Referring to Figure B.2.1:

$$\text{normal moveout} = \Delta t = \sqrt{\frac{x^2}{V^2} + T_0^2} - T_0, \quad (\text{B.2.1})$$

where V is rms velocity, x is shot-to-detector distance, and T_0 is reflection time for normal incidence.

As an example, in current marine shooting for industrial purposes, 48-channel recording with a 2400-m streamer is common. Detector groups are 50 m in length and for 48-fold stacking a shot is recorded every 25 m along the line; for 24-fold stacking a shot is recorded every 50 m along the line. This is illustrated in Figure B.2.2.

The power and contribution of the multichannel seismic technique lie in two areas: *velocity extraction* and *noise suppression*.

VELOCITY EXTRACTION

Root-mean-square velocity is obtained from Eq. (B.2.1) by inserting measured (x) and observed quantities (T_0 and Δt). High multiplicity of common-depth-point recording gives high statistical accuracy in the determination of velocity. William A. Schneider, in *Geophysics* (Vol. 36, No. 6, December 1971) gives an excellent description of the factors affecting the extraction of velocity. He provides the data given in Table B.2.1. He also enumerates the principal factors affecting accuracy as follows:

1. Random moveout estimation error,
2. Random static correction error,
3. Near-surface and distributed-velocity anomalies,
4. Multiple reflection interference,
5. Offset-dependent wavelet shape (complex reflector),
6. Curved-ray path,
7. Complex structure (in three dimensions),
8. Survey geometry,
9. Selection of intervals for interpretation.

Figure B.2.3 is taken from Schneider's paper. It shows the combined effects of statistical averaging, signal-to-noise

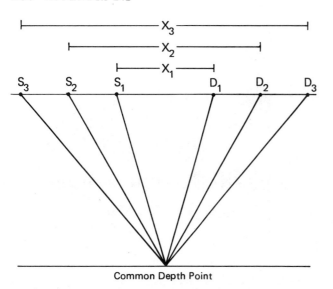

FIGURE B.2.1 Geometric layout for common-depth-point recording. (Courtesy of Gulf Science and Technology Company.)

NOISE SUPPRESSION

In the first applications of CDP, shooting noise suppression was visualized solely as the reduction of random noise according to the square root of the number of recordings added.

In 1958, stacking was conceived as a modification of the spatial filtering techniques used in detector and shot-point arrays. As such, it was applied to the reduction of systematic noise with recording distances and multiplicities tuned to achieve optimum rejection of the most troublesome noise events in any area of operation. Figure B.2.5 illustrates the idea.

The surface shot and detector configuration for fourfold

ratios, and spread length. The dashed curves show spread-length extensions equal to reflector depths.

Velocities derived from CDP shooting have been compared with velocity surveys in wells. Figure B.2.4. shows typical results for the Gulf Coast. This subject is discussed in depth by A. F. Woeber and J. O. Penhollow in *Geophysics* (Vol. 40, No. 3, pp. 388–398). An accuracy of 1 to 2 percent is common for reflection velocities in the depth range 5000 to 12,000.

Table B.2.1 Factors Affecting Accuracy of Velocity Determination

Use of Velocity	Acceptable Error	
	rms Velocity	Interval Velocity
NMO corrections for CDP stack as currently practiced	2–10 percent	—
Structural anomaly detection: 100-ft anomaly at 10,000-ft depth	0.5 percent	—
Gross lithologic identification: 1000-ft interval at 10,000-ft depth	0.7 percent	10 percent
Stratigraphic detailing: 400-ft interval at 10,000-ft depth	0.1 percent	3 percent

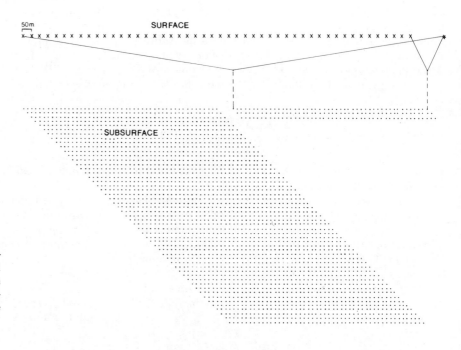

FIGURE B.2.2 Diagrammatic representation of marine shooting for industrial purposes using 48-channel recording with a 2400-m streamer in the common-depth-point mode. (Courtesy of Gulf Science and Technology Company.)

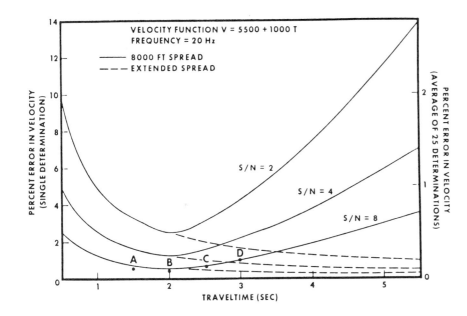

FIGURE B.2.3 Root-mean-square velocity statistical accuracy versus travel time and signal-to-noise ratio (S/N). [From W. A. Schneider (1971). Developments in seismic data processing and analysis, *Geophysics* 36(6):1043–1073.]

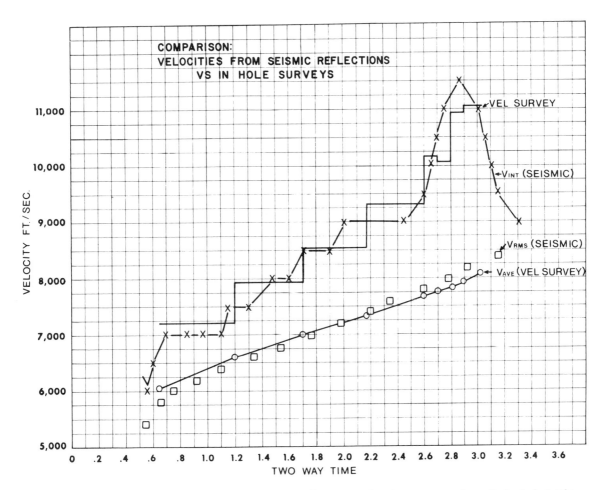

FIGURE B.2.4 Comparison of velocities from seismic reflections with velocity surveys in wells (or holes). This figure shows typical results for the U.S. Gulf Coast. (Courtesy of Gulf Science and Technology Company.)

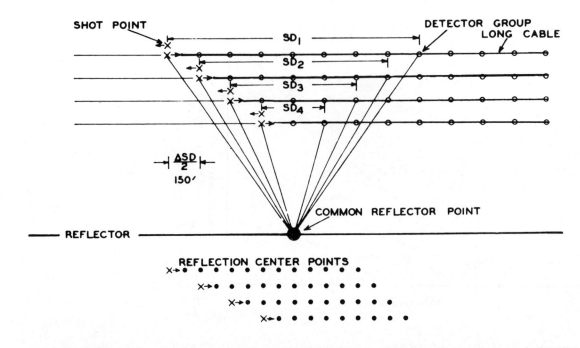

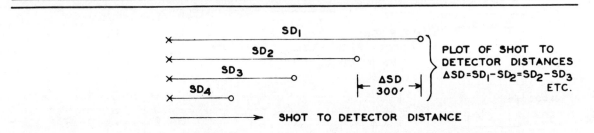

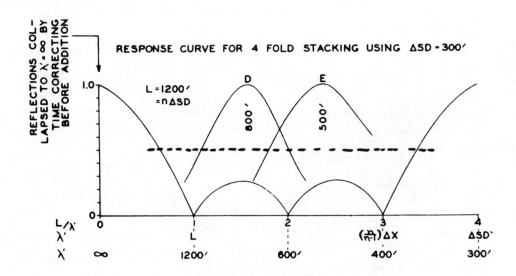

FIGURE B.2.5 Spatial array principles applied to common-depth-point (stack) shooting. (Courtesy of Gulf Science and Technology Company.)

stacking is shown with the corresponding reflection centerpoints displayed below.

If one visualizes shot-to-detector distances (SD$_1$, SD$_2$, SD$_3$, and SD$_4$) as measured from a common point, the detector positions are equivalent to a detector array with element separation equal to the difference between successive shot-to-detector distances. This is illustrated in the center panel of Figure B.2.5.

The bottom panel shows the corresponding array response designed to optimize the rejection of noise events with wavelengths ranging from 350 to 1400 ft. As compared with random noise rejection as shown by the dashed line, the use of CDP stacking as an array yields far more efficient rejection.

Because individual traces are corrected for NMO before summation, the dimensional limitations of conventional arrays are overcome and arrays can be as long as the maximum shot-to-detector distance employed.

One of the first applications of the array concept in CDP recording was to the suppression of formational multiples in the Gulf Coast. The upper part of Figure B.2.6 shows recordings obtained by the "expanded-spread" technique, wherein a shooting boat approaches a recording boat, meets above a common reflection point, then continues shooting as the two vessels separate. In this illustration, ranges in excess of 11,000 ft were recorded. The procedure can be considered the marine analogy to land noise spreads. In this case, multiples (A, C, E, and F) are distinguished from reflections (B and D) by the excess moveout of the former.

The lower panel of Figure B.2.6 shows the data after removal of normal moveout for the reflections at section velocity. Note that reflections occur at a common time, while multiples retain residual moveout. These multiples can be characterized by their residual wavelengths (after correction for reflection NMO) and a stacking array devised to optimize their suppression.

Refinements to the array application have been developed to account for the curvature of the multiple wave fronts over the long differential in shot-to-detector distances involved in CDP stacking.

APPLICATION OF THE CDP TECHNIQUES TO PROBLEMS IN MARINE GEOLOGY

CDP systems as developed for the industry have been applied to some problems in marine geology. A long multichannel traverse along a flow line transect in the North Atlantic was commissioned jointly by the U.S. Geological Survey and the International Program for Ocean Drilling Site-Survey Panel. Interesting intracrustal reflection events were recorded. One event is postulated to be the Moho interface.

To help appreciate some of the requirements for deep penetration and velocity resolution, some sample calculations have been made:

1. For a "typical" passive margin with a velocity distribution as shown in Figure B.2.7;
2. For a "typical" active margin with a velocity distribution as shown in Figure B.2.8.

The calculations revealed the following:

(1)	B in	*For Passive Margin:*	A in
	Figure B.2.7	*To Moho Depths*	Figure B.2.7
	11.9K	Length of cable for 10% accuracy in rms velocity	6.1K
	15.9K	Length of cable required for multiple rejection	8.1K
	20.9 sec	Approximate recording times	9.1 sec
	138K	Shot-to-detector distance for critical angle	29K
(2)	B in	*For Active Margin:*	A in
	Figure B.2.8	*To Moho Depths*	Figure B.2.8
	8.2K	Length of cable for 10% accuracy in rms velocity	5K
	10.9K	Length of cable for multiple rejection	6.7K
	8.9 sec	Approximate recording times	6.15 sec
	13.9K	Shot-to-detector distance for critical angle	20.5K
(3)	C in	*Oceanic Crust: To Top*	
	Figure B.2.8	*Asthenosphere*	
	18.7K	Length of cable for 10% accuracy in rms velocity	
	24.9K	Length of cable for multiple rejection	
	32.3 sec	Approximate recording time	
	7.5K	Shot-to-detector distance for critical-angle reflection from Moho	

Accuracy for velocity determination from methods of inverse moveout is primarily dependent on the signal-to-noise ratio and accurate distance determination. Solving the normal moveout equation explicitly for V_{rms} yields:

$$V_{rms} = \frac{x}{\Delta T \, (2T_0 + \Delta T)^{1/2}}. \qquad (B.2.1)$$

Hence, the derived velocity is directly proportional to the offset distance, x, and inversely proportional to the square root of ΔT. Any percentage error in x will cause a like percentage error in V_{rms}. The error induced in ΔT is different, however. It is more likely to be a fixed, additive error caused by noise than a percentage error. Figures B.2.9–B.2.13 illustrate the magnitude of the V_{rms} error that would be obtained using various streamer lengths, assuming no error in x and a given range of ΔT errors. These figures illustrate that cable configurations that are currently available cannot alone reduce the error to a desired 10 percent interval velocity error. An increase in multiplicity over that yielding the postulated ΔT errors would, however, reduce the effective ΔT error by the

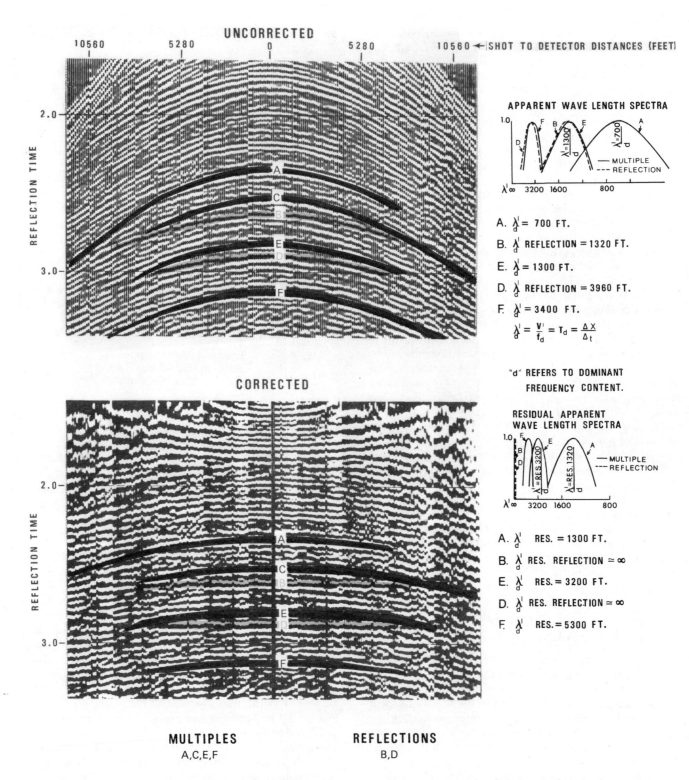

FIGURE B.2.6 Long-spread recording; an early application of the array concept. (Courtesy of Gulf Science and Technology Company.)

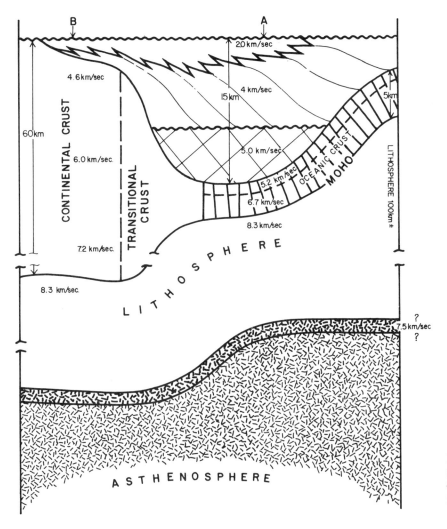

FIGURE B.2.7 A "typical" passive margin, with velocity distribution. (Courtesy of Gulf Science and Technology Company.)

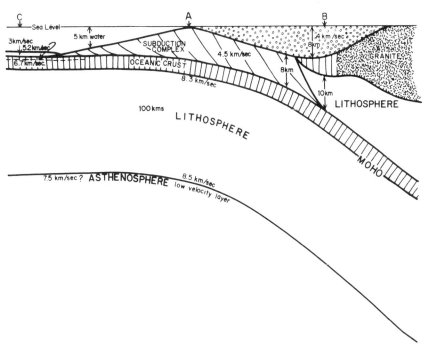

FIGURE B.2.8 A "typical" active margin, with velocity distribution. (Courtesy of Gulf Science and Technology Company.)

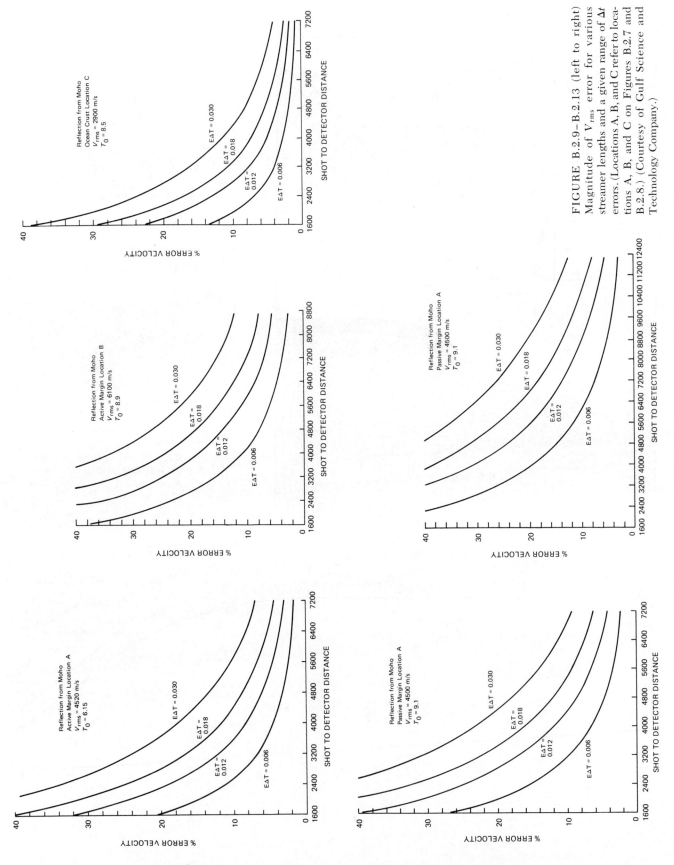

FIGURE B.2.9–B.2.13 (left to right) Magnitude of V_{rms} error for various streamer lengths and a given range of Δt errors. (Locations A, B, and C refer to locations A, B, and C on Figures B.2.7 and B.2.8.) (Courtesy of Gulf Science and Technology Company.)

236

square root of the increase. For example, Figure B.2.9 shows a 100 percent error reduction going from a 3200-m cable to a 4800-m cable. This same reduction could be accomplished by going from 12-fold CDP stacking to 48-fold CDP stacking. A high multiplicity is also indicated, because, with enough redundancy, coherency measures can be made to determine proper NMO alignment in signal-to-noise ratios that are far too poor for a velocity determination by the single-fold x^2-T^2 method.

The method for determining appropriate cable length was to assume that 100-msec ΔT on the far trace would be sufficient for a 10 percent interval velocity tolerance given sufficient CDP-fold. Also, distances to yield 100-msec moveout of differentials between primary and multiple times were computed by assuming that the multiple velocity was 0.8 times the primary velocity. This is a good rule of thumb assumption and yields the simple result that the cable length required is 1 1/3 times the cable length needed for an equivalent amount of primary ΔT.

B.3 | Seismic Data Processing

P. G. MATHIEU

INTRODUCTION

Reflection seismic data recorded by the petroleum industry were subjected to some kind of mathematical manipulations from the earliest days of seismic exploration. In the 1930's and 1940's when the data were recorded on paper, hardwire processes were being applied for such functions as frequency filtering; constant gain expansion; and automatic gain control, multichannel composites (mixing).

With the advent of analog recording of seismic data in the 1950's, there was an expansion of processing capability, with increased sophistication being applied. Innovative designs for single-channel filters were made; these included integration for low-frequency enhancement, differentiation for high-frequency enhancement, and, by extension, primitive attempts at amplitude and phase deconvolution; also gain normalization and static and dynamic time-shifts. Multichannel operations were also implemented, and these included such processes as common-depth-point stack, common-offset stack, velocity filtering. With this expanded processing capability, the industry was able to produce a variety of seismic displays useful to the exploration interpreter, whose effectiveness was increased.

It was with the advent of the "Digital Revolution" of the 1960's that seismic processing really bloomed. Scientists from diverse disciplines such as mathematics, physics, electrical engineering, and computer sciences, joined the geophysicists in writing a plethora of seismic programs. Whereas the geophysicists had used essentially deterministic models for seismic phenomena and seismic processes, the "new wave" adapted, for seismic application, the statistical and probabilistic techniques used in communication theory.

During the past 10–15 years, literally hundreds of seismic programs have been written. A very large number of them were later abandoned because they either proved to be mathematically unsound, or presumed an overly restricted model, or proved to be impractical in their application to real-life data. Yet, there are today hundreds of seismic digital processes in daily use.

COMPUTER ENVIRONMENT

Digital processing of seismic data probably dates back to the 1940's, when geophysical researchers would laboriously digitize by hand seismic responses which were then subjected to computer processes.

In the 1950's, larger and faster computers came on the market and the pace of geophysical computer experimentation accelerated. However, routine application of computer processing was uneconomical. The computers were designed for simple mathematical operations on fairly large commercial and financial data bases or for complex mathematical operations on relatively small data bases. For seismic processing, fairly complex operations had to be applied to very large data bases. The traditional seismic record contained 24 traces, and the typical recording was 6 seconds; with a 1-msec sampling rate, a seismic record contained 144,000 values. Even a simple operation such as filtering with a 100-msec operator would require some 15 million multiplications and additions for each seismic record. Even though computers were considered marvels of speed in those days, they were very slow by today's standards, and computer logic was not at all optimum for seismic processing. Since 10 or more records would be obtained each day on a land crew and several hundred records each day on a marine crew, seismic computer processing was very time-consuming and very costly.

Starting in the early 1960's, there were clear demonstrations of the benefits to be accrued from digital processing and there was a rapid acceleration of seismic processing caused by the following factors: (a) computer speeds increased, and over the next 15 years arithmetic unit costs decreased dramatically; (b) certain mathematical algorithms (e.g., multiply–add) specific to seismic processes were hard-wired into very high-speed "array processors" (the first AP's were convolvers); (c) mathematicians–computer scientists devised clever algorithms that took advantage of computer logic so as to achieve very substantial speedups over conventional algorithms. At that time, several small, special-purpose computers were designed with seismic processing as a specific objective, and a number of them were put into use.

In the late 1960's, the large computer manufacturers became aware that they were losing some of the the potential business in seismic processing to the special-purpose computers and they designed special peripherals to their large computers, thus regaining the seismic processing business.

In the early 1970's, a new breed of minicomputers appeared, and these were designed for processing at remote sites. These have been used either for preliminary processing or as an adjunct to a main data center or as a principal data center by itself.

The chronology of computer usage for seismic processing has been as follows:

1. *1950's:* Large-data-center computers used mostly by the majors and some contractors for R&D.
2. *Early 1960's:* Large-data-center computers and small, special-purpose data center computers used by majors, not-so-majors, and contractors for routine processing.
3. *Late 1960's:* Mostly large-data-center computers; some remote-job-entry (RJE) facilities to the data centers for processing from district locations.
4. *Early 1970's:* Large-data-center computers; minicomputers at remote sites.

COMPUTER PROCESSES

There are three principal categories of seismic processes:

    I. Data-arrangement processes
   II. Data-enhancement processes
  III. Analytical or interpretive processes

The first two categories are invariably applied by all oil explorationists, while the third category is applied more selectively.

Although most seismic processing has for an objective the enhancement of the data, this type of processing cannot be applied until the data are properly conditioned or arranged.

I. DATA-ARRANGEMENT PROCESSES

This stage is normally called "preprocessing," and its purpose is to take the data as received from field recordings and organize them in the fashion expected by the mathematical algorithms used in data enhancement.

Some of the principal processes in this stage are as follows:

1. Demultiplex

Data are recorded in the field from a number of channels simultaneously onto a single digital tape. Since this operation is in real time, at a given time sample, the seismic response of all channels has to be written before the response at the next time sample can be written. In other words, the data are multiplexed on tape: $11, 21, 31, 41, \cdots n1, 12, 22, 32, 42, \cdots n2, 13, 23, 33, 43, \cdots n3, \cdots nm$, where the first index represents channel number and the second index represents time sample.

The demultiplexing operation is equivalent to ordering the data into a matrix:

$$11, 21, 31, 41, \cdots n1$$
$$12, 22, 32, 42, \cdots n2$$
$$13, 23, 33, 43, \cdots n3$$
$$\cdot$$
$$\cdot$$
$$\cdot$$
$$1m, 2m, 3m, 4m, \cdots nm$$

then transposing it:

$$11, 12, 13, 14, \cdots 1m$$
$$21, 22, 23, 24, \cdots 2m$$
$$31, 32, 33, 34, \cdots 3m$$
$$\cdot$$
$$\cdot$$
$$\cdot$$
$$n1, n2, n3, n4, \cdots 4m$$

Now all time samples of channel 1 are streamed together, followed by all time samples of channel 2, and so on. The data are now in trace sequential mode and ready for further processing.

2. Gain Recovery

Since seismic data, which decay with time (distance traveled) and with frequency, have a very wide dynamic range, the recording system is designed to retain the same significance for the smallest amplitudes as for the largest, without using unwieldingly large numbers. In order to achieve this, every amplitude has a scalar applied so that it can be represented by the full number of available digits. The scalar, which is a multidigit number, is logged simultaneously with the amplitude value. This arrangement of writing on tape a scalar (called "gain bit" or "exponent") followed by an amplitude value (called "mantissa") results in a compact binary number of fixed format (typically 15 bits: sign bit + 4 exponent bits + 10 mantissa bits).

In the gain recovery process, each sampled amplitude is rewritten in true relative amplitude and the scalar is removed. The data-center computer can accept a wider range of numbers without the restriction of 15 bits.

After this recovery process, the data display the full time-dependent decay of the seismic response.

Some processors chose to apply, at this stage, an inverse to the decay curve so that amplitudes are reasonably balanced in time; i.e., amplitudes at 5.0 sec have the same order of magnitude as amplitudes of 1.0 sec. One scalar is applied for one time to all channels and/or to all recordings of one line or of one prospect. There are many variations of this approach.

Some processors also chose to apply, at this stage, scalars that will balance the amplitude distribution in space, so as to remove nongeological spatial variations that are caused by source or receiver characteristics or by near-surface effects. One typical approach is to compute the power (sum of squares) of a trace and to apply one scalar to all trace values such that the power will be a constant. The result of this process is that all traces have the same power. There are many variations of this approach.

3. Merging of Location Data

For a number of processes, there is a need to locate the seismic traces: sources and receivers need to be located relative to some coordinate origin and relative to each other. This information is external to the seismic recording: surveyor's notes for land surveys, various types of navigation data for marine surveys. For either type of survey, the location data have to be subjected to a data-reduction step before they are merged with the seismic data. In the merging step, each seismic trace on digital tape is tagged with a "header" containing all pertinent geographic information.

Besides the location information, water depth may be added to the trace header in the case of marine surveys; elevation and depth of weathering may be added for land surveys.

4. Gathers

Seismic data are recorded in groups of channels, say 24 or 48 channels. As stated earlier, the recording is in multiplex form, but after the demultiplex process, the data are written on tape in a trace sequential mode, with trace 1 of the first group being followed by trace 2 of the first group, and so on to trace 24; then the first trace of the second group is written on tape, followed by the second trace of the second group, etc. This follows the sequence of the field recording where a source is energized and the response is recorded by 24 receivers, then another source is energized and this is recorded by 24 receivers, etc.

For single-channel enhancement processes, the trace sequence is unimportant. However, for multichannel processes, such as CDP-stack or velocity analysis, which will be described later, a certain trace sequence is required. For example, the process may require that trace 1 of record 1 be followed by trace 3 of group 2 and by trace 5 of group 3, etc. There are many different sequences that might be required depending on the multichannel processes and according to the characteristics of the survey.

This gathering of traces is generally done in the preprocessing stage.

5. Time Shifts

Fixed time shifts, which are not dependent on the seismic responses but are due to predetermined causes, are applied in preprocessing. These would include instrumental shifts, datum shifts, elevation corrections, etc.

6. Vibroseis Correlation

This is a special process used for special types of source—the vibroseis technique in which vibratory waves are sent into the ground. The vibratory waves have unique characteristics, and by cross-correlating the resulting recording with a pilot vibration, the seismic sequence is restored. For a whole survey, the same vibration is used and the same process is applied to all recorded traces.

There are other, less prevalent, data-arrangement processes that are applied ahead of the data-enhancement processes. The main feature of all of these preprocesses is that, although they move data around and change the digital formats, the seismic response remains essentially unchanged and the signal-to-noise ratio or the resolution of the data is unaffected.

II. DATA-ENHANCEMENT PROCESSES

It is this stage of processing that has been of greater interest to the geophysicist, who wants the optimum data quality for interpretations: (a) classical structural interpretation based on spatial correlations of seismic events and (b) interpretations of rock and fluid properties from seismic response unpolluted by noise.

The geophysical objectives of processing can be summarized as follows:

(a) Improve signal-to-noise ratio (S/N).
(b) Improve temporal resolution.
(c) Improve spatial resolution.

There are two types of processes available:

(A) Single-channel.
(B) Multichannel.

Table B.3.1 lists some of the current processes.

As mentioned earlier, the processes involve either deterministic or probabilistic techniques or both.

(a) *Improve S/N*

Seismic noise has great variety, whereas signal, to the geophysicist, is very restrictively defined: it is the sequence of reflection coefficients convolved with a simple, uniform pulse (preferably a spike). Noise can be ambient, i.e., it exists before and after a recording is made; it can be random or coherent, and it can include wind, wave action, culture, tides, microseisms, power lines, and traffic. Noise is also an integral part of the seismic model and of the survey technique, and it is, in the true sense, signal generated by the experimenter. This source-generated noise is generally more severe than the ambient noise and in-

cludes all seismic events that are not primary reflections: Rayleigh waves, shear waves, refractions, reflected refractions, diffractions, ghosts, multiples, and others.

Processes that enhance S/N depend on different characteristics between signal and noise. Differences can exist in frequency, wavelength, or statistics. If no difference exists, no enhancement is possible and the recording technique has to be designed to create differences between signal and noise.

(a.A) *Single-channel processes* take advantage of the wavelet differences between signal and noise.

(a.A.1) *Frequency filters* are effective because the signal is reasonably broadbanded (10–70 cps); certain types of noise have a restricted bandwidth. Shear waves and Rayleigh waves peak in the low frequencies, below 10 cps, and a high-pass or bandpass filter can provide strong attenuation. Certain types of ambient noise, such as wind or culture noise, peak in the higher frequencies and can be attenuated by low-pass or bandpass filters.

(a.A.2 and a.A.3) *"Bubbles" and "surface multiples"* are called noise although they are an integral part of the seismic wavelet. They both involve a long tail attached to the wavelet, and the tail interferes with the separation between events. The attenuation of these tails is as much a problem of improved signal resolution as of S/N enhancement. They are included here because of their length (up to 1.0 sec or more) when compared with the typical seismic wavelet (less than 200 msec).

In marine recordings, the explosive source (dynamite, air gun) quickly creates a large gas bubble, which collapses and then expands again as it moves upward to the surface, where it vents. This expansion–collapse–expansion–etc. creates a long wave.

TABLE B.3.1   Data-Enhancement Processes

Geophysical Objectives of Processing	Process Type	
	(A) Single-Channel	(B) Multichannel
(a) Improve S/N	1. Frequency filter	1. CDP-stack
		2. Coherence-stack
		3. Automatic static corrections
		4. Velocity filter
	2. Debubble	5. Phase filter
	3. Demultiply	6. Radial arm
(b) Improve temporal resolution	1. Deconvolution	1. Multichannel deconvolution
	2. Deghosting	
	3. Frequency filter	2. Multichannel deghosting
(c) Improve spatial resolution	1. Normal-move-out correction	1. Two-dimensional migration
		2. Three-dimensional migration

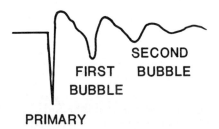

Where there is a strong shallow reflector (e.g., water bottom), a multiple reflection system is created between the reflector and surface:

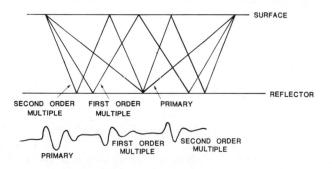

For either type of wave train, inverse filters have to be designed that will attenuate the tail, but leave the primary wavelet undistorted:

There are two principal methods of designing an inverse filter: (1) from an exact model of the tail and (2) from an estimate obtained from the autocorrelation of the seismic trace. With either approach, there are several mathematical methods of inversions, and all result in a collapsed wavelet that makes discrete seismic events separable.

(a.B) *Multichannel processes* take advantage of spatial recording differences between signal and noise: signal is the same, or can be made the same on several channels, while noise is different on the several channels; compositing of the channels produces S/N enhancement. Compositing can be achieved with various degrees of mathematical sophistication that optimize the S/N enhancement. In general, sophisticated processes producing optimum enhancement require exact modeling of signal and noise; since this is rarely available in the real world, the more effective processes are those that are robust.

(a.B.1) One of the most widely used composites is the *common-depth-point* (CDP) *stack*. The signals recorded from the same depth point have the same response after a suitable normal-move-out correction (mentioned in c.A.1):

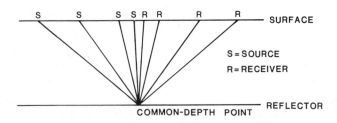

The signals add linearly, and if the amplitude on one channel is a, then the amplitude of an n-channel composite is na.

The compositing of random noise increases the amplitude of the noise as a function of the square root of the number of channels. If the average noise amplitude on one channel is b, then the average amplitude of an n-channel composite is $\sqrt{n}b$. With the S/N on one channel being a/b, the S/N of an n-channel composite is $na/\sqrt{n}b$, an improvement of $n/\sqrt{n} = \sqrt{n}$. The greater the number of channels, the greater the improvement.

The compositing of coherent noise will have an amplitude response that is dependent on the wave number between channels; with equally spaced channels, the response is as follows:

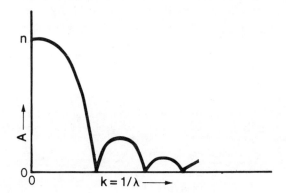

With signals being in perfect coincidence, the wave number between channels is zero and the signal composite equals n. Noises with linear coherence (constant phase between channels) will have amplitude response that will go to zero at certain frequencies. The larger the number of channels, the greater the attenuation.

Noises that have nonlinear coherence between channels have different responses to a composite, e.g.,

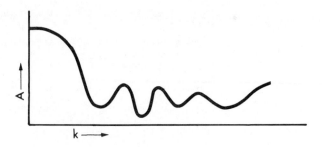

There are other responses. In every case, the signal composite has a higher response than the noise.

This type of composite or stack presumes each channel to have unit weight and to be undistorted.

Figure B.3.1 is an example of the enhancement capability of CDP stack.

(a.B.2) *Coherence stack.* Signal coherence between channels can be measured. One commonly used measure is the cross-correlation $\phi gr = \int_{-T}^{T} g(t) r(t + \tau) \, dt$ or, for the sampled case $\phi xy = \sum_{i=-N}^{N} X_i Y_{i+j}, j = -m \ldots -2, -1, 0, 1, 2, \ldots m$. g and r, x and y represent the two functions being cross-correlated; t and i represent time; τ and j represent lags; T and N represent maximum lags. The maximum value of the cross-correlation is a measure of coherence. There are other measures available. One can show that in order to optimize the S/N of the multichannel composite, one should give greater weight to those channels that have high signal coherence and lower weight to those with low coherence (it can also be shown that it is better to use a channel, even with the lowest coherence weight, than not to use the extra channel at all).

One technique is to cross-correlate each channel to a reference (e.g., previous composite, first-round composite of the n or $n - 1$ channels, etc.) and to weigh each channel in the stack by the cross-correlation value:

$$\text{Stack} = \phi_1 a_1 + \phi_2 a_2 + \ldots \phi_n a_n.$$

This stack produces a better S/N than the straight stack $= \sum_{i=1}^{n} a_i$.

Figure B.3.2 compares a straight stack to a coherence stack.

(a.B.3) *Automatic static corrections.* Optimum S/N in a stack is obtained when all signal channels are coincident. However, if there are time shifts between channels, there will be suboptimum stacking. Common-depth-point recordings should have coincident times, after normal-move-out corrections. However, there are surface and near-surface spatial variations that affect reflection times.

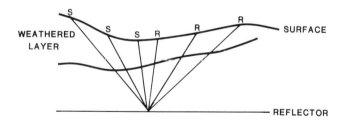

Elevation and weathered layer variations affect the reflection times in a static manner, i.e., all reflection times are affected equally at a particular receiver location. To a great extent, these time variations can be corrected for with suitable recordings. However, exact corrections are almost impossible, and, in some cases, substantial residual static shifts remain.

To bring the channels in signal time-coincidence, the residual static shifts must be measured. One means of measuring the shift is through cross-correlation, $\phi xy = \sum_{i=1}^{n} X_i Y_{i-j}, j = -2, -1, 0, 1, 2, \ldots m$. The lag j at which the cross-correlation response peaks is the shift needed to bring channels X and Y in coincidence. There are other mathematical means of measuring shifts.

The shifts from all channels of a CDP can be measured and statics can be applied, thus optimizing S/N of the stack.

Any technique for measuring time-shifts can fail at low S/N's. To ensure that the "true static" has been measured, the redundancies of the CDP recordings are used:

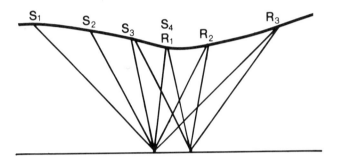

The static under R_2 must be the same for the S_2R_2 recording as it is for the S_4R_2 recording; the static under R_3 must be the same for S_1R_3 as for S_3R_3; and so on. This is ensured by clustering, least-square solutions, and/or other mathematical approaches.

It is only after the statistical determination of the statics that corrections are applied.

Figure B.3.3 shows the comparison between a conventional CDP stack and a stack with automatic statics.

(a.B.4) *Velocity filters.* Seismic reflections will have certain dips that are geologically plausible; larger dips would be attributed to noise. Since seismic dips are measured by time increments over space, an apparent velocity results. In the same manner, multiples have more dip/lower apparent velocity than primaries. For this reason, the filtering of noise on the basis of dip has been called "velocity filtering." It has also been called "fan filtering."

These are n-channel spatial filters that will pass events with certain wavelengths (apparent velocity) between channels and attenuate all other events. This is achieved by applying a convolusive filter, deterministically derived, to each channel and summing the results:

Stack $f(t) = a_1*f_1 + a_2*f_2 + \ldots a_n*f_n$, where a_i denotes channel response, f_i denotes filter, and $*$ denotes convolution.

The desired response of the velocity filter can be shown schematically:

MULTI - CHANNEL PROCESS

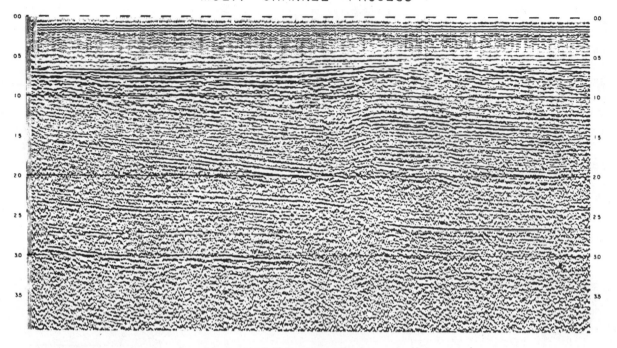

SINGLE CHANNEL

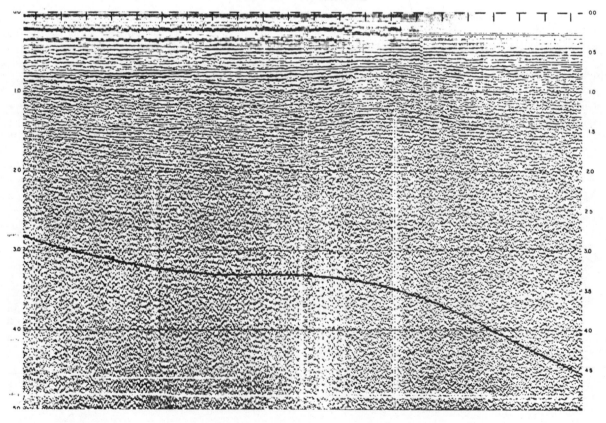

FIGURE B.3.1 Signal-to-noise enhancement. (Courtesy of Gulf Science and Technology Company.)

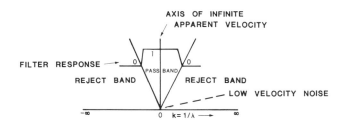

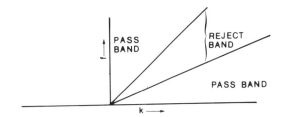

The common velocity filter is a passband filter, where a small portion of the f–k plane (frequency–wave number) has the pass response and the remainder of the plane is rejected.

A variation is the reject velocity filter, where rejection occupies a small portion of the f–k plane, and the pass response occupies the remainder:

There are many mathematical variations of these filters. These types of filters can be applied to "stacked" data or to CDP gathers, where all primary reflections have infinite apparent velocity after NMO correction (see c.A.1).

(a.B.5) *Phase filters.* One feature of the seismic model is that the signals on n-channels are time-coincident (no phase between channels), while noise has phase between channels and the channel composite has a finite S/N:

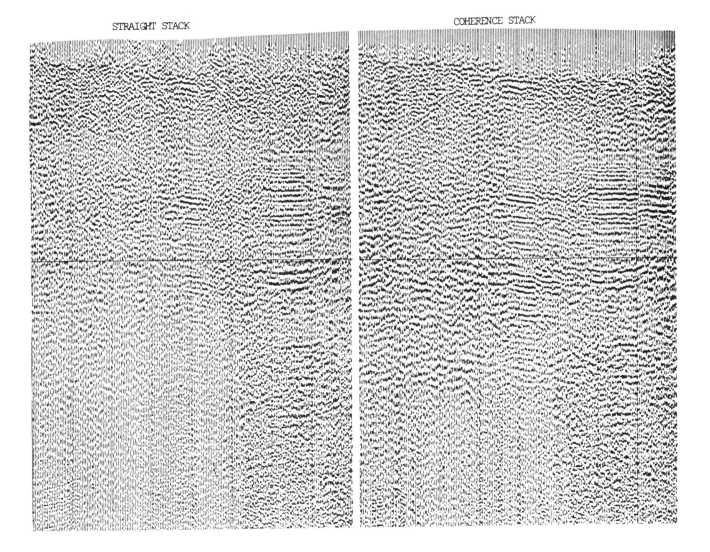

FIGURE B.3.2 Comparison of a straight stack with a coherence stack. (Courtesy of Gulf Science and Technology Company.)

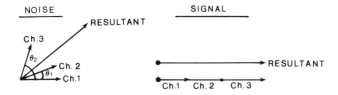

Filters are designed to rotate the phase so as to produce a zero resultant on the noise:

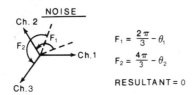

$$F_1 = \frac{2\pi}{3} - \theta_1$$

$$F_2 = \frac{4\pi}{3} - \theta_2$$

RESULTANT = 0

The application of these filters modifies the signals:

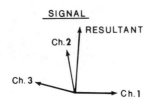

The filters are modified (phase is rotated and scalars are applied) so that the output signal is undistorted.

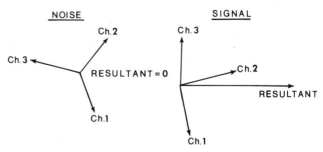

This operation is performed for every frequency in the seismic passband (say, 10–70 cps).

These are powerful filters in cases where the channel-to-channel time delay (phase) of the noise is accurately known and the signal is properly aligned between channels. These types of filters have been used to attenuate multiples.

(a.B.6) *Radial arm.* This is a multichannel filter especially designed for the attenuation of deepwater (>500 feet) multiples, which are troublesome in marine recordings.

This technique takes advantage of the fact that the primary response for one source–receiver geometry is exactly the same as the response of the multiple for another source–receiver geometry:

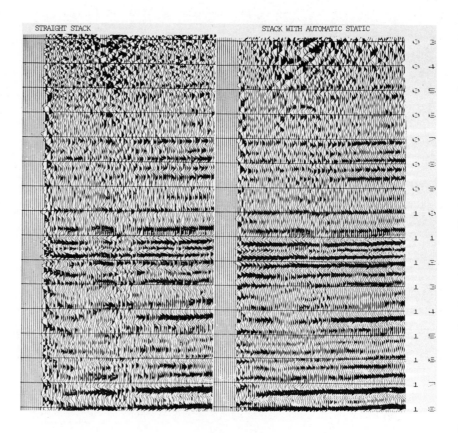

FIGURE B.3.3 Comparison of a conventional CDP stack with a stack with automatic static. (Courtesy of Gulf Science and Technology Company.)

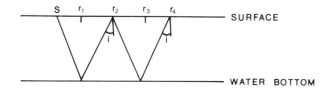

The angle of incidence i for the primary at r_2 is the same as for the multiple at r_4. There is no distortion of the multiple relative to the primary except for an amplitude decay caused by distance traveled and the reflection coefficient R at the water bottom.

The scalar difference between the primary and multiple is determined by cross-correlating the primary on channel r_2 with the multiple on channel r_4. The scaled version of the primary is then subtracted from the multiple.

This operation is performed for all channels and has proved to be quite effective for the attenuation of first- and higher-order surface multiples.

There have been many, many more multichannel operations designed for data enhancement. Each has utilized some presumed special properties of the signal and the noise. Included are such multichannel filters as maximum likelihood, minimum power, and cross-equalization; also included are filters with exotic names such as APE, SOUSTON, WIPER, DOVE, and HAWK.

Figure B.3.4 is an example of a multichannel process that attenuates multiples.

(b)   *Improve Temporal Resolution*

The seismic trace is assumed to be the result of convolving the reflection coefficient function with a wavelet:

$$S(t) = R(t)*W(t).$$

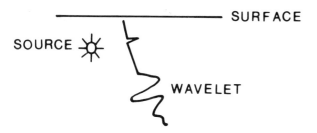

$R(t)$ is the function that is of geological interest to the explorationist. It has high resolution, and discrete geological events are separable. The wavelet $W(t)$ limits the resolution of the data because the sequence of events is replaced by overlapping wavelets. The shape of the wavelet is determined soon after it enters the ground and is affected by reverberations near the source and by the elastic characteristics of the medium.

The wavelet acts as a frequency filter, reducing the bandwidth of the reflection coefficient function:

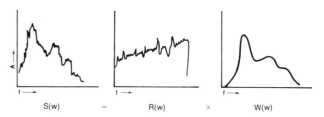

(b.A.1) *Deconvolution*. To improve the resolution, we would like to recover the reflection coefficient function by undoing the convolution with the wavelet, i.e., we want to "deconvolve" the wavelet or replace with a spike, ___⋀___, which does not affect the response of the reflection coefficients.

A simple operation would be to filter the seismic trace $S(t)$ with an operator $f(t)$ such that the result is $R(t)$. In other words, the convolution of the filter $f(t)$ with the wavelet $W(t)$ would produce a spike. In polynomial notation:

$$(W_0 + W_1 + \cdots W_n)(f_0 + f_1 + \cdots f_n)$$
$$= (1 + 0 + 0 + 0 + \cdots).$$

One method for determining $f(t)$ is through polynomial division $(1 + 0 + 0 \cdots)/(W_0 + W_1 + \cdots W_n)$. The quotient is of infinite length, and finite truncation produces an error. Furthermore, exact knowledge of the wavelet, $W(t)$, is required.

The preferred approach is to require the sum of the square of the errors resulting from the application of the filter be a minimum. This means that when we apply the filter:

$$(f_0 + f_1 + f_2)(W_0 + W_1 + W_2) = f_0W_0 + (f_0W_1 + f_1W_0)$$
$$+ (f_0W_2 + f_1W_1 + f_2W_0) + (f_1W_2 + f_2W_1) + f_2W_2$$

we want a least-squares error solution for our desired result of $1 + 0 + 0$. We want

$$(1 - f_0W_0)^2 + (0 - f_0W_1 - f_1W_0)^2 + (0 - f_0W_2 - f_1W_1$$
$$- f_2W_0)^2 + (0 - f_1W_2 - f_2W_1)^2 + (0 - f_2W_1)^2 = \text{minimum}.$$

The solution takes the following form:

$$\begin{bmatrix} W_0^2 + W_1^2 + W_2^2, & W_0W_1 + W_1W_2, & W_0W_2 \\ W_0W_1 + W_1W_2, & W_0^2 + W_1^2 + W_2^2, & \\ & & W_0W_1 + W_1W_2 \\ W_0W_2, & W_0W_1 + W_1W_2, & W_0^2 + W_1^2 + W_2^2 \end{bmatrix} \begin{bmatrix} f_0 \\ f_1 \\ f_2 \end{bmatrix} = \begin{bmatrix} W_0 \\ 0 \\ 0 \end{bmatrix}$$

We recognize, on the left-hand side, the autocorrelation of the wavelet, AC_w. The solution for the filter is

$$F = (AC_w)^{-1} w_0.$$

Since w_0 is a scalar in this instance, we can say $F = (AC_w)^{-1}$.

This shows that, rather than the wavelet, we need to

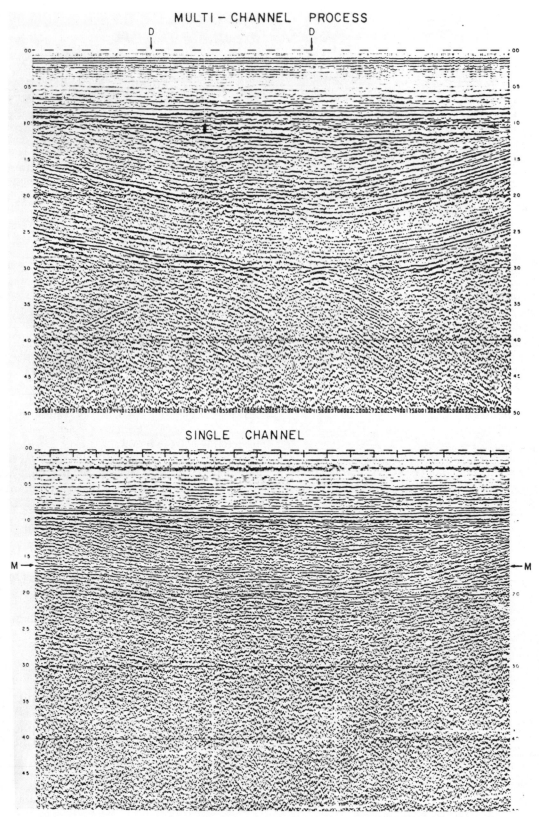

FIGURE B.3.4 Example of a multichannel process that attenuates multiples. (Courtesy of Gulf Science and Technology Company.)

know the autocorrelation of the wavelet. If we state that the autocorrelation of the reflection coefficient function, $R(t)$, is a spike $(\cdot \ \cdot \ \cdot, 0, 0, 0, 1, 0, 0, 0, \cdot \ \cdot \ \cdot)$, then the autocorrelation of the trace, for finite length, $S(t)$, is the autocorrelation of the wavelet. This is an important result, since we always have the seismic trace, but it is difficult to record the source wavelet independently.

The usual deconvolution is performed by convolving the seismic trace with the inverse of the matrix of the autocorrelation of the seismic trace. The autocorrelation is of finite length, typically 200 msec; this represents the expected length of the wavelet.

The same operation can be achieved in the complex Fourier transform by inverting the amplitude and phase spectra of the wavelet obtained from the power spectrum of the trace. The phase spectrum is obtained through Hilbert transformation.

There are other methods to separate the wavelet from the reflection coefficient function. One method that has been used occasionally is the homomorphic deconvolution in which the spectrum of a trace is used.

The deconvolution can be performed so that the wavelet, instead of being replaced by a spike, is replaced by a collapsed wavelet of specified shape:

$$(W_0 + W_1 + W_2 + \cdots W_n)(f_0 + f_1 + f_2 + \cdots f_n) = (a_0 + a_1 + \cdots a_n).$$

This is called "shaped deconvolution" or "predictive deconvolution" and is generally more robust than "spiking deconvolution."

An approximate deconvolution, which is computationally fast because it avoids the matrix inversion, is the polynomial division of the trace autocorrelation.

The deconvolution can be applied in a time-varying mode by defining different operators for different trace segments.

Figure B.3.5 shows an example of the effectiveness of

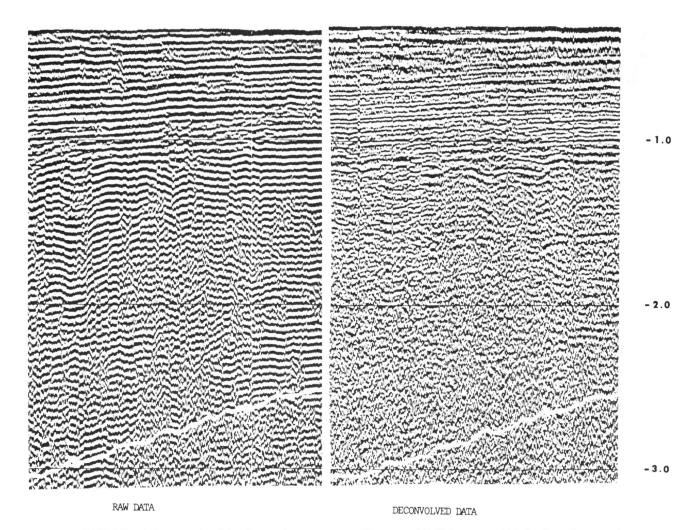

RAW DATA   DECONVOLVED DATA

FIGURE B.3.5   Example of the deconvolution process. (Courtesy of Gulf Science and Technology Company.)

the deconvolution process; resolution and event separation have been achieved.

(b.A.2) *Deghosting.* When a source is energized below the surface, energy is reflected from the surface and trails the primary pulse as a ghost:

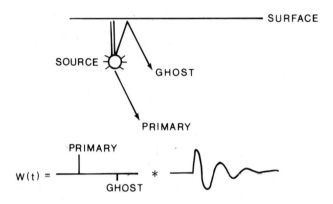

$$W(t) = \quad * \quad$$

Ghosting is a special wavelet problem similar to the "bubble" and "multiple" mentioned earlier.

The ghost can be detected from the autocorrelation of the trace or from special recordings, and a simple inverse operator can be designed. Although the exact inverse is of infinite length, an operator of finite length produces adequate results:

$$1/(1 - g) = 1 - g + g^2 - g^3 + g^4 - \cdots .$$

(b.A.3) *Frequency filters.* By judicious selection of the wavelet spectrum, one can enhance the resolving power.

Wavelets that have most of their energy in a narrow frequency band, say three fourths of an octave or less, are quite "ringy," i.e., they have long tails:

MINIMUM PHASE WAVELET (20-35 CPS)

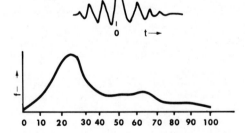

ZERO-PHASE WAVELET (20-35 CPS)

One could select the band 30 to 100 cps, filtering out the low-frequency portion of the spectrum and have a sharper wavelet with dampened tail:

Since it has less power than the wavelet peaking at 25 cps, it has lower S/N even though it has greater resolving power.

Coupled with the filter, simple deterministic operations can be applied. Integration raises the low frequencies at the rate of 6 dB per octave:

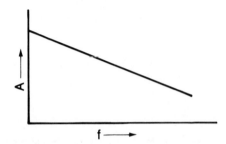

Differentiation raises the high frequencies at the rate of 6 dB per octave:

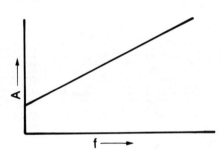

These operations also produce 90° of phase shift.

(b.B.1) *Multichannel deconvolution* is a direct extension of single-channel operation. It is assumed that each channel is convolved with an operator, and all channels are then summed to produce the desired result. The problem is set to produce a least-squares error solution to the following equation:

$$(W_{10} + W_{11} + \cdots W_{1n}) (f_{10} + f_{11} + \cdots f_{1n})$$
$$+ (W_{20} + W_{21} + \cdots W_{2n}) (f_{20} + f_{21} + \cdots f_{2n}) + \cdots$$
$$+ (W_{m0} + W_{m1} + \cdots W_{mn}) (f_{m0} + f_{m1} + \cdots f_{mn})$$
$$= (1 + 0 + 0 + 0).$$

The advantage of the multichannel operation over the single-channel deconvolution is that, to the extent that power spectra are different on different channels, optimum S/N is used from each channel in constructing the desired result.

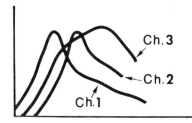

The difficulty with these filters is that they are very sensitive to the lack of time-coincidence between all channels and are computer-intensive (large matrices need to be inverted).

Those filters are seldom used, but they have been used as the foundation to other special-purpose multichannel filters (maximum likelihood, cross-equalization, etc.).

(b.B.2) *Multichannel deghosting.* If two or more recordings are made at the same location with the source at different depths, ghosts occur with different time delays to the primary for the different channels.

The deghosting operation can be performed as described above or as described in a.B.5 for phase filters.

(c) *Improve Spatial Resolution*

Seismic recordings do not measure depth vertically below the receiver but rather time along a seismic ray that obeys Fermat's principle (shortest travel time) and Snell's laws (continuity across boundaries). The distortions caused by the travel times must be corrected before the geological shapes can be resolved by the explorationist.

There are two types of structural distortions: one caused by the recording geometry, the other caused by the geological features themselves.

(c.A.1) *Normal-move-out correction.* If source and receiver were coincident, times along a seismic event would in general represent the true time relationship of the geological feature. However, seismic recordings are made with variable source–receiver distance, and all data need to be corrected to a source–receiver coincident position before the data can be interpreted or processed through the various multichannel filters.

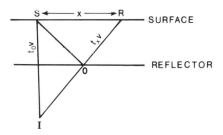

The vertical ray *SI* represents the two-way travel path to the reflector; this is the travel path that would be recorded if the receiver was located at S. With the receiver offset from S by a distance *R*, the travel path *SOR* is equal

in distance to *IOR*. The travel time for *SI* is t_0 and for *SOR* it is t_x. We can compute t_x:

$$t_x{}^2 V^2 = t_0{}^2 V^2 + x^2$$
$$t_x = \sqrt{t_0{}^2 + x^2/V^2}.$$

The correction to be applied to t_x, $\Delta t = t_x - t_0$, is called the normal-move-out correction (NMO). To apply NMO, a correct estimate of velocity is required. Current technology provides velocities accurate to within 1–2 percent, which is more than adequate for the purpose of correcting NMO.

(c.B.1) *Two-dimensional migration.* The seismic profile is viewed as representing the structural shapes vertically below the surface. This is only true when we deal with horizontal layers:

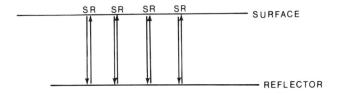

Dips, curvatures, and discontinuities distort the ray path, and the resultant time-data can be misinterpreted. A dipping layer shows the simplest distortion:

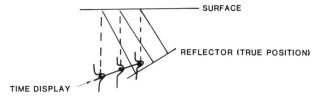

It can be seen that the data need to be shifted sideways and up ("migrated") in order to be in true position.

Discontinuities (point sources) cause similar distortions:

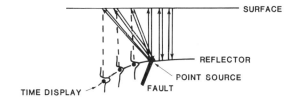

This particular distortion of a point source is called a "diffraction." Curved surfaces have similar distortions.

If one considers every reflecting point as a point source generating a diffraction pattern (similar to the NMO curve), then a migration technique becomes obvious: sum all data values along a diffraction curve and place the result at the apex of the curve:

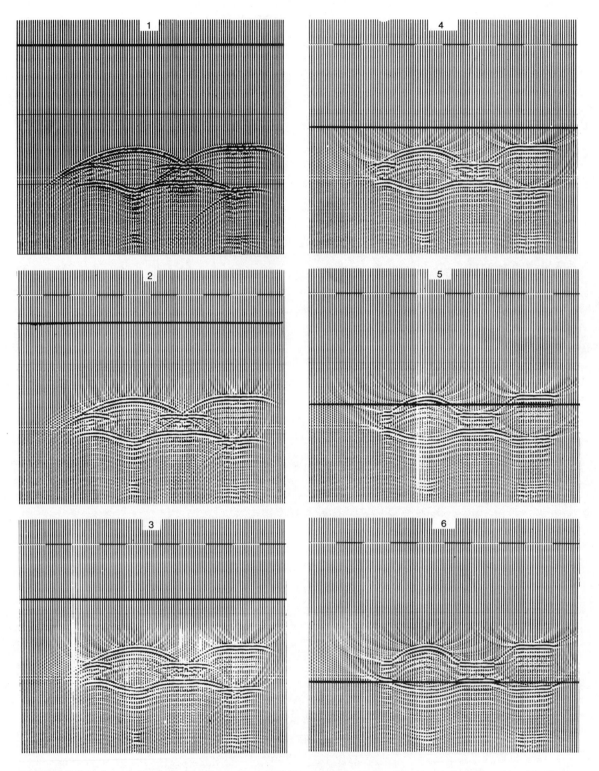

FIGURE B.3.6   Snapshots of migration in progress. Solid lines indicate the imaginary geophone horizons. (Courtesy of Gulf Science and Technology Company.)

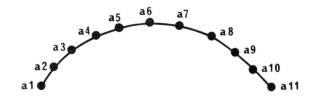

Migrated value = $\sum a_i$ at location a_6. This process is commonly called "ray-path migration." If there is reflector energy, then the migrated value at a_6 will be large. If there is no energy along the diffraction curve, the sum will be small.

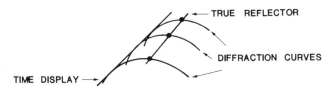

The length of the diffraction curve along which the summation is performed (the "aperture") depends on the expected dips; steeper dips require a larger aperture in order to be accurately migrated. A typical aperture is 144 traces (about 3 miles) with 72 traces on either side of the apex.

One of the flaws of the ray-path migration is that it can create continuities or pseudo-reflection out of noise. In recent times a more powerful imaging technique has been developed. It is called "wave-equation migration," and it uses finite differences rather than sums. It takes a deterministic approach to solve the migration problem, while the conventional ray-path migration utilizes a statistical approach that is based on the Huygens-Fresnel principle.

Wave-equation migration involves solving a scalar wave equation to synthesize seismic-wave fields that would have been observed with imaginary receivers at various depths. Based on the concept of "downward continuation," the receivers are thought to move downward into the earth in progressive steps, while the corresponding wave fields are computed using finite-difference techniques. Thus the closer the receivers are to a reflector, the sharper the image becomes in the computed wave field. When the receivers reach the location of the structure, the wave field yields the correct image, which is free from diffractions.

Figure B.3.6 (parts 1–6) shows a series of snapshots of a migration of tank model data in progress. As the imaginary geophone horizon approaches the depth of the structure, the diffractions are progressively contracted and the image comes into focus. The section above the receiver line is fully migrated, and only the data below will be changed in the following iteration. Part 6 is the true image.

On field data, wave-equation migration demonstrates its superior performance in preserving original character, amplitude, and frequency content. It is particularly effective on data with low signal-to-noise ratios where ray-path migration schemes tend to produce significant distortion in frequency and amplitude.

There are other mathematical techniques for 2-D imaging of seismic data.

Since a very large number of input traces is used for every migrated output, migration is a time-consuming process. However, because the imaging provides a much clearer picture of the subsurface structures, the process is becoming widely used, even though costly.

(c.B.2) *Three-dimensional migration.* Since geological features are seldom two dimensional, the 2-D imaging is often inadequate. In such cases, the structural picture seen along a profile is not the accurate structure vertically below the line of profile. The location of the reflection energy is not along a diffraction curve, but rather along a diffraction surface.

An extension of 2-D ray-path migration to a 3-D process is straightforward. Of course, the field recording must be so designed that spatial coverage is obtained; rather than a profile line, the survey must provide a carpet of locations, and the dimensions must be adequate to satisfy aperture requirements.

In addition to the correct imaging along one vertical profile, the 3-D technique also provides good sampling, in all directions, of the geological features. The resolution power is thus enhanced.

Three-dimensional migration requires a costly surveying technique and a costly computer process. Although the technique is not often used, it is gaining in popularity.

Figure B.3.7 compares the results of 2-D and 3-D migration. For the 2-D migration, one profile was processed; for 3-D migration, 60 profiles were processed to produce Figure B.3.7. The salt dome is better imaged by the 3-D process.

III. INTERPRETIVE PROCESSING

When the data have been optimally enhanced through the processes described earlier, the explorationist is ready to interpret the data in geological terms. He can use the data as they are, or he can seek the help of the computer to provide transformation or displays that have geological meaning. These processes modify the data or abstract subsets of the data, but they do not make any basic change to the S/N or to the frequency bandwidth of the data. The S/N and bandwidth have been optimized by the enhancement processes.

A large number of processes are useful to the interpreter. They have been designed to extract from the seismic data information that is correlated to structural or lithologic properties. Many processes were designed to test geological hypotheses of the interpreter. We will discuss briefly the following processes, which are in fairly common use:

1. Time–Depth Conversion
2. Velocity Analysis
3. Pseudo-velocity Log
4. Anelastic Attenuation
5. Parameter Extraction and Multivariate Analysis
6. Displays

1. Time–Depth Conversion

The seismic time can be converted to depth through multiplication of the time by the average medium velocity from surface to the seismic event. This puts the data into conventional geological units (feet and meters) instead of seconds and milliseconds. It makes it possible for the interpreter to incorporate other subsurface data such as well logs.

If the velocity does not vary spatially, the time–depth conversion produces convenience to the interpreter but no change in structural shape. However, if velocities vary spatially, which they do, then the shapes of the time structures are altered, sometimes substantially. For example, this time feature:

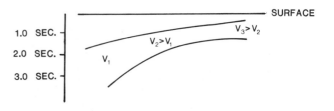

could be converted to this feature:

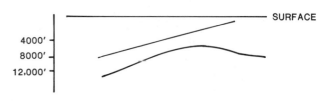

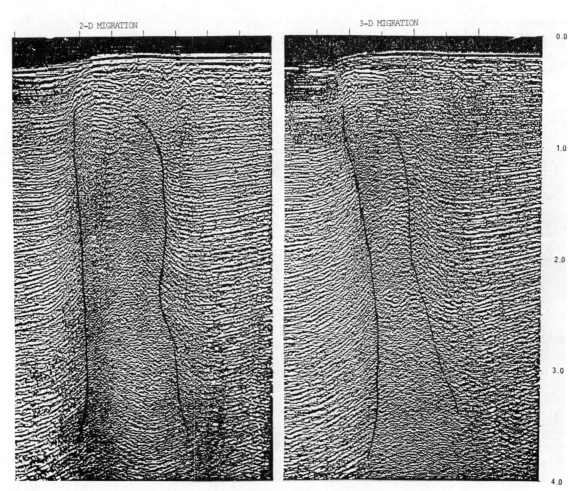

FIGURE B.3.7   Comparison of results of 2-D (left) and 3-D (right) migration. (Courtesy of Gulf Science and Technology Company.)

Whenever there are substantial spatial variations of velocities, time–depth conversion is essential.

Velocities can be obtained from well surveys or from the reflection data themselves or from both.

Time–depth conversion is very fast, computationally.

2. Velocity Analysis

As mentioned earlier, velocity information is needed for normal-move-out correction and for time–depth conversion. Velocity is also a parameter that can be correlated to rock and fluid properties. To make discrete interpretation of the variability of rock properties (porosity, for example) or fluid (oil, gas, water) properties, the interpreter needs very extensive velocity data. These are available from the reflection data, particularly the common-depth-point recordings.

In the normal geometry of seismic recordings, the velocity can be computed by using the arrival times at two different offset distances (source-to-receiver distances), say zero and x.

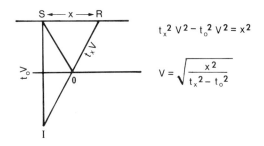

$$t_x^2 V^2 - t_o^2 V^2 = x^2$$

$$V = \sqrt{\frac{x^2}{t_x^2 - t_o^2}}$$

Any two offset recordings x and y will provide velocity information through the same medium,

$$V = [(y^2 - x^2)/(ty^2 - tx^2)]^{1/2}.$$

Since the current CDP surveys involve 24 or 48 offsets (occasionally as many as 96, 120, or 250 offsets), there is a great deal of redundancy available for statistical averaging, which increases the S/N of the velocity data. There are two principal methods of computing velocities: (a) track a seismic event through some cross-correlation scheme and measure the redundant velocities:

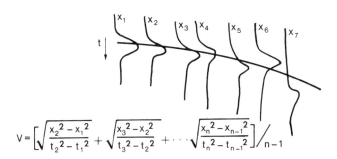

$$V = \left[\sqrt{\frac{x_2^2 - x_1^2}{t_2^2 - t_1^2}} + \sqrt{\frac{x_3^2 - x_2^2}{t_3^2 - t_2^2}} + \cdots \sqrt{\frac{x_n^2 - x_{n-1}^2}{t_n^2 - t_{n-1}^2}}\right]/n-1$$

(b) use several test velocities to apply normal-move-out to the traces, and for each velocity measure the cross-coherence; the largest coherence represents the true stacking velocity. This is the method that is most widely used, and a typical approach is to sum algebraically in space x along an NMO hyperbola, $\sum_{x=1}^{n} a_{t,x}$, the amplitude values for discrete times t within a time gate (say, 25 samples), then to sum the squared values of the results within the gate:

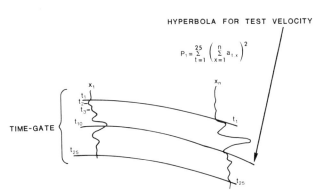

HYPERBOLA FOR TEST VELOCITY

$$P_t = \sum_{t=1}^{25}\left(\sum_{x=1}^{n} a_{t,x}\right)^2$$

The powers, Pj, for j test velocities are examined, and the highest power is related to the best estimate of the velocity of that location. There are many mathematical variations of this approach; each is designed to improve the reliability of the results.

Figure B.3.8 shows an example of continuous velocities determined from reflection data.

Very fast algorithms have been developed, and, with the use of array processors, computers can produce these analyses very fast even when they are computed at every recording location.

The velocity obtained in this fashion is not the true velocity of the medium, and it is called "stacking" velocity.

When velocities are obtained at two geological interfaces, the interval velocity can be computed:

AVERAGE VELOCITY V_{a2}

AVERAGE VELOCITY V_{a1}

$VI_{1,2}$

$$VI_{1,2} = \frac{t_2 V_{a2} - t_1 V_{a1}}{t_3 - t_1}$$

This equation is correct for true average velocities. However, in the case of "stacking velocities" a good approximation is

$$\hat{VI}_{1,2} = \sqrt{\frac{t_2 V_{s_2}^2 - t_1 V_{s_1}^2}{t_2 - t_1}},$$

where V_s is the stacking velocity.

These interval velocity results can be subjected to two-dimensional filters that improve the reliability of the

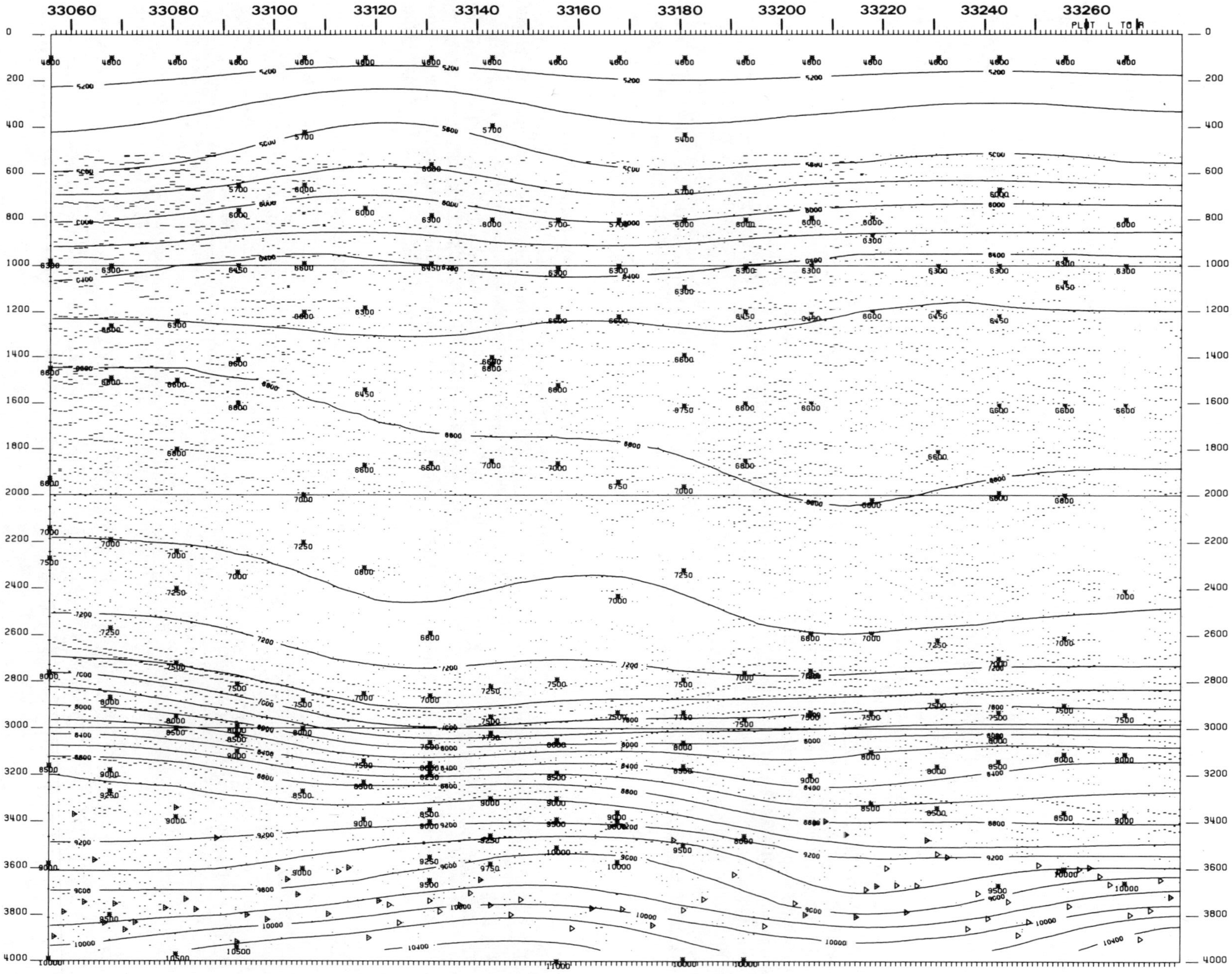

FIGURE B.3.8 Example of continuous velocities determined from reflection data. (Courtesy of Gulf Science and Technology Company.)

256

results. The explorationist then uses the results for his interpretation. For example, gas zones can be predicted from velocity behavior.

3. Pseudo-velocity Log

Synthetic seismograms have been constructed from velocity logs for the past 30 years. The log is converted to reflection coefficients, and these are filtered to the seismic band.

The logs contain more information than the seismic trace, particularly information about rock and fluid properties. There has been a strong incentive to construct the inverse, that is, synthetic or pseudo-logs from seismic traces, and many schemes have been attempted over the years. The difficulty in achieving the transformation is that logs are broadband, containing essential velocity information at the very-low-frequency end, below 3 cps, whereas the seismic traces have very little usable data below 8 cps and they do not have the higher frequencies (above 80 cps) found in the logs.

In the past few years, some effective schemes have been developed for producing pseudo-logs, and the general approach is as follows: the seismic trace is deconvolved and very high frequencies are retained while the low-frequency end, say below 8 cps, is filtered out; a velocity curve is obtained, either from well-logs in the vicinity or from velocity analyses of the reflection data; the velocity curve is filtered to the band 0–8 cps. The deconvolved seismic trace is considered to be a reflection coefficient function, and it is combined with the velocity curve. The pseudo-logs can be interpreted by the explorationist as if he had a string of wells along the surface.

4. Anelastic Attenuation

Seismic amplitudes attenuate as a function of distance traveled, and they are proportional to $1/R$, where R is the radial distance from the source. This attenuation is the same for all seismic frequencies.

There is an additional attenuation that involves solid friction and that is called "anelastic attenuation." This is a frequency-related phenomenon in which higher frequencies are more strongly attenuated than low frequencies:

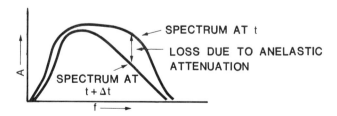

It is the spectral ratio between the wavelet at t and $t + \Delta t$ which defines the attenuation.

The anelastic attenuation is related to rock properties: hard rocks such as carbonates have low attenuation; softer

rocks such as clays have higher attenuation; gas-filled reservoirs have high attenuation, etc. It is a useful parameter to the explorationist.

To measure the anelastic attenuation across a geological unit, one needs to examine the seismic wavelet before it enters the unit and after it leaves the unit. To obtain the wavelet is a difficult task, and many estimation methods have been used: least-squares and homomorphic deconvolution and gated autocorrelation, to name but three. By whatever method, estimating the wavelet from a small sequence of data is always imperfect, and anelastic attenuation computations are very sensitive to noise.

5. Parameter Extraction and Multivariate Analysis

In addition to specific parameters such as velocity analysis and anelastic attenuation, explorationists have found it useful to extract systematically a large number of parameters available in the processed seismic data. Parameters are extracted from specified seismic events or from arbitrary time gates (say every 100 msec). The list of parameters would include the following:

Highest positive amplitude
Highest negative amplitude
Ratio of positive/negative amplitude
Power $\sum (a_i)^2$ within specified gate
Ratio of highest amplitude to power
Peak frequency
Slopes on the amplitude spectrum
Coherence
Velocity
Anelastic attenuation

These parameters are examined by the interpreter, who seeks to correlate parameter behavior to geological condition. He may try to relate the change of water to oil to a change in amplitude or frequency content or coherence or to a combination of parameters.

Combinatorial techniques exist that can study, statistically, the relationship of a large number of parameters to specified conditions. For example, factor analysis is a method of determining the relative importance of several parameters in identifying a specific condition, say oil.

Another method is discriminatory analysis, in which parameters are manipulated to optimize the discrimination between conditions of interest. For example, conditions Za and Zb (e.g., oil and water) can be described as follows:

$$Za_1 = \lambda_1 X_{11} + \lambda_2 X_{12} + \cdots \lambda_n X_{1n} + C_1,$$
$$Za_2 = \lambda_1 X_{21} + \lambda_2 X_{22} + \cdots \lambda_n X_{2n} + C_1,$$
$$\cdot$$
$$\cdot$$
$$\cdot$$
$$Zb_1 = \gamma_1 Y_{11} + \gamma_2 Y_{12} + \cdots \gamma_n Y_{1n} + C_2,$$
$$Zb_2 = \gamma_1 Y_{21} + \gamma_2 Y_{22} + \cdots \gamma_n Y_{2n} + C_2,$$

where the X's and Y's are the parameters (amplitude, power, velocity, etc.) for conditions a and b, respectively, and a's and b's are statistical samples for each condition; λ's and γ's are discriminants computed to satisfy the following equation:

$$\frac{(\overline{Z}a - \overline{Z}b)^2}{(\overline{Z}a - Za_i)^2 + (\overline{Z}b - Zb_i)^2} = \text{maximum}.$$

If discrimination is found at a significant level, sets of parameters from locations of unknown conditions can be tested to determine whether they belong to one condition or another (e.g., water or oil).

Other computer manipulations are also possible.

6. Displays

Finally, we come to the subject of displays. After all the computer processing to enhance and massage the data has been completed, the explorationist wants to study the results in a format that is comprehensible to him. He seldom looks at lists of digits, but rather he wants displays that emphasize the characteristics that are of interest to him, e.g., phase alignment, amplitude variation.

The bulk of processed seismic data is displayed in the form of an amplitude–time section. The section can be displayed as variable amplitude (also called "wiggletrace"), variable density, or variable area mode or combinations of these modes. Variable area and variable area-wiggle are the most popular forms of displays.

Certain seismic parameters are displayed as contoured line drawings. For example, velocity data along seismic profiles are displayed in this manner.

Finally, data can also be displayed in color. Besides the fact that variability can be enhanced through color, there is the additional advantage that more than one parameter can be examined on a single display, e.g., amplitude in black and white, velocity in color.

Most modern display units (plotter) are computer controlled. Great plotting speeds have been achieved, and there is great flexibility.

CONCLUSIONS

Geophysicists being quite inventive, there are many, many more digital processes than have been described in this paper. The list submitted here represents the more important and commonly used processes.

B.4

Integrated Approaches to the Interpretation of Gravity and Magnetic Data in Petroleum Exploration: Applications to Continental-Margin Tectonic Studies

J. G. SMITH

SUMMARY

During the past decade, petroleum explorationists have evolved an integrative approach to the interpretation of gravity and magnetic data in which inferences drawn from these data are considered in full awareness of other types of exploration information. Previous studies were generally restricted to interpretations based on independent, isolated studies of potential field data, seismic data (reflection or refraction), and stratigraphic information obtained through drilling, coring, and dredging. These more recent integrative studies, although very simple in concept, have greatly expanded insights into continental-margin structure, composition, and origin.

Similar integrative approaches to the interpretation of gravity and magnetic data need to be adopted by academic, institutional, and government research agencies, if public knowledge of the nature and development of continental margins is to increase substantially. The cost of acquiring high-quality regional gravity and magnetic surveys is the main obstacle to this type of application. Nevertheless, extensive, detailed aeromagnetic surveys should be the first step in any thorough investigation of a continental-margin segment. Research vessels carrying out subsequent seismic surveys should be equipped to collect gravity data as well.

The utility of gravity and magnetic data in integrated continental-margin studies lies in that (a) they often provide the only clue to the deep structural framework of the continental margin lying beyond the resolution of seismic data; (b) they complement seismic data in indicating the composition, size, and shape of structural elements within even the shallow sedimentary mantle of the margin; and (c) they provide a means for mapping these structural and compositional features on a regional basis—all at a cost that is quite low compared with that for seismic surveying and for drilling.

INTRODUCTION

Interpretations of gravity and magnetic data have always been an integral part of the continental-margin studies carried out by academic institutional and governmental research groups. Certainly the discovery and study of magnetic lineations within the ocean floors have supported the concepts of seafloor spreading and have greatly increased our insight into some of the general processes involved in continental-margin development. Gravity data have also provided means for making approximations of the gross structure of the continent–ocean interface. For the most part, however, these studies have been based on small areal surveys, or widely spaced or isolated profile surveys; as a result, the regional significance of the interpretations has been limited. Furthermore, these observations commonly do not overlap or cannot be related to other sources of information such as seismic surveys, deep drilling, or onshore geology, limiting their value even further.

The utilization of gravity and magnetic data in the search for oil and gas has been quite different and offers some approaches that should be more actively employed by continental-margin research groups. In general, industry-conducted potential field surveys have been more extensive and comprise more intensive coverage of higher-quality data. Even in petroleum exploration, these data sets were, at first, considered independently of each other, of seismic data, and of other geological information. Gradually, however, explorationists became aware that much could be learned if gravity and magnetic interpretations were considered in conjunction with or integrated into the structural and stratigraphic patterns that emerged through seismic mapping, drilling, and other geological investigation. This awareness was greatly accelerated during the past decade by the advent of multisensor survey vessels simultaneously collecting seismic-reflection, magnetic, and gravity data.

Integrated approaches to the utilization of gravity and magnetic data applied in petroleum exploration and examples of their application to three principal types of continental-margin problems follow. These approaches are quite unsophisticated; they do require, however, extensive, overlapping networks of high-quality gravity, magnetic, and seismic data, anchored if possible to at least a few stratigraphic control points. Finally, such integrated interpretations require special attitudes and a great deal of coordination among the specialized geologists and geophysicists required to accomplish this type of study. The first requirement can be bought (at great cost), the second requirement is more elusive and needs to be nurtured.

DIFFERENTIATION AND RESOLUTION OF COMPOSITIONAL AND STRUCTURAL ELEMENTS WITHIN THE SHALLOW SEDIMENTARY MANTLE OF THE CONTINENTAL MARGIN

Geophysical data collected several years ago off the western coast of Africa have provided an outstanding example of the role of gravity and magnetics as a critical supplement to modern seismic surveying of continental margins.

The seismic-reflection profile seen in the central panel of Figure B.4.1 clearly indicates two high-standing structural masses of the same apparent size, shape, and origin. Both features appear to penetrate the sedimentary veneer of the continental margin. If the interpretation of this traverse had been limited to seismic data, it is unlikely that even the most experienced observer would have differentiated between the two masses, i.e., one would have mapped two salt or shale piercements, two basement uplifts, or perhaps two intrusive plutons. Fortunately, however, gravity and magnetic data were also available, which demonstrated that the two masses are quite dissimilar—that is, a salt or shale piercement on the left and probably a fault-defined block of dense, magnetic crystalline basement rock on the right. The significance of this distinction to a proper understanding of the nature and evolution of this segment of the west African continental margin is easily apparent.

The African profile illustrates an extreme example of the utility of potential field data; it is hardly unique. During the past few years, gravity and magnetics have played a key role in the recognition of crystalline rock masses (horsts, stocks, dikes, sills, flows) penetrating and/or lying within the sedimentary mantle of the continental margins of Brazil, the eastern United States, and Alaska, to mention only a few areas. These features are important clues to a fuller understanding of continental-margin development and would have been unresolved, if not undetected, by seismic methods alone.

In the west African example the seismic data did, however, provide an adequate representation of at least uppermost outlines of the two positive structural blocks. There are sometimes instances when quality of the seismic data is such that this is not possible. Figure B.4.2 illustrates two versions of the outlines of a high-pressure shale mass that penetrates a sequence of shales and petroliferous sandstones. An unambiguous and accurate representation of this shale mass is, of course, critical to exploratory drilling in the area. The initial seismic interpretation was checked against the observed gravity expression of the shale mass using standard modeling techniques and assumed density partitions and gradients based on drilling results elsewhere in the region. The left side of Figure B.4.2 indicates that the density distribution required by the seismically defined shale-mass geometry would not produce an approximation of the observed gravity anomaly. However, the right side illustrates that if the seismic data are re-examined with an awareness of the geometric constraints imposed by the gravity data, an integrated interpretation can be developed that satisfies both sets of information.

Similar modeling approaches can be used to supplement seismic data in the resolution of a wide range of complicated structural situations or, for that matter, for defining the gross distribution of major facies within the sedimentary mantle of continental margins.

RESOLUTION OF THE DEEP STRUCTURAL FRAMEWORK OF CONTINENTAL MARGINS

We have seen that even within the relatively shallow sedimentary veneer of the continental margin, gravity and magnetic data are essential adjuncts to seismic information. Deeper within the margin—perhaps beneath a highly reflective layer that retards the penetration of seismic energy or simply beyond the limits of seismic resolution—gravity and magnetic data provide the only clues to margin infrastructure and composition.

Another geophysical profile across the continental mar-

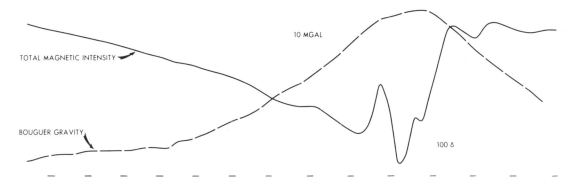

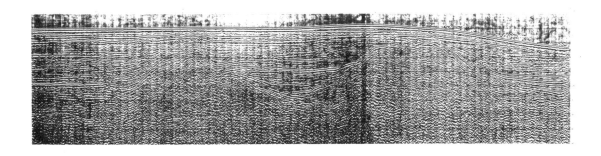

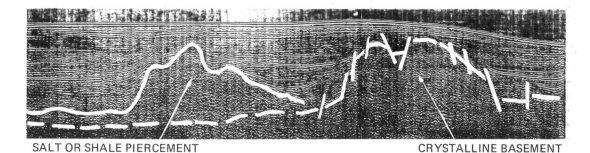

SALT OR SHALE PIERCEMENT   CRYSTALLINE BASEMENT

FIGURE B.4.1   Integrated interpretation of geophysical information (example from western continental margin of Central Africa). *Top*: Outlines of structural uplifts based on seismic data; *middle and bottom*: compositional interpretation based on gravity and magnetics. (Courtesy of Gulf Science and Technology Company.)

gin of west-central Africa (Figure B.4.3) illustrates such a situation. One can see on the seismic profile that there are few coherent seismic events below about 3.5 sec, a level that corresponds to a salt layer indicated by drilling and mapped as the deepest, most coherent seismic reflection. Petroleum explorationists were fortunate in finding oil in sandstones and limestones in early wells drilled into the presalt sedimentary sequence. It was, therefore, critical to discover the nature and extent of the presalt basins, the thickness and composition of sediments, the size and attitude of the faults that bound them, and the presence of anticlines or other structures within the basin sediments.

One would also like to know whether the floor of the basin is composed of basaltic, nearly oceanic rocks or sialic crystalline rocks or whether the basin sediments have been penetrated by magmatic rocks. Such inferences, of course, also bear on developing a full insight into the nature and evolution of the continental margin and are not limited to hydrocarbon prospecting. Gravity and magnetic data provide many of these answers, and in the specific existence shown in Figure B.4.3, gravity modeling indicates the presence and also the shape of a number of sediment-filled depressions (graben) lying beneath the salt. As we will see below, regional gravity and

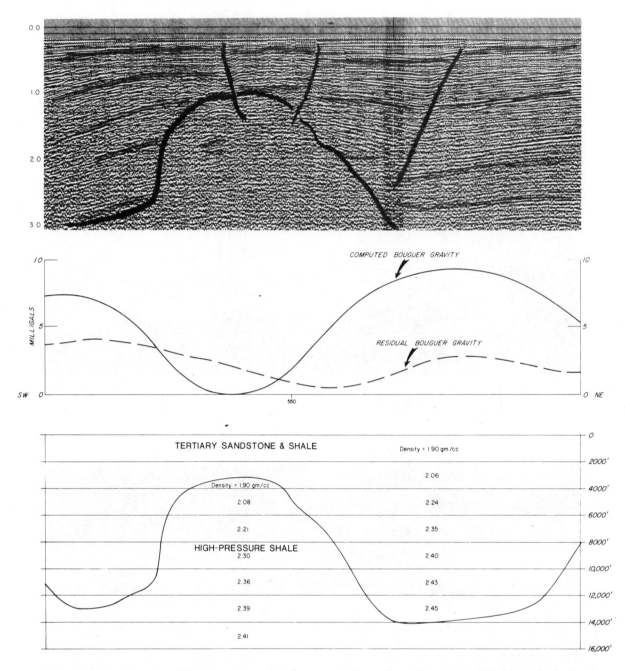

FIGURE B.4.2 *Left*: Outline of shale piercement based on seismic data and resulting mismatch between observed and computed gravity.

magnetic data provide the only insight into the areal extent of these grabens.

This modeling approach, although not fundamentally different from gravity modeling done by academic institutions to approximate the gross nature of the continent–ocean interface, does require high-quality gravity and magnetic coverage, which can be integrated with well and seismic data.

REGIONAL MAPPING OF STRUCTURAL AND COMPOSITIONAL ELEMENTS OF CONTINENTAL MARGINS

Despite the great utility of gravity and magnetic data in the examples cited, the most significant application has been to the recognition and mapping, on a regional scale, of the complex of structural and compositional elements

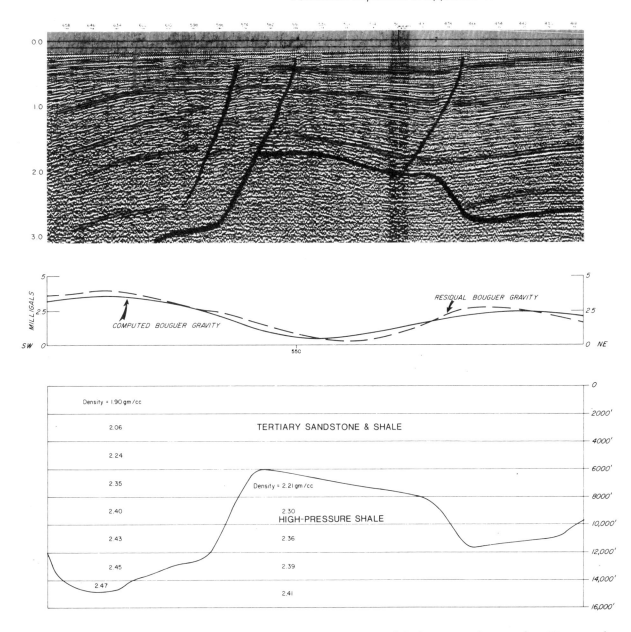

FIGURE B.4.2. *Right*: Revised outline of shale piercement consistent with both gravity and seismic data. (Courtesy of Gulf Science and Technology Company.)

that constitute the continental margin, i.e., structural depressions or basins, high-standing blocks and platforms, volcanic belts, plutonic bodies. This application has been the traditional one for aeromagnetic data, which, because of its relatively low cost, has provided an ideal reconnaissance tool for searching out prospective sedimentary basins, prior to more intensive and expensive seismic surveying. More recently, with the advent of shipborne gravimeters, networks of gravity data that are generally collected simultaneously with seismic data have become available for integrative interpretation on a regional scale.

The approach to regional integrated interpretation of gravity and magnetic data depends on the amount of other overlapping data available—on whether the continental-margin study is at an early or more mature stage. Aspects of these approaches are illustrated on Figure B.4.4, a portion of a U.S. government magnetic survey of the Atlantic continental margin. This was probably the first nonproprietary survey that even approached the extent and quality of magnetic surveys carried out in petroleum exploration. The contour map was based on widely spaced profiles, which have missed many smaller fea-

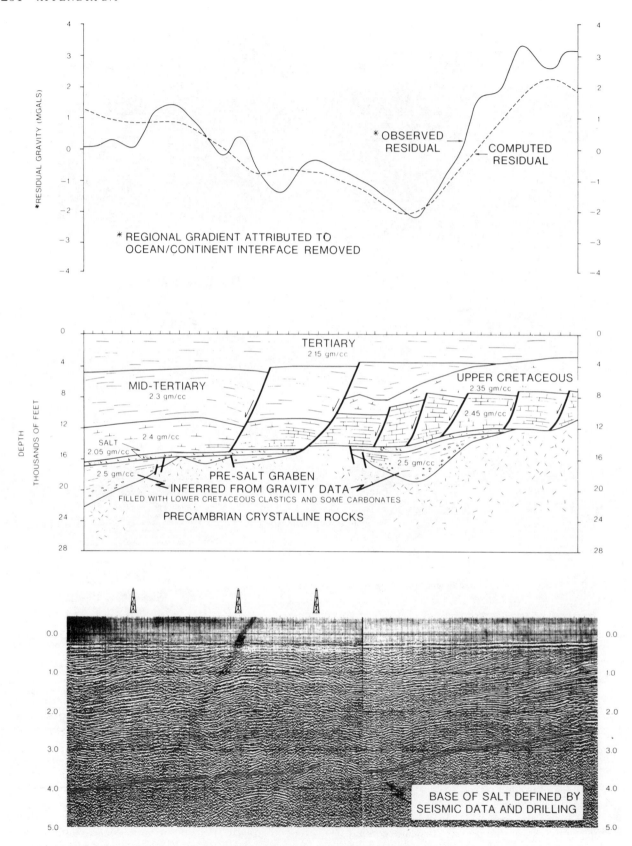

FIGURE B.4.3 Deep structure of continental margin unresolved by seismic data, but inferred from gravity modeling (West Africa example). (Courtesy of Gulf Science and Technology Company.)

tures, but still the map can serve to illustrate, although somewhat simplistically, the value of aeromagnetic surveying in regional continental-margin studies.

The magnetic data covering the Blake Plateau are immediately encouraging from an exploration point of view. They indicate that the Plateau is underlain by a great thickness of ostensibly prospective sedimentary rocks uncomplicated by igneous activity of any type. The magnetic data, when integrated with gravity information, indicate that the floor of the basin comprises crystalline rocks, which are continental in character at the north and become intermediate between continental and oceanic types to the south. Moreover, it is possible to relate the onshore magnetic anomaly character with the pre-Cretaceous compositional and structural elements of Florida that are known from drilling and more detailed geophysical work. These elements can then be projected beneath the Plateau on the basis of magnetic trends and offer a provisional understanding of deep structure that certainly provides objectives for subsequent detailed geophysical surveys, deep drilling, and interpretation.

Similar regional approaches are relevant even in in-

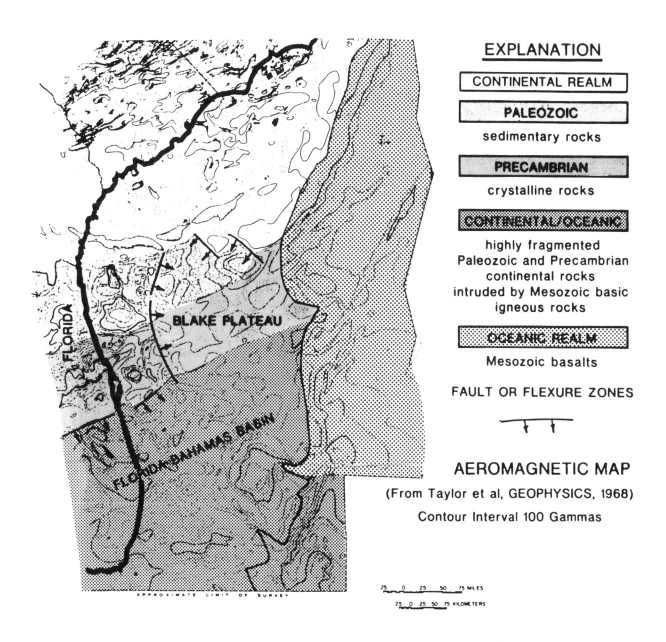

EXPLANATION

CONTINENTAL REALM

PALEOZOIC
sedimentary rocks

PRECAMBRIAN
crystalline rocks

CONTINENTAL/OCEANIC
highly fragmented
Paleozoic and Precambrian
continental rocks
intruded by Mesozoic basic
igneous rocks

OCEANIC REALM
Mesozoic basalts

FAULT OR FLEXURE ZONES

AEROMAGNETIC MAP
(From Taylor et al, GEOPHYSICS, 1968)
Contour Interval 100 Gammas

FIGURE B.4.4   Pre-Cretaceous tectonic elements of the Blake Plateau. Highly generalized interpretation based on association and extrapolations of onshore geology and magnetic data. [From P. T. Taylor, I. Zietz, and L. S. Dennis (1968). Geologic implications of aeromagnetic data for the eastern continental margin of the United States. *Geophysics* 33(5):755–780.]

tensely explored continental margins such as the Gulf of Mexico. Despite the fact that it has been crisscrossed with tens of thousands of miles of seismic surveys and penetrated by thousands of deep wells, knowledge of the nature and origin of the deep structural framework of the Gulf and its development is still fragmentary, partly because of the limitations of the data. Much of the infrastructure of the Gulf is beyond the resolution of seismic-reflection data, whereas refraction seismic data as well as deep-drilling data provide quite isolated points of reference from which only very generalized connections and extrapolations can be made. On the other hand, regional gravity and magnetic surveys now provide not only an additional means for resolving deep structure but more significantly an excellent regional framework into which

to fit fragmentary, previously isolated observations into a consistent picture revealing the infrastructure of the margin; viz., detailed connections among similar structural elements can be drawn, as well as the limits between dissimilar units formerly recognized only in isolated localities by drilling or from seismic data.

This is not to say, however, that the utility of regional gravity and magnetic observations need be limited to the deeper portions of continental margins. Studies in long-explored onshore basins have shown remarkable associations between basement structure and composition inferred from gravity and magnetic data and the development of regional structural and stratigraphic trends in the shallow sedimentary veneer of the basins.

Bright-Spot Interpretation

ROSS A. DEEGAN

INTRODUCTION

In the late 1960's and early 1970's it was recognized that a large class of hydrocarbon reservoirs have anomalous seismic response that can be employed to infer the presence of hydrocarbons. Figures B.5.1A and B.5.1B show such response on a strike line and dip line, respectively, intersecting over a known offshore gas accumulation in a sandstone reservoir of Miocene age.

The example shown in Figure B.5.1 is representative of bright-spot technology as it is most commonly used today. The presence of gas in poorly consolidated sediments (often geologically young and of shallow to medium depth of burial) alters the acoustic properties by dramatically lowering the acoustic impedance; the effect is most evident on high-quality data (often collected offshore).

This paper summarizes briefly our understanding of how the sediment acoustic properties are altered by hydrocarbon presence and discusses some details of interpreting seismic bright-spot anomalies, especially with regard to verifying that the anomalous event is associated with low impedance and estimating the thickness of this low-impedance sedimentary unit.

EFFECT OF HYDROCARBONS ON ACOUSTIC PROPERTIES

For reservoir rocks with approximately equidimensional (as opposed to elongated) pores, the Geertsma-Gassmann formulation[1-4] provides a convenient mechanism for conceptual as well as quantitative analysis of fluid effects on rock properties. In this formulation, the p-wave velocity of a fluid-filled porous rock is given in terms of the elastic moduli of the rock frame when the pores are empty and the elastic properties of the pore fluid or fluids.

Consider a rock in which the pores contain water and a second fluid, such as gas or oil. The Geertsma formulation of the Gassmann equation describes the p-wave velocity of the rock as a function of eight parameters:

$$V = V(\phi, c_p, \sigma, s_w, \rho_w, c_w, \rho, c),$$

where ϕ is porosity; c_p, σ are the "pore-volume" compressibility and the Poisson ratio of the dry rock frame, respectively, which is just a convenient way to express the bulk and shear elastic moduli of the rock when the pores are empty; s_w is the volume of pore space that is water-saturated; ρ_w and c_w are the density and compressibility of the formation water; and ρ and c are the density and compressibility of the second formation fluid.

Figure B.5.2 shows some calculations using the Geertsma equation. The velocity in the fluid-saturated rock is shown as a function of water saturation, from 0 (100 percent gas saturation) to 1 (100 percent water saturation). Both curves in Figure B.5.2 result from using properties for water and dry gas (methane) corresponding to normal pressure at a depth of 6000 ft in a young deltaic environment.[5] For each curve, the porosity is 33 percent. For the curve labeled "unconsolidated," the elastic moduli of the

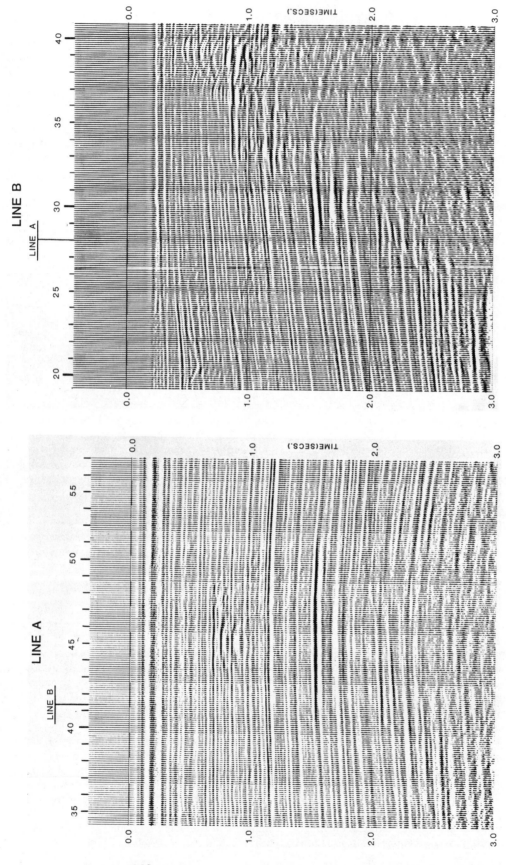

FIGURE B.5.1 (A) Strike line over known gas-sand at about 1.6 sec, extending approximately from shotpoint 40 to 50. Gas pay thickness equals 80 ft at SP 44. Tie with line A is indicated. (B) Dip line over same gas-sand as in (A). Tie with line B is indicated. (Courtesy of Gulf Science and Technology Company.)

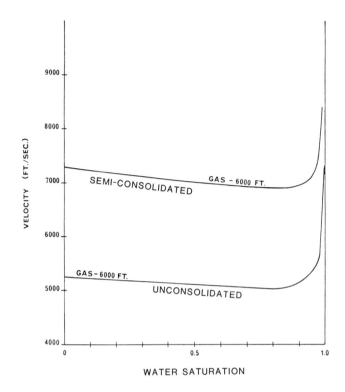

FIGURE B.5.2 Calculated velocity versus water saturation in two reservoirs differing only in degree of consolidation of rock frame. Porosity = 33 percent in both cases. (Courtesy of Gulf Science and Technology Company.)

TABLE B.5.1 Typical Theoretical Velocity Anomalies in ft/sec

Gas	Unconsolidated	Well Cemented
Shallow	3000	750
Normal	2200	750
Deep	1700	750

water-sand, taking into account the density effect in addition to the velocity decreases of Table B.5.1.

Figure B.5.3 further demonstrates the magnitude of the reflection anomaly resulting from the above theory. The synthetic seismogram labeled A was constructed from a well in the vicinity of the profiles of Figure B.5.1, with water-sand velocities at the position of the reservoir sand of interest, near 1.6 sec; the synthetic labeled B has been constructed by replacing this water-sand velocity by an appropriately calculated gas-sand velocity. Note that the corresponding reflection amplitude is the largest event on the seismogram.

rock frame correspond to completely unconsolidated sand packings (after Gardner and Harris[6]). For the curve labeled "semi-consolidated," the rock-frame elastic moduli are taken from measurements on a Mississippi Gulf Coast sandstone (Frio of Oligocene age[7]).

Note from Figure B.5.2 that the effect of gas in the unconsolidated reservoir is to reduce the velocity from 7300 ft/sec at 100 percent water saturation to 5200 ft/sec for 20 percent water (80 percent gas) saturation. This dramatic drop in velocity is primarily responsible for the anomalous strength of seismic reflections from the top and base of the gas-saturated reservoir. For the semiconsolidated reservoir, Figure B.5.2 shows that the velocity drops from 8300 ft/sec for 100 percent water saturation to 7200 ft/sec for 20 percent water (80 percent gas) saturation. This latter effect may be detectable, although other lithologic variations of comparable magnitude are expected in a sedimentary sequence. Also remember that, in addition to the velocity effect, the density of the fluid-saturated rock decreases linearly with increasing gas saturation, contributing to the lowering of acoustic impedance.

Table B.5.1 lists representative calculations of decrease in reservoir velocity due to the presence of gas, for the extreme limits of consolidated and fully consolidated (well-cemented) sandstone. Table B.5.2 shows the corresponding reflection coefficients between gas-sand and

INTERPRETATION OF BRIGHT-SPOT RESPONSE

In interpreting bright-spot response on seismic profiles, we will make the assumption that the processing has done an adequate job of recovering true amplitude primary reflection coefficients with known phase. Seismic processing is covered in detail by Mathieu in this Appendix (paper B). Processing features that are of particular concern to bright-spot interpretation include:

Spatial and temporal gain adjustments, to recover relative reflection strengths;
Deconvolution, to estimate and remove the effective seismic wavelet;
CDP compositing to enhance primary reflection events relative to other arrivals;
Final bandpass filtering with a wavelet of known shape (e.g., a zero-phase bandpass filter).

Commonly, a potential hydrocarbon-related seismic event is first detected on the basis of anomalously strong amplitude response. This detection is normally qualita-

TABLE B.5.2 Typical Theoretical Reflection Coefficients (Water Sand–Pay Sand)

Gas	Unconsolidated	Well Cemented
Shallow	0.31	0.06
Normal	0.22	0.05
Deep	0.16	0.05

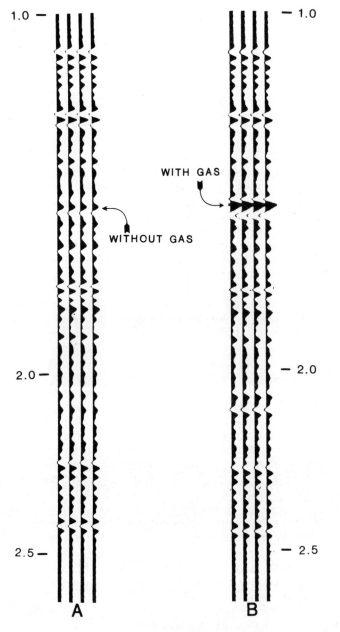

high-impedance layer (the latter may be associated with carbonate, basalt, calcareous sandstone, etc.); Estimation of the thickness of the low-impedance zone; Search for confirming evidence, such as "flat-spots," phasing at termination of events, and anomalously low seismic velocity.

Polarity and thickness estimates normally involve direct or inverse seismic modeling. To a lowest approximation, a single gas-saturated, poorly consolidated reservoir is simply a zone of constant, low impedance, in a medium of uniform higher impedance. To this approximation, the expected seismic response can be modeled simply. Figure B.5.4 shows the synthetic response of a series of isolated low-impedance zones of thickness ranging from 4 to 40 msec two-way time (12- to 120-ft thickness at a gas-sand velocity of 6000 ft/sec); the response modeled consists of primary reflection coefficients only, with a zero-phase filter having a bandpass from 6 to 40 Hz. Both polarity responses are shown.

If the seismic data acquisition and processing are suffi-

GAS-SAND THICKNESS

2-WAY TIME (MSEC.)	THICKNESS FOR VELOCITY = 6000 FT./SEC. (FEET)	IMPEDANCE	POSITIVE POLARITY	NEGATIVE POLARITY
4	12			
8	24			
12	36			
16	48			
20	60			
24	72			
28	84			
32	96			
36	108			
40	120			

FIGURE B.5.3 Synthetic seismograms for a well near the bright spot of Figure B.5.1. Seismogram A is calculated with water-sand velocity at sand of interest near 1.6 sec; seismogram B is with insertion of calculated gas-sand velocity at position of sand of interest. (Courtesy of Gulf Science and Technology Company.)

tive from true-amplitude seismic displays but may involve special postprocessing displays or quantitative data analysis. Further high grading beyond amplitude analysis may involve:

Verification that the reflection is of the expected polarity, from an anomalously low-, rather than anomalously

FIGURE B.5.4 Modeled synthetic response to low-impedance zones, simulating gas-sands of varying thicknesses. Response is primary reflections only, zero-phase filtered in the band 6–40 Hz, both polarities displayed. (Courtesy of Gulf Science and Technology Company.)

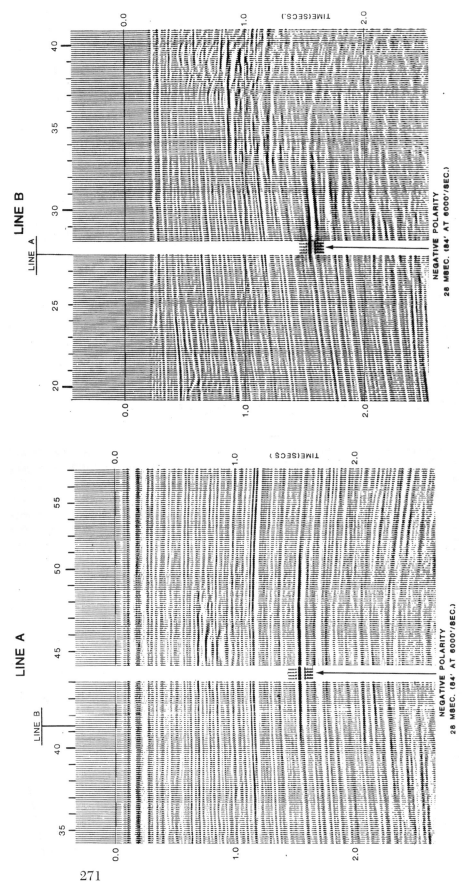

FIGURE B.5.5 (A) The appropriate thickness model of Figure B.5.4 is displayed at location of well penetrating 80-ft gas-sand. Seismic data are assumed to be negative-polarity convention. (B) The appropriate thickness model of Figure B.5.4 is displayed at projection of well penetrating 80-ft gas-sand. Seismic data are assumed to be negative-polarity convention. (Courtesy of Gulf Science and Technology Company.)

271

ciently well understood to predict the polarity of the processed seismic section, then the above models can be used to discriminate low- from high-impedance zones. For example, if the processed data are negative-polarity convention, zero-phase filtered 6–40 Hz, then the expected gas-sand response is as shown in the "negative polarity" displays of Figure B.5.4 (i.e., strong black on top of strong white reflection); high-velocity units would have opposite polarity (white above black, just as for the "positive-polarity" displays).

The models of Figure B.5.4 can also be used for thickness estimation by direct comparison with the observed bright spot. Figure B.5.5 shows that the bright spot of Figure B.5.1 compares closely to the 28-msec (84-ft-thickness) model; the position at which the comparison is shown in Figure B.5.5 corresponds to a well that penetrated an 80-ft gas-sand at that horizon.

If the reservoir has a fluid contact (gas-oil or gas-water) we might expect a corresponding event that is constant in depth, and therefore approximately constant in time—a "flat spot." The dip line of Figure B.5.1B shows some indication of a flat spot at the bright spot of interest, and a gas-water contact was found in drilling.

If, for the geological section under study, the acoustic impedance of the water-saturated leg of the sand is expected to be higher than the impedance of the encasing shales, then the bright spot should exhibit a lateral phase reversal at each terminating edge; this results from the lateral impedance change from very low to slightly high, relative to the shales. There is some evidence of such phasing at the termination of the bright spot of interest in Figure B.5.1A.

If the gas-sand is thick, we might expect an observable anomaly in seismically derived interval velocity. With dense velocity analysis in good data areas, and careful analysis of velocity resolution accuracy relative to the size of the expected anomaly, the seismic velocity can be an additional high-grading tool; it is of particular importance as an amplitude-independent indicator.

DISCUSSION

The brief exposition given above is intended to emphasize aspects of bright-spot interpretation that are now in everyday use by explorationists working in young Cenozoic clastic basins. It should be mentioned, however, that much current effort by petroleum companies and geophysical contractors is directed toward continued advances in this state of the art. Data-acquisition and -processing advances include high-resolution work, especially to obtain higher-frequency response; in terms of the thickness models discussed above, this is directly translatable into improved stratigraphic resolution. Interpretative processing advances include inverse seismic modeling as an alternative to the trial-and-error direct modeling discussed above. Advanced analysis attempts to detect effects of oil, hydrocarbon effects in better-consolidated sediments, and lithologic discrimination.

The seismic processing goal, to recover primary reflection coefficients with correct amplitude and phase, was mentioned only in passing. However, the bulk of bright-spot research during the past five years has been directed toward this end. There is still much improvement to be made in data-acquisition technology and in postprocessing data analysis, and these cannot be separated from data processing; however, perhaps our greatest current restrictions in the area of direct detection of hydrocarbons and lithology stem from our limited ability to turn recorded seismic data into subsurface mapping of acoustic impedance.

REFERENCES

[1]F. Gassmann (1951a). Über die Elastizität poröser Medien, Vierteljahrsschrift der Naturforschenden Gesellschaft in Zurich 96, 1–23.
[2]F. Gassmann (1951b). Elastic waves through a packing of spheres, Geophysics 16, 673–685; 18, 269.
[3]J. Geertsma (1961). Soc. Petrol. Eng. J. 1, 235.
[4]M. J. Geertsma and D. C. Smit (1961). Geophysics 26, 169.
[5]S. N. Domenico (1974). Geophysics 39, 759.
[6]G. H. F. Gardner and M. H. Harris (1968). SPWLA Ninth Annual Logging Symposium.
[7]W. Van Der Knapp (1959). Trans. AIME 126, 179.

B.6 | Use of Shear Waves in the Offshore Environment

M. D. COCHRAN

Present-day seismic exploration makes use of only part of the available information. Only the compressional waves (P-waves) are analyzed. All other energy is considered noise, and great efforts are made to eliminate it. Part of this so-called noise does, however, contain additional independent information that might yield additional parameters to be used in defining the geology of the subsurface. A P-wave impinging on an interface between elastic media will not only produce reflected and transmitted P-waves, it will also produce reflected and transmitted shear waves (S-waves). While the P-wave data yields information as to the behavior of the media under a compressional stress, the S-wave data will give information as to the behavior of the media under shear stress. By combining these two sets of data, one should be able to define fully the elastic behavior of the media and, hence, be able to better predict the lithology, fluid content, and porosity of the rocks in the subsurface. In addition, in certain engineering applications—drilling and foundation stability studies—the shear properties of the medium are more important than the compressional properties.

In the late 1950's and early 1960's there was a great deal of interest in use of shear waves for oil exploration. It was hoped that, as they traveled at about one half the speed of P-waves, S-waves would allow one to obtain higher resolution (i.e., at a given frequency, the S-waves would have a shorter wavelength and, hence, be able to resolve thin-

ner beds). Unfortunately this was not the case; because energy lost as a result of anelastic attenuation in the subsurface is inversely proportional to wavelength, the resolution was no better than that for P-waves. Because of additional difficulties in generating, recording, and analyzing S-waves, research in this area was dropped (Cherry and Waters, 1968; Erickson *et al.*, 1968).

In the last ten years, technological improvements have changed the picture. Seismic data are no longer used only to obtain the geometrical configuration (i.e., structure) of the subsurface; they are now being used to predict the lithology in the subsurface—rock type, fluid content, porosity. This change in focus has rekindled interest in S-waves, as they can supply additional parameters for use in lithologic prediction. One should note that most of the information as to the nature of the interior of the earth is based on shear-wave data from earthquakes. An indication of the renewed interest in S-waves is Conoco's proposed shear-wave group shoot, which was begun in early 1977; most major oil companies have joined.

From P-wave seismic data one can estimate the P-wave velocity distribution in the subsurface and hence part of the elastic behavior of the material in question, i.e., its behavior in compressional stress. The P-wave velocity data do not, however, give any insight as to the behavior of the material under shear stress; this is given by the shear-wave velocity of the medium. While the P-wave velocity can be used to discriminate lithologies, it is

273

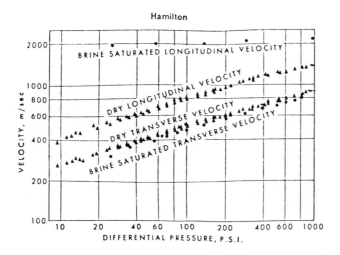

FIGURE B.6.1 Shear velocities of unconsolidated sands—brine saturated and dry. (After Hamilton, 1976.)

convert some of its energy at the sea bottom to S-waves; this wave energy will be transmitted down into the substrate, be reflected back to the sea bottom, converted to P-waves, and recorded. In addition, at each reflecting interface in the substrate, the impinging P-waves will also convert some of their energy to S-waves, both reflected and transmitted. Because of this, much of the energy recorded as P-waves at the surface even in the marine environment are converted waves (i.e., for part of their reflection path they traveled as S-waves) and contain information as to the shear properties of the subsurface. Heretofore, we have considered these converted waves to be noise and have assiduously attempted to attenuate them. One can, however, as Tatham and Stoffa (1976) point out, design the recording and processing to emphasize these waves and extract the relevant information—S-wave interval velocity and S-wave reflection amplitudes—for use in lithologic prediction.

In summary, recent experimental work has suggested that by using S-wave data to augment P-wave data one can improve one's prediction of the subsurface lithology, type fluid content, and porosity. While the generation and recording of primary shear waves are relatively easy on land

highly dependent on the fluid content of the rock in question (Figure B.6.1); in fact this is the basis of the bright-spot technique, i.e., presence of hydrocarbons lowers the P-wave velocity. As a result, it is often difficult to separate lithologic changes from fluid changes. The shear velocity on the other hand, while dependent on lithology, has the added advantage of being relatively independent of fluid content (Figure B.6.1). As such, in addition to giving one another lithologically dependent parameter with which to discriminate lithology (and, hence, improve our discrimination by a factor on the order of 40 percent), it will allow one to predict both lithologic changes and fluid changes. While the experimental evidence is not conclusive (extensive measurements have been made only on sandstones and carbonates), it bears this out (Figure B.6.2); the Vp/Vs ratio appears to be fairly constant for a given lithology. More measurements on other sedimentary rocks of interest (salt, rock, and shale) are, however, required.

In addition, to describe fully the elastic behavior of the substrate for engineering studies (e.g., earthquake stability of structures and drilling) the shear properties of the substrate are of paramount importance both onshore and offshore. The drillability of the substrate is also dependent on the shear properties of media. These shear properties would be easily estimable from shear-wave velocity data.

While it is reasonably easy to generate and record S-waves on land, the problem is more difficult offshore. Water does not transmit shear stress; hence one cannot generate and record pure shear waves in a marine seismic survey. One can, however, record waves that have traveled through to substrate as S-waves. The P-waves will convert part of their energy to S-waves upon reflection and transmission at an interface between elastic media. Therefore, a P-wave generated in the water will

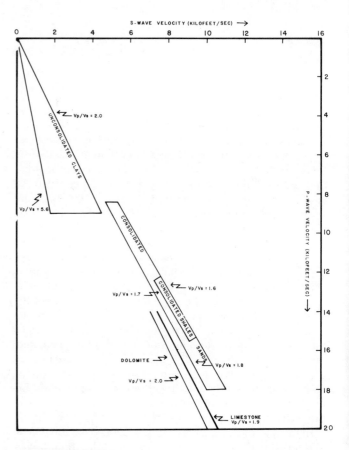

FIGURE B.6.2 Vp/Vs ratios for various common sedimentary rocks (based on a collation of various laboratory and field measurements). (Courtesy of Gulf Science and Technology Company.)

and impossible in the marine environment, one can obtain shear-wave information from marine seismic surveys by making use of converted waves. In the past, these converted waves have been considered noise and attenuated in recording and processing; by changing the recording and processing parameters one can emphasize these converted waves and extract the S-wave velocities and reflection amplitudes. The improved lithologic prediction should allow one to better reconstruct the geological history of the continental shelves and improve our ability to estimate their economic resources.

In addition, improving lithologic prediction, the knowledge of the shear properties of the substrate should aid in engineering studies for onshore structures and in improving drilling techniques.

REFERENCES

Cherry, J. T., and K. H. Waters (1968). Shear wave recording using continuous signal methods, Part I—early development, *Geophysics 33*, 229–239.

Erickson, E. L., D. E. Miller, and K. H. Waters (1968). Shear wave recording using continuous signal methods, Part II—later experimentation, *Geophysics 33*, 240–254.

Hamilton, E. L. (1976). Shear wave velocity versus depth in marine sediments: A review, *Geophysics 41*, 985–996.

Tatham, R. H., and P. L. Stoffa (1976). Vp/Vs—A potential hydrocarbon indicator, *Geophysics 41*, 837–849.

APPENDIX

B.7 | Well Logging

L. L. RAYMER

Most present-day well logs are run in boreholes drilled for the purpose of finding oil or gas in commercial amounts. From the dozens of logging services now available, the most appropriate ones are selected by the operator to furnish the information needed about the subsurface formations penetrated by the borehole. The well logs and wellsite interpretations allow the oil company to make rapid and reliable short-range completion decisions. However, the detection of potential oil and gas production is only one aspect of log interpretation. Well logs and the answers that can be obtained from them through appropriate interpretation techniques are often the most efficient means of reservoir evaluation.

The evaluation of a reservoir requires knowledge of such things as the structural relationship of the strata from one well to another, the thicknesses of the various strata, the identification of permeable beds, the pore volumes of the rock, the type of fluids contained within the pores, the producibility of these fluids, and the reservoir pressure. Although these parameters cannot yet be measured directly, they can be obtained by quantitative log interpretation, that is, the transformation of log data (resistivity, density, sonic transit time, etc.) into the desired formation parameters through the use of appropriate petrophysical relationships.

For rapid and reliable short-range decisions, numerous logs and wellsite (or quick-look) interpretation techniques are available. Each is tailored for a particular set of drilling and formation conditions. They are primarily designed for manual use to obtain lithology identification, average porosity values, hydrocarbon detection, average water saturation values, and oil–water and/or gas–oil contacts.

However, when using log data for reservoir evaluation on a foot-by-foot basis, the amount of data to be processed, as well as the speed required, necessitates the use of computer-oriented techniques.

In this discussion we will be able to touch on only a few of the uses of well logs and interpretation techniques. Anyone who wishes to go into the subject in greater detail is referred to the bibliography listed at the end of this paper.

WELLSITE TECHNIQUES

For decisions that usually are made at the wellsite, information needed includes porosity, lithology, and fluid content of the formation.

POROSITY–LITHOLOGY

Porosity from well logs is usually determined using sonic, density, and/or neutron logs. The readings of these logs, however, depend not only on porosity but also on formation lithology and the type of fluid in the pore space.

276

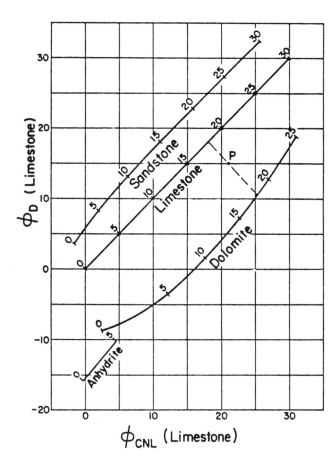

FIGURE B.7.1 Neutron–density crossplot, showing "standard" matrix lines. (Courtesy of Schlumberger-Doll Research Center.)

For more complex lithologies, up to four unknowns (including porosity as an unknown) can frequently be resolved by use of all three porosity logging devices.

Crossplots are convenient to display both porosity and lithology information when two porosity logs are available. Figure B.7.1 is an example in which density and neutron porosities are crossplotted on linear scales. When the matrix lithology is a binary mixture (e.g., quartz–lime or lime–dolomite), the point plotted from the log readings will fall between the corresponding lithology lines, thus allowing one to determine porosity as well as the proportions of the two minerals making up the rock matrix.

When the makeup of the formation is more complex, indications of lithology, gas, and secondary porosity can often be seen by use of the MID (matrix identification) plot. In this technique, "apparent matrix parameters," $(\rho_{ma})_a$ and $(\Delta t_{ma})_a$, are obtained by use of appropriate neutron–density or sonic–neutron crossplots such as those shown as Figures B.7.2 and B.7.3. These values are then used to enter the MID-plot chart of Figure B.7.4. On this example chart, the most common matrix minerals

Thus, when the matrix-lithology parameters (transit time of the matrix material, Δt_{ma}, the bulk density of the matrix material, ρ_{ma}, and the neutron index for the matrix material, ϕ_{Nma}) are known, correct porosity values can be derived from these logs in clean water-filled formations. Under these conditions a single log, either neutron, density, or (if there is no secondary porosity) sonic, should suffice to determine porosity, ϕ. When the matrix lithology is unknown or consists of two or more minerals in unknown proportions, accurate porosity determination becomes more difficult. The interpretation is further complicated when the pore fluids in the portion of the formation investigated by the tools differ appreciably from water.

Sonic, density, and neutron logs respond differently to the different matrix compositions and to the presence of gas and light oil. Thus, combinations of these tools furnish more information about the formation and its contents than can be derived from a single log. For example, if a formation consists of only two known minerals, a pair of porosity logs, one of which is usually a neutron, will suffice to determine the proportion of the minerals in the rock matrix and to determine a better value of porosity.

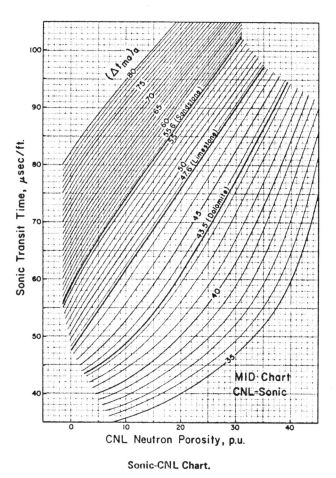

Sonic-CNL Chart.

FIGURE B.7.2 Neutron–sonic MID chart, for determination of "apparent" matrix transit time $(\Delta t_{ma})_a$. (Courtesy of Schlumberger-Doll Research Center.)

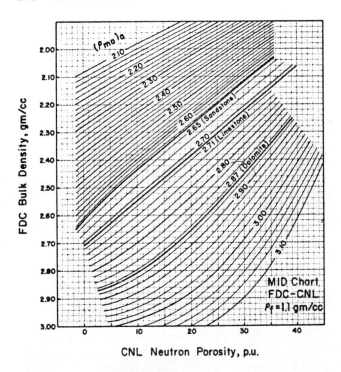

FIGURE B.7.3 Neutron–density MID chart, for determination of "apparent" matrix bulk density $(\rho_{ma})_a$. (Courtesy of Schlumberger-Doll Research Center.)

also related to porosity and can be defined as $F = 1/\phi^m$, where the cementation exponent m can usually be taken as equal to 2. Therefore, F may be obtained by the use of the appropriate $F-\phi$ relationship, taking the value of porosity from neutron, density, or sonic logs.

Other wellsite procedures can be classified according to the type of logs used, as:

1. Methods comparing a deep-investigation resistivity log with a porosity log. These can provide values of R_w and apparent water saturation, S_{wa}.
2. Methods comparing resistivity logs of different depths of investigation. These provide S_{wa} values and indications of hydrocarbon movability.
3. A method comparing a resistivity ratio (from resistivity logs of different depths of investigation) with the SP. This method provides an indication of movable hydrocarbons that is little affected by shaliness or changes in R_w.
4. Methods of comparing two or three resistivity logs with a porosity log. These provide movable oil plots.

To illustrate, consider the interpretation of a clean-sand formation, containing oil and saltwater, that was drilled using a mud that is less saline than the formation water. For this sample case, an adequate logging program should include deep resistivity, shallow resistivity, spontaneous-potential (SP), gamma-ray (GR), and sonic transit-time measurements. Since we will be looking at a number

(quartz, calcite, dolomite, anhydrite) plot at the positions shown. Mixtures of these would plot at locations between the corresponding pure mineral points. And other minerals of known characteristics can be plotted.

Lithologic trends may be seen by plotting many levels over a zone and observing how they are grouped on the chart with respect to the mineral points. A typical example is shown as Figure B.7.5. From the points plotted, it is obvious that the zone is primarily a carbonate with secondary porosity, but it also contains some salt and anhydrite.

SATURATION AND HYDROCARBON MOVABILITY

A wide diversity of procedures has been developed to (1) show which permeable zones have potential interest; (2) derive a fairly accurate value of water saturation, S_w (and thereby hydrocarbon saturation); and (3) give some valuable information about hydrocarbon movability.

Saturation evaluation is based eventually in one form or another on the Archie saturation formula, which for clean formations is $S_w{}^n = FR_w/R_t$. S_w is the water saturation of the formation, the saturation exponent n is nearly always taken equal to 2, F is the formation resistivity factor, R_w is the resistivity of the formation water, and R_t is the resistivity of the virgin formation. The formation factor, F, is defined as R_o/R_w, when R_o is the resistivity of the formation 100 percent saturated with water of resistivity R_w. F is

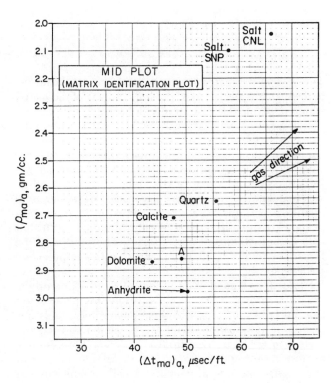

FIGURE B.7.4 MID plot, for lithology identification. (Courtesy of Schlumberger-Doll Research Center.)

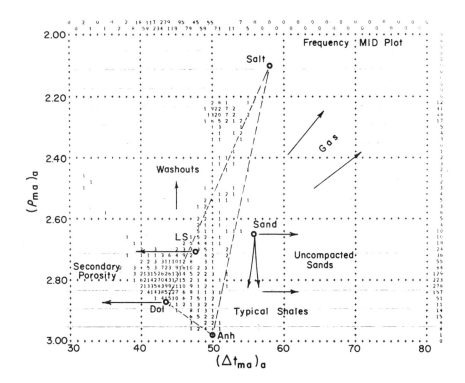

FIGURE B.7.5 Example of a machine-made MID plot, showing the trends produced by various formation and borehole factors. (Courtesy of Schlumberger-Doll Research Center.)

of zones, it is convenient to use a resistivity–porosity crossplot (Figure B.7.6).

For our clean-sand formation, Δt values from the sonic have been plotted versus corresponding values of R_t from the deep-investigating resistivity log. Values of S_w are readily obtained for all the zones plotted.

Logarithmic scaling of resistivity logs is also useful for wellsite quick-look presentations, because of the properties of logarithms.

Instead of computing an apparent water saturation (S_{wa}) from deep resistivity and porosity data, the formation factor, F (obtained from the porosity log), can be recorded with logarithmic scaling, directly on the resistivity log. S_{wa} would then be determined by the separation on logarithmic scale between F_R (F from resistivity-log values) and F (from the porosity log) as follows:

F_R is defined as R_{deep}/R_w; thus log F_R = log R_{deep} − log R_w. Then the log F_R curve is found by shifting the logarithmic R_{deep} curve by a distance R_w. This is shown on Figure B.7.7. The meaning of F_R in terms of S_w is seen by replacing R_t/R_w (by use of Archie's relation) to get $F_R = F/S_w^2$, or, $2 \log S_w = \log F - \log F_R$. Thus, S_{wa} (apparent water saturation) can be found from the separation on the log F curve from the log F_R curve, using the scaler shown.

The interpretation procedures touched on thus far have been mainly for use in clean, nonshaly formations and for formations whose lithology is relatively simple. As the complexity increases, the interpretation becomes more difficult. With the availability of computers, more sophis-

ticated interpretation techniques are possible and have been developed for shaly sands and other complex lithologies.

COMPREHENSIVE SHALY-SAND INTERPRETATION

The effect of shaliness on a log reading depends on the proportion of shale, the physical properties of the shale, and the way it is distributed in the formation. Shaly material may be distributed in formations in three possible ways:

1. Shaliness exists in the form of *laminae*, between which are layers of sand. Laminar shale does not affect the porosity or permeability of the sand streaks.

2. Shale may exist as grains or nodules in the formation matrix. This matrix shale is termed *structural* shale and is considered to have properties similar to those of laminar shale.

3. The shaly material may be dispersed throughout the sand, partially filling the intergranular interstices. The *dispersed shale* may be in the form of accumulations adhering to or coating the sand grains, or it may partially fill the smaller pore channels. Dispersed shale in the pores markedly reduces the permeability of the formation.

All three forms may occur simultaneously in the same formation.

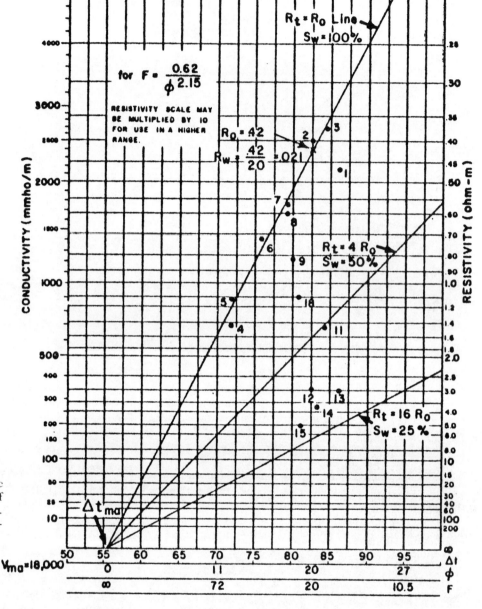

FIGURE B.7.6 Resistivity–sonic crossplot, showing typical scatter of oil- and water-bearing intervals. (Courtesy of Schlumberger-Doll Research Center.)

The responses of the radioactivity tools (gamma ray, neutron, density) are not affected by the way the shale is distributed in the formation (whether in laminated, dispersed, or structural form). However, the responses of other logs (resistivity, sonic, SP) are variously affected by the shale distribution.

One shaly-sand interpretation technique uses a silty-shaly-sand model, in which the shales are assumed to be of the laminated, dispersed, and structural types. The basic model is suggested by the groupings of the plotted points shown in Figure B.7.8. This figure is a typical neutron–density frequency crossplot through an interval

of a sand-shale sequence. Most of the data belong to two groups: Group A identified as sands and shaly sands and Group B identified as shales.

The spread of the shale points in Group B along the line from Point Q through Point Sh_o and Point Cl is explained by the fact that shales are considered to be essentially mixtures of clay minerals, water, and silt in various proportions. Thus Point Cl corresponds to shales that are relatively silt-free, while Point Sh_o corresponds to shales containing a maximum amount of silt.

The shaly sands in Group A grade from shales on Line $\overline{Sh_oCl}$ to clean sands at Point Sd. The shale in the shaly

Well Logging 281

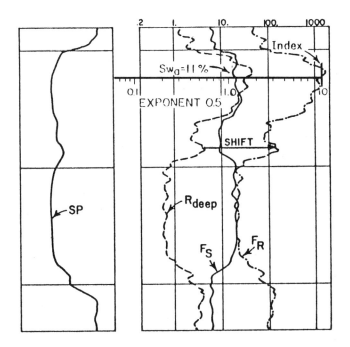

FIGURE B.7.7 Use of logarithmic scaling on resistivity log, to simplify log calculations. (Courtesy of Schlumberger-Doll Research Center.)

sands may be distributed in various ways. When all the shale present is laminar shale, the points will fall on the $Sd–Sh_o$ line. Dispersed shale causes the points to plot to the left of the line. Structural shale causes the points to plot to the right of the line.

A computer program has been developed to perform the required processing of the method. The technique provides a complete continuous analysis of both clean and shaly sands. The results are presented as a computed log (Figure B.7.9) and as a listing (Figure B.7.10). Outputs include the following:

1. In Track 1, the shale fraction (V_{sh}).
2. In Track 2, a hydrocarbon analysis consisting of the following curves: water saturation (S_w), residual hydrocarbon volume ($\phi \cdot S_{hr}$), and residual hydrocarbon weight ($\phi \cdot S_{hr} \cdot \rho_h$). (Hydrocarbon type (gas or oil) is usually identifiable by comparison of hydrocarbon volume and hydrocarbon weight.)
3. In Tracks 2 and 3, a differential caliper (caliper minus bit size).
4. In Track 3, a porosity analysis showing: porosity (ϕ), "core porosity" (ϕ_{core}), bulk-volume water in the invaded zone (ϕS_{xo}), shown by the outline of the left edge of the dotted area, and bulk-volume water in the noninvaded zone (ϕS_w), shown by the outline of the left edge of the

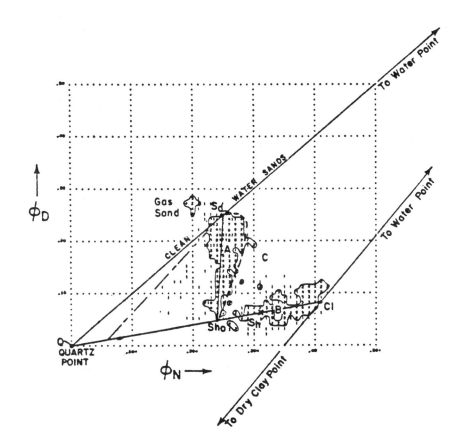

FIGURE B.7.8 Neutron–density cross-plot, showing development of shaly-sand interpretation model. (Courtesy of Schlumberger-Doll Research Center.)

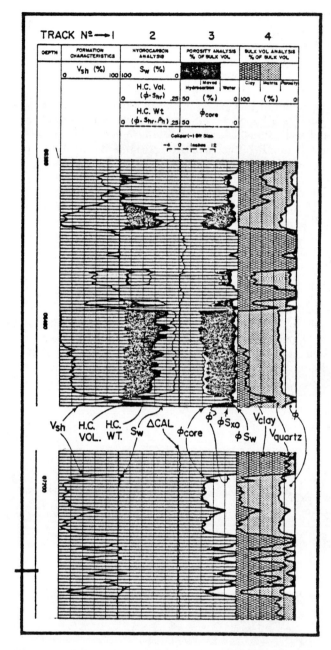

FIGURE B.7.9 Example answer log produced by the shaly-sand computer program. (Courtesy of Schlumberger-Doll Research Center.)

sents the unmoved (residual) hydrocarbon in the invaded zone.

5. In Track 4, a bulk-volume analysis showing the clay fraction (V_{clay}), matrix-solids fraction, and the porosity (ϕ). The matrix-solids fraction includes both the actual matrix and the nonclay materials (e.g., silt) in the shales.

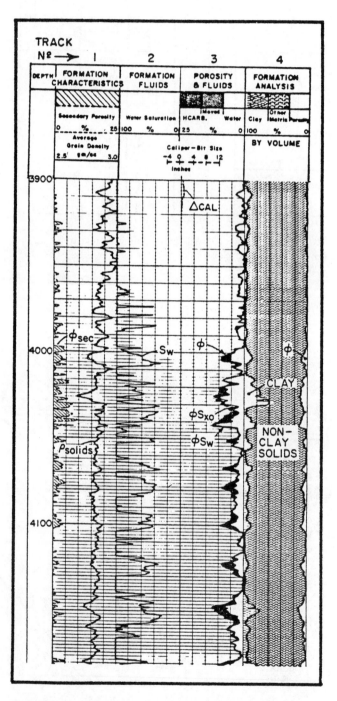

FIGURE B.7.10 Example answer log produced by the complex-lithology program. (Courtesy of Schlumberger-Doll Research Center.)

white area. "Core porosity" is a computed value of porosity, which usually represents more closely the value of porosity from core analysis; it is obtained by adding to ϕ the bulk volume of the water bound to those clays that are not laminated. The dotted area in Figure B.7.9 is the difference, $\phi S_{xo} - \phi S_w$, which represents "moved hydrocarbon" (hydrocarbon displaced by mud invasion). The darkened area is the difference, $\phi - \phi S_{xo}$, which repre-

Other outputs shown on the listing, Table B.7.1, are permeability index, hydrocarbon density (when computable), cumulative porosity feet, and cumulative hydrocarbon feet.

INTERPRETATION IN COMPLEX LITHOLOGIES

A technique has also been developed that is a general interpretation method for complex lithologies. In concept, the technique uses a neutron–density crossplot, as discussed previously. However, the procedure is much more general and includes, among other things, techniques to handle the presence of shale and hydrocarbons and other nonstandard but well-defined minerals.

A computer program is available to perform the required processing of the method. The results are presented as a continuous computed log shown as Figure B.7.10. Outputs shown in the figure include:

1. In Track 1, the average density of the rock matrix (including dry clay), ρ_{Ga}. From this, lithology can be inferred. And if a sonic log is available, a secondary porosity index is presented.
2. In Track 2, the hydrocarbon–water saturations analysis.
3. In Tracks 2 and 3, a differential caliper (caliper minus bit size).
4. In Track 3, a porosity–fluid analysis showing porosity (ϕ), bulk-volume water in the invaded zone (ϕS_{xo}) if an R_{xo} measurement is available, and bulk-volume water in the noninvaded zone (ϕS_w). Total hydrocarbons are indicated by the separation, $\phi - \phi S_w$, moved hydrocarbons by the separation, $\phi S_{xo} - \phi S_w$, and water by ϕS_w.
5. In Track 4, a formation analysis showing the clay fraction (V_{clay}), the matrix-solids fraction, and the porosity (ϕ).

COMPUTERIZED LOGGING

The techniques of well logging have grown up around analog measurements, recorded on analog equipment. As the number of log types grew, each type tended to develop into a modular system using its own unique hardware. The only commonality among systems was in the hoisting machinery and cable and in the recorder.

As the advantages of digital data-processing methods emerged, a need became evident for a new, integral logging system. The objectives were defined:

Reduce the operator burden caused by dozens of complex, dissimilar operating procedures.
Take full advantage of digital methods for log-data acquisition, storage, processing, transmission, and presentation.
Remove the logging-speed limitation imposed by analog recording.
Provide increased wellsite log computation capability.

In Schlumberger, the consummation of this effort is a completely new logging system called CSU. Figure B.7.11 is a functional block diagram, minus many redundant functions and special features that contribute to reliability.

The CSU system is built around a high-speed minicomputer, which uses preprogrammed operating routines, stored on magnetic tape, to simplify and expedite logging

TABLE B.7.1 Interpretive Output of Well-Log Analyses

Depth (ft)	Permeability Index (md)	Porosity (%)	Water Saturation (%)	Hydrocarbon Density (g/cc)	Clay (%)	Cumulative Porosity (ft)	Cumulative Hydrocarbon (ft)
6321.0	0.0	8.5	96		42	217.19	97.06
6322.0	0.6	9.4	67		38	217.10	97.04
6323.0	0.1	9.4	96		39	217.00	97.02
6324.0	0.1	9.6	96		46	216.91	97.01
6326.0	0.1	9.1	94		44	216.75	97.01
6327.0	5	11.5	34		38	216.65	96.98
6328.0	100	17.9	21		26	216.52	96.89
6329.0	300	22.0	21	0.1	28	216.32	96.73
6330.0	100	21.8	28		19	216.10	96.56
6331.0	200	20.2	22		11	215.89	96.40
6332.0	1000	22.8	11		6	215.68	96.24
6333.0	3000	26.9	10	0.2	4	215.44	96.02
6334.0	3000	29.6	12	0.2	0	215.16	95.77
6335.0	4000	28.0	9	0.2	0	214.87	95.51
6336.0	3000	26.6	9		1	214.59	95.26

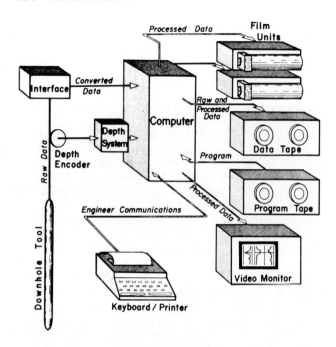

FIGURE B.7.11   Functional schematic diagram of the CSU logging system. (Courtesy of Schlumberger-Doll Research Center.)

operations (CSU is a trademark of the Schlumberger-Doll Research Center).

The various signal conditioning and processing functions are accomplished in common for all tools by a general electronics unit. Only the minimum possible circuitry, contained in a small semipermanently mounted module, is specific to a given service. Careful programming allows the operator to interact with or override the programs by means of the keyboard printer. A storage CRT provides a display of selected log intervals, data tables, or parameter options as desired. And log-analysis programs

permit on-site log computations of unprecedented speed and thoroughness.

The more important advantages of the CSU system are the following:

Reduction of rig time needed for logging.
More accurate and consistent logs.
Wellsite log playback on different scales, options, etc., using no rig time.
Real-time video log display.
Wellsite computed logs.

BIBLIOGRAPHY

1. Archie, G. E. (1942). "The electrical resistivity log as an aid in determining some reservoir characteristics," *Petrol. Technol.* 5(1), Jan.
2. Doll, H. G. (1948). "The SP log: theoretical analysis and principles of interpretation," *Petrol. Technol. 11,* Sept.
3. Poupon, A., C. Clavier, J. Dumanoir, R. Gaymard, and A. Misk (1970). "Log analysis of sand-shale sequences—a systematic approach," *J. Petrol. Technol.* July.
4. Poupon, A., W. R. Hoyle, and A. W. Schmidt (1971). "Log analysis in formations with complex lithologies," *J. Petrol. Technol.* Aug.
5. Schlumberger, C., and M. Schlumberger (1934). "Electrical coring—a method of determining bottom-hole data by electrical measurements," *Transactions AIME 110.*
6. *Schlumberger Log Interpretation: Vol. I—Principles,* Schlumberger Limited, New York, 1972.
7. *Schlumberger Log Interpretation: Vol. II—Applications,* Schlumberger Limited, New York, 1974.
8. *Schlumberger Log Interpretation Charts,* Schlumberger Limited, New York, 1977.
9. Tixier, M. P. (1949). "Electric log analysis in the Rocky Mountains," *Oil and Gas J.* June 23.
10. Wyllie, M. R. J. (1957). *The Fundamentals of Electric Log Interpretation,* Academic Press, New York, second ed.

APPENDIX C

Research Vessel *Hollis Hedberg*: Exploration Capabilities

GULF OIL CORPORATION

INTRODUCTION

Although Gulf Oil Corporation has been exploring for and producing oil from water-covered areas for nearly 50 years, it was not until late 1966 that Gulf's management approved a long-range research project designed to improve exploration techniques in the world's oceans.

Providing the impetus for this new thrust in exploration was Hollis D. Hedberg, formerly Vice President for exploration and currently Gulf Exploration Advisor. The principal component of the new program, as envisioned by Dr. Hedberg, was a ship-mounted exploration system capable of generating technically optimal data for making exploration decisions.

The program began to unfold in early 1967, when Gulf signed a contract to operate what was said to be the world's largest private oceanographic ship. A converted fishing trawler, the 220-foot-long ship was christened in October 1967 as the R/V *Gulfrex* (representing Gulf Research and Exploration) and set sail from West Palm Beach, Florida, for South America waters.

Foremost in the minds of *Gulfrex* planners was the creation of a completely mobile, self-contained exploration system, capable of traveling to all marine areas of the world and, in addition, serving as a laboratory for both basic and development research in marine geology and exploration systems (including data gathering, processing, and interpretation).

When the christening of the first research vessel was held in 1967, it was hailed as the first endeavor of its kind ever undertaken by a private corporation. The *Gulfrex* had the distinction of being equipped with the first commercial satellite navigation system and was the first fully integrated multisensor marine geophysical gathering and processing vessel.

During its seven-year history, the *Gulfrex* program has led to the discovery of several successful wells and has brought to the attention of Gulf and host governments numerous new potentially petroliferous basins. Its usefulness has not been limited to Gulf scientists alone, as data from the *Gulfrex* program have aided Joint Oceanographic Institutions Deep Earth Sampling (JOIDES) program in its Deep Sea Drilling Project, provided computer processing techniques to Lamont-Doherty Geological Observatory, and advised the National Science Foundation on ways to improve the effectiveness of government oceanographic surveys. Data gained through the *Gulfrex* program have also been presented to various government agencies and have been instrumental in producing a more realistic government stance as to the desired national jurisdiction over subsea mineral resources.

Based on the overwhelming success of the *Gulfrex* and

because of the impending worldwide petroleum shortage, Gulf's management decided in 1972 to construct an entirely new research vessel. As a result of this decision, the R/V *Hollis Hedberg* was launched in May 1974. It represents a $6 million investment. The vessel was designed from the keel up to be the most modern marine geophysical research platform in the world and is the only vessel of its kind equipped with a complete onboard seismic data-processing capability.

The R/V *Hollis Hedberg* is patterned after the R/V *Gulfrex*, but several desired improvements and modifications have been incorporated into its design. It is wider and more stable with better handling characteristics. Hull reinforcement permits operation in higher-latitude areas. Laboratory space has been more than doubled in size, and working space on the fantail has also increased. For increased low-frequency energy content and concomitant deeper penetration, air guns are used on the R/V *Hollis Hedberg* instead of the gas exploders as on the R/V *Gulfrex*. The magnetic gradiometer should greatly improve magnetic interpretation by enabling the removal of the diurnal variations in the earth's field. The addition of Loran C and provision for other radio-navigation instruments to be readily integrated into the positioning system increase the accuracy of the navigation system, which before was regarded as the best commercially available. An improved and expanded hydrocarbon analyzer permits digital logging and a more sophisticated geochemical interpretation.

The R/V *Hollis Hedberg* is a self-contained marine research laboratory, capable of operating 24 hours a day for periods up to 60 days. The vessel is operated by the Exploration and Production Division of Gulf Research & Development Company, under the direction of Gulf Oil Exploration and Production Company.

Since the inception of the *Gulfrex* program, the ships have logged more than 300,000 nautical miles and have collected more than 175,000 nautical miles of multichannel seismic data.

SYSTEM CAPABILITY

SHIP CAPABILITY

Figure C.1 is a photograph of the R/V *Hollis Hedberg* as it departed Vancouver, Canada, in May 1974 on its maiden cruise. Table C.1 lists operational and exploration capabilities. Vessel length is 61.6 m (202 ft), beam is 12.8 m (42 ft), and its gross tonnage is 1360 tons. The ship has a cruising range of 9000 miles with an endurance of up to 60 days.

The main laboratory has over 100 sq m (1100 sq ft) of area. Additional work space is also provided for data analysis, preparation of maps (Figure C.2), scientific equipment storage, etc.

Figure C.3 is a photograph of the ship's wheel house, and Figure C.4 shows the radio room equipped with the latest, most modern radio communication equipment.

Provision has been made for a crew of 16 and a technical staff of up to 34. A Gulf Chief Scientist and at least two Gulf geophysicists are onboard at all times to provide almost real-time interpretation of the data. In addition, room is available for visiting scientists.

The worldwide operating range of R/V *Hollis Hedberg* is shown in Figure C.5.

DATA ACQUISITION AND PROCESSING

Figure C.6 schematically illustrates the basic onfiguration of the vessel and the exploration systems. A block diagram of the complete data-acquisition and -processing system is shown in Figure C.7. Table C.2 gives the manufacturer and model number of the scientific instruments.

Seismic Energy Source and Sensors

The seismic energy source consists of two air guns, whose volume can be varied from 13,100 to 32,775 cu cm (800 to 2000 cu in.), which operate at a maximum pressure of

FIGURE C.1 Research Vessel *Hollis Hedberg*.

FIGURE C.2 Data interpretation office.

13.78 MPa (2000 psi). Four 8.5 cu m/min (300 cfm) compressors supply sufficient high-pressure air to fire the guns at a maximum 8-sec rate.

One of these air guns (at 32,775-cu-cm volume) generates a peak pressure of approximately 7 bar-meters and at a 9.1-m (30-ft) tow depth is equivalent to 0.8 kg (1¾ lb) of 60 percent dynamite. Figure C.8 is a photograph of the R/V *Hollis Hedberg* showing the bubbles rising from the air guns. Figure C.9 shows the air guns being launched.

The two-mile-long seismic streamer consists of 48 sections each 67 m (220 ft) in length containing 26 acceleration cancelling hydrophones per group in a tapered array with a frequency passband from 4.5 to over 200 Hz. Compliant decoupler sections are used in front of and behind the active sections to mechanically isolate the cable from the ship and attenuate towing noise. Up to 12 automatic depth controllable birds are used on the streamer to establish any desired towing depth down to 30.5 m (100 ft).

FIGURE C.3 Wheel house.

FIGURE C.4 Radio room.

A buoy is employed at the end of the cable to determine the average azimuth of the streamer with respect to the ship's course.

For high resolution of a shallow geological section, the two streamer sections closest to the ship can be replaced by six groups of shorter length. Each of these short sections contains 20 acceleration cancelling hydrophones and is 10.7 m (35 ft) in length. The recorded data can be used either for single-channel coverage or 200 percent coverage. A change back to 48-channel recording can be made with a minimum of operational difficulty any time the recording of high-resolution data is not required.

Refraction seismology is accomplished through use of expendable sonobuoys with the data telemetered to a receiver onboard the vessel. A range of up to 26 km (14 nm) is possible using two air guns. Using the streamer, reflection data are recorded concurrently with the refraction information. The sonobuoy data also provide normal moveout information for use in one of the methods available for determining average reflection velocities.

Seismic Amplifiers and Digital Magnetic Tape Recorders

A Globe Universal Sciences (GUS) seismic system employing binary gain ranging amplifiers is used in digital recording. In Figure C.10, the GUS electronics are shown in the middle. The high-density digital recorder (HDDR) is shown on the right. This tape unit records data at 8000 bits per inch of tape in a unique 14-track sequential format. Only one track is recorded at a time and thus avoids the skew problems that obviously limit the packing density of 9-track tape recorders. An oscillograph camera (shown on the left in Figure C.10) presents raw seismic data from each of the 48 channels and is also used to monitor the stacked (up to 48-fold) data and trace gathers.

TABLE C.1 Data Sheet—R/V Hollis Hedberg[a]

Vessel Characteristics

Length	61.57 m (202 ft)
Beam	12.80 m (42 ft)
Mean draft	4.27 m (14 ft)
Gross tonnage	1360.20 tons
Registered tonnage	893.91 tons
Engines	2-GM 1454-kW (1950 SHP), 800 rpm
Generators	2-250 kW
	2-40 kW (regulated)
Laboratory air conditioning	2-20 ton, air cooled
Main laboratory space	106.1 sq m (1142 sq ft)
Sewage disposal unit	Northern Purification Limited
Twin props and rudders	
Autopilot	
Ice capability	ABS-Class 1A
Weather map	Facsimile receiver and recorder

Operational Characteristics

Cruises	Up to 60 days
Range	9000 miles
Speed	Up to 14 knots
Personnel (Crew)	50
(Scientists and technicians)	16
	Up to 34
Navigation equipment	Radar (2 units)
	Celestial
	EM Log
	Satellite[b]
	Doppler Sonar[b]
	Inertial Platform[b]
	Forward Scan Sonar
	Gyro Compass[b]
	Loran C with Atomic Frequency Standard
	Shoran
	Lorac

Exploration Capabilities

Two Geometrics magnetometers
LaCoste-Romberg stable platform gravimeter
LaCoste-Romberg portable gravimeter
Fathometers
Full ocean depth EDO UQN-4
Shallow water (bridge and lab)
Marine seep detectors
Hull
61-m (200-ft) depth tow
183-m (600-ft) depth tow
Globe Digital Seismic System
48-Channel multidyne ACH streamer 3200 m
6-Channel high-resolution "ministreamer"
Automatic streamer depth controllers
Binary gain ranging amplifiers
Two 13,100–32,774 cu cm (800–2000 cu in) air guns
Four 8.5 cu m/min (300 cfm) compressors
Multisonobuoy FM refraction seismic system
EMR 6130 data-processing system for Edit-SUM and demultiplex
48-fold stack
Trace gather—velocity control
Tape output, single and multichannel
PDP-11 communication
32K word core
Two Varian line printer plotters
Two PFR's for onboard geophysical display
PDP-11 data processor for logging potential field and navigation data
PDP-11/10 programmable data logger for logging potential, navigation, and geochemical data
Western inertial navigation control
Bottom grab sampler
Piston corer

Radio Transmitting Equipment

Make	Model		Frequencies	Power	
Marconi	Conqueror SD Main and SSB Transmitter	cw	405–525 Hz	(cw)	500 W
				(mcw)	320 W
		cw/AM	1.6–3.8 Hz (cw/mcw/R.T.)		400 W
		SSB/AM/cw	4–25 Hz (cw/R.T.)		1800 W
Marconi	Salvor III Reserve MF WT transmitter	cw	25.070–25.110 Hz (cw)		350 W
		cw/mcw	404–525 Hz		140 W
Svenska	ME-42 VHF Transmitter	FM	156.0–157.5 Hz		20 W
			156.0–156.9 Hz		
			160.0–162.1 Hz		
Marconi	Survivor Type 610 Lifeboat transceiver	cw/AM/cw	500/2182/8364		10 W
Motorola	Walkie talkies	FM	150/160 Hz		5 W

[a] Port of Registry: Georgetown, Grand Cayman, W.I.; Owner: Cayman Island Vessels Ltd.; Official Number: 356550, Radio Call Sign: ZCFS.
[b] Dedicated computer.

288

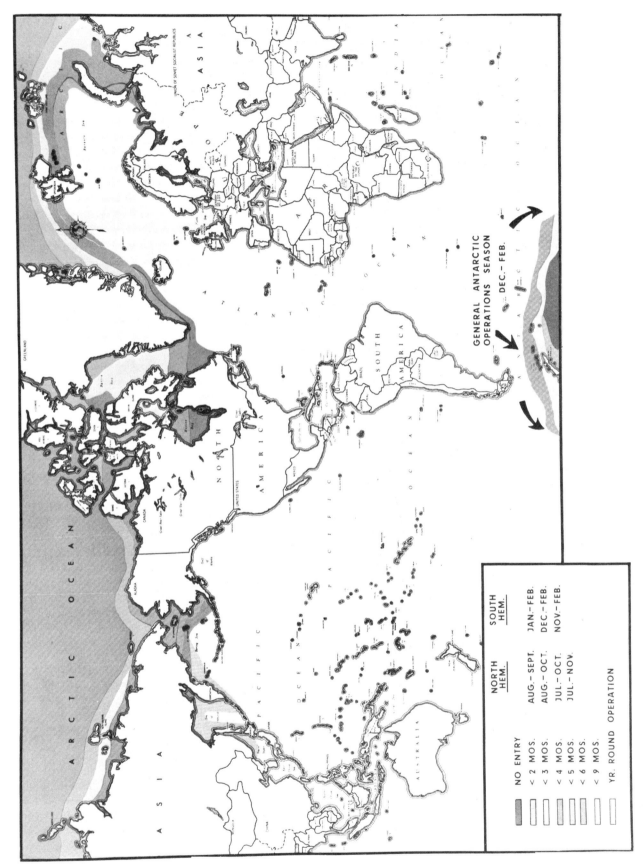

FIGURE C.5 Seasonal range of R/V *Hollis Hedberg* relative to broken-ice conditions.

289

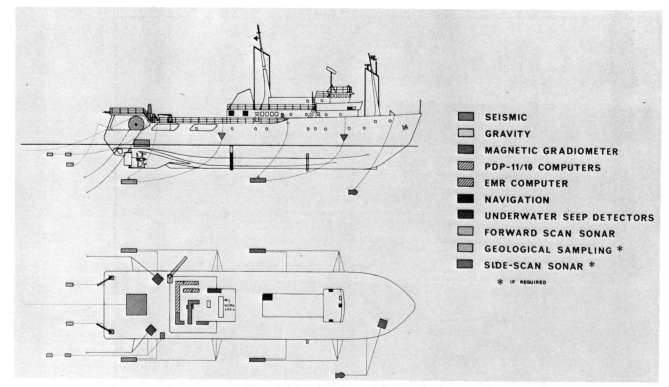

FIGURE C.6   R/V *Hollis Hedberg* exploration systems.

EMR Computer and Peripherals

The EMR computer is used to process the seismic data collected by the GUS amplifiers. It is a 16-bit word machine, which has 32,000 words of core storage. The memory cycle time is 670 nsec. The central processing unit contains two arithmetic registers, three index registers, and provides for multiple levels of indirect addressing. Peripheral devices include a card reader, teletype, an 850,000-word disk with a 17-msec access time, an 8000-bpi high-density magnetic tape drive (HDDR), and three 1600-bpi magnetic tape drives. In addition, a cathode-ray-tube display and keyboard and a switch register are used for control functions and input for velocity calculations. The switch register, card reader, and control unit are shown in the center of Figure C.11, and the HDDR unit on the right.

The processing performed by the EMR computer is limited to a 6-sec record length, exclusive of a deepwater delay and includes the following:

Demultiplexing of 48 channels of seismic data
Stacking of the near channels (up to 48-fold)
Gather of 48 traces of a common-depth-point group
Deconvolution of single-channel, stack, and/or gather data
Production of single-channel and stacked seismic profiles

Production of a single-channel/stacked seismic data tape
Production of a demultiplexed seismic data tape
Calculation of velocities based on residual normal moveout

Deconvolution and bandpass filtering may be applied to any combination of single-channel, stacked, and trace-gather data. The calculations are accomplished through use of the polynomial division method and are performed simultaneously using one operator. The operator sample interval is selectable at either 4 or 8 msec. The operator length is also selectable, up to a maximum of 64 operator points. The operator is calculated over a window beginning at water bottom and extending for one fourth of the record length.

The horizontal stacking algorithm assumes that the shot spacing is one half the group interval. The near channels are used, producing a stacked trace (up to 24-fold). Trace gathers may be either the near 24 traces or the even 24 traces of a common-depth-point group.

Velocities are calculated from residual normal moveout data and may be updated at any time. The results of the velocity calculations are displayed on the cathode-ray tube (one in laboratory, one in scientific office) and may be entered/changed through use of the switch register (middle of Figure C.11). The display includes a corrected record and a velocity versus travel-time graph.

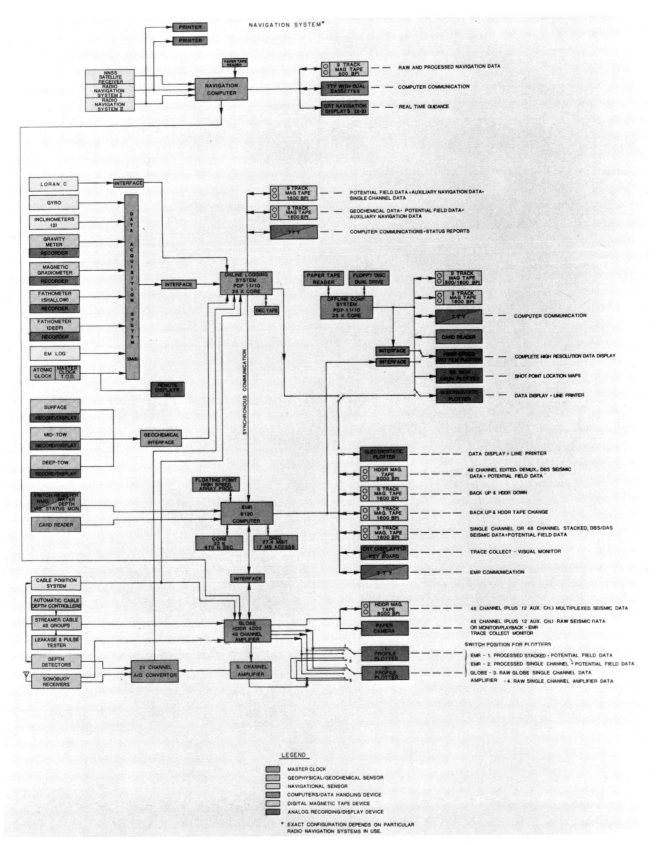

FIGURE C.7   R/V *Hollis Hedberg* instrumentation.

TABLE C.2   Manufacturers and Models of Geophysical, Geochemical, and Auxiliary Instruments Onboard the R/V *Hollis Hedberg*

Seismic System

Source: Bolt PAR Air Gun Model 800C; 2 each 13,100 to 32,775 cu cm (800 to 2000 cu in.) at 13.78 MPa (2000 psi)

Hydrophone Cable: Seismic Engineering Multidyne Streamer, 3200 m (10,500 ft) long, 48 groups each 67 m (220 ft) in length, 26 acceleration cancelling hydrophones per group, ACDC automatic depth controllable birds. Western Geophysical High-Resolution "Ministreamer," 133.8 m (439 ft) long, 6 groups each 22.3 m (73.3 ft) in length, 20 acceleration cancelling hydrophones per group (replaces nearest 2 groups of main cable)

Sonobuoy System: Aquatronics Telseis Model STR70-2F Receiver; SM44C Sonobuoys

Digital Seismic System: GUS HDDR Model 4000, 62 channels, 2-msec sampling rate

Single-Channel Amplifier: SIE Model PT 100 (Modified)

Seismic Processing Computer and Peripherals: EMR Model 6130 (32K 16-bit word core, 13.7-mbit disk), Globe HDDR Model 2000 high-density magnetic tape drive (2 each), Pertec Model 6640 magnetic tape drive (3 each), Computer Model 300 CRT data terminal, Mohawk 200 cpm card reader, Gulf R&D switch register, Texas Instruments Model 733KSR silent data terminal

Display Plotters: Varian Statos Model 31 printer/plotter (2 each), Raytheon Model UGR-196A plotter (2 each), Dresser SIE Model ERC-10 electrostatic oscillograph

Other Geophysical and Geochemical Systems

Gravity Meter (sea): LaCoste & Romberg. Sea gravity meter Model S (stable platform)

Gravity Meter (land): LaCoste & Romberg Model G geodetic meter

Magnetometer/Gradiometer: Geometrics Model G-802G marine gradiometer

Fathometer (shallow, 0–2400 ft): Simrad Model EN echosounder (38 kHz)

Fathometer (deep ocean): Edo Western Model 9057 (AN/UQN-4) sonar sounding set (12 kHz)

Geochemical System: Gulf R&D three-inlet hydrocarbon analyzer; hull inlet, 61-m (200-ft) inlet, 183-m (600-ft) inlet

Data Acquisition System: Western Geophysical DAS

Data Logging Computer and Peripherals: Digital Equipment Model PDP 11/10 (28K 16-bit word core), Pertec Model 6640 magnetic tape drive (1 each), Texas Instruments Model 733 KSR silent data terminal

Data Processing Computer and Peripherals: Digital Equipment Model PDP 11/10 (28K 16-bit word core), Pertec Model 6640 magnetic tape drive (1 each), Digital Equipment Model CR-11 card reader, Texas Instruments Model 733KSR silent data terminal

Navigation System

Western Geophysical Integrated Inertial Navigation System (WINS) consisting of
Satellite System: ITT Model 5001 Doppler satellite receiver
Doppler Sonar: Marquardt Model 2020A (300 kHz)
Velocimeter: NUS Model 1020-102
Gyrocompass: Sperry Mark 227 gyrocompass

Inertial Platform: Litton Model LN-15G

Computer and Peripherals: Litton Model LC-728 computer (12K 28-bit word core), Potter Model 152 magnetic tape drive, Digitronics Model 2540EP high-speed paper tape reader, Mohawk Data Systems Model 1200 printer

Loran C System: Teledyne Model TDL 601G Receiver, Hewlett-Packard Model HP-2100A computer (24K 16-bit word core), DI/AN 9030 Terminal

Interface for Additional Radionavigation Sensors: Western Geophysical NSI

Other Navigational Sensors:

EM Log: Sagem Model LH

Inclinometer: Gap Instruments Model S2000-208-5A level sensor

Clock and Frequency Standard

Master Clock: Chronolog Model S7121-1

Atomic Frequency Standard: Hewlett-Packard Model 5061-A cesium-beam frequency standard

Other Equipment

Weather Map Recorder: Alden Model 519 marine radiofacsimile recorder

Forward Scanning Sonar: Simrad Model SK3

Data Displays

Seismic data are displayed on four different plotters in addition to the oscillograph camera. Two Raytheon PFR plotters (Figure C.12) and two Varian Statos 31 electrostatic plotters are used to display:

FIGURE C.8   Air-gun detonation.

FIGURE C.9   Launching of air guns.

FIGURE C.11   Fiber optic camera (in foreground), EMR computer control unit, switch register, and CRT display.

Single-channel data
Single-channel deconvolved and filtered data
Stacked data
Stacked deconvolved and filtered data
Ships course, speed, potential field curves, time of day, and geochemical curves can be superimposed

The Varian Statos units use a medium for plotting that is reproducible onboard, thus allowing multiple copies of processed data to be available at any time.

The vertical scale of the Varian units are switch-

selectable to enable the seismic sections to be plotted at one half of the paper width (mini-section) or at full width.

In addition, the Varian units can be used off-line to display fully processed potential field data (e.g., two-dimensional Bouguer gravity, magnetics with the International Geomagnetic Reference Field removed), water depth, and a suite of geochemical information from two different inlets. These displays are plotted at a horizontal scale that can be selected to correspond to the seismic displays. The number of scans per trace and timing line spacing can also be varied.

Magnetics

The R/V *Hollis Hedberg* tows a magnetometer/gradiometer system consisting of two Geometrics G801

FIGURE C.10   Seismic system electronics.

FIGURE C.12   Seismic profile plotters.

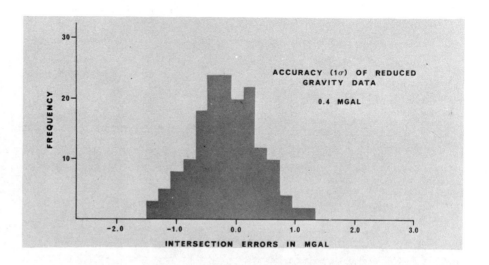

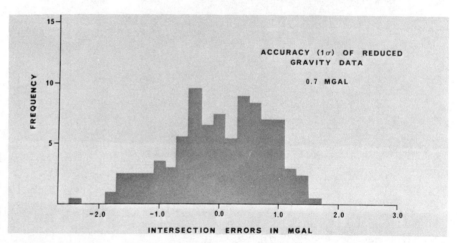

FIGURE C.13 Accuracies of reduced shipborne gravity data at line intersections. *Top*: using high-precision electronic positioning system to reduce gravity data and *bottom*: using integrated satellite positioning to reduce gravity data (in shallow water).

marine proton magnetometers. These magnetometers have a maximum sensitivity of one eighth of one gamma.

Superimposed on the earth's total magnetic field are time-varying perturbations. Assuming that short-term time variations of the total magnetic field affect both sensors equally at the same time, the difference between the responses of the two magnetometers is undisturbed and can be integrated to obtain total magnetic-field profiles free of time variations. Gulf's research division is currently developing an effective gradiometer reduction technique based on an extensive amount of gradiometer data obtained from a specific recent test survey of the R/V *Hollis Hedberg*.

Gravity

Within recent years a major breakthrough has been made in improving the accuracy of shipboard gravity meters. Gyro stabilization and reduction of errors due to wave action and to cross-coupling were paramount factors in this improvement. In effect, one-half milligal variations in the earth's gravity field are now detected in a noise back-

ground due to ship's acceleration of up to 100,000 milligals. Accurate navigation is mandatory, however, to correct the gravity data adequately. Currently achievable accuracy for the reduced gravity data is illustrated in Figure C.13 and involves the following errors:

Using high-precision radio location system half a milligal or less.
Using WINS* in shallow water 1 milligal or less.
Using WINS in deep water usually between 1 and 2 milligals.

The gravity-meter control electronics and analog recorders are shown on the left-hand side of Figure C.14.

Geochemical Seep Detector

The seep detector system was originally assembled by Gulf for research purposes and has since been developed

*WINS is a registered trademark of Western Geophysical Company of America.

FIGURE C.14 Gravity-meter control electronics and WINS navigation system.

into a practical and highly valuable new supplement to other exploration tools.

The major components of the system are shown in Figure C.15. Seawater from three inlets at different depths (hull, 61 m, 183 m) is continuously circulated through a gas stripper chamber. A preselected volume of gas is trapped at regular intervals and analyzed by a gas chromatographic analyzer with flame ionization detector. The analyzer is adjusted to examine hydrocarbons in the C_1 to C_4 range. Both saturated and unsaturated hydrocarbons are determined. The results of analysis of each sample are displayed on a time-referenced recorder. The analyzer and recorder are shown in Figure C.16. The system has sufficient resolution to detect light hydrocarbon components in seawater at the level of 10 parts per billion.

A major contribution of the present system is continuous recording whereby it has been possible to establish a normal background from thousands of miles of recording and to characterize different types of anomalies.

Figure C.17 shows the principles of seep detector interpretation. Gases are eluted in the time sequence shown from right to left. Ethylene and propylene, the unsaturated C_2 and C_3 hydrocarbons, do not occur in petroleum. Interpretation of significant anomalies is based on differences between concentrations of background related hydrocarbons and assemblages of petroleum related hydrocarbons. These include methane, ethane, propane, *i*-butane and *n*-butane.

Geochemical data are displayed for analysis in a variety of ways as briefly mentioned under Data Displays above. More specifically, one of the Varian plotters can display off-line, in addition to potential field information, geochemical data from two different inlets:

Propane concentration	Water temperature
Methane concentration	Inlet depth
Water salinity	Water depth

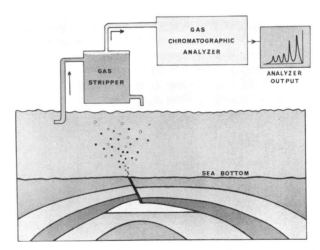

FIGURE C.15 Underwater seep detector.

FIGURE C.16 Geochemical equipment.

Propane and methane concentration curves are displayed also on the bottom of the "mini-section" (described under Data Displays) along with potential field data.

Data Logger and Data Processor

There are two Digital Equipment Corporation Model PDP-11/10 computers with associated peripheral equipment installed onboard and operating as two functional systems. One PDP-11 system is referred to as the logger and the other is referred to as the processor. The PDP-11/10 is a byte-addressable 16-bit word size computer with such features as eight general-purpose registers, hardware stack, automatic power fail/restart, vectored multilevel interrupt structure, asynchronous operation

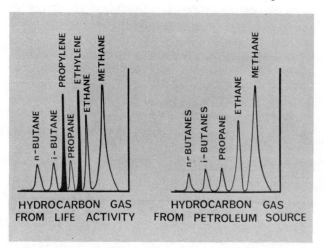

FIGURE C.17 Interpretation of seep detector data.

along a single communications path (UNIBUS) for highest speed of components.

The PDP-11 logger has a 28K word-core memory, hardware-extended arithmetic element, TI Silent 700 terminal, Pertec 1600 bpi 9-track tape unit, interfaces to Western's DAS (Data Acquisition System) and to the real-time clock, synchronous communications link to the PDP-11 processor and to the EMR computer. The logger gathers, smooths, reports, records on magnetic tape, and transmits to the processor basic information such as ship's speed and heading, water depth, gravity and magnetics data, raw navigation system data, and calculated information from the WINS system. In addition the logger gathers, reports, and records information from navigation satellites as they pass within range.

The PDP-11 processor has a 28K word-core memory, hardware-extended arithmetic element, TI terminal, two Pertec 1600 bpi 9-track tape units (Figure C.18), card reader, high-speed paper tape reader, DEC tape magnetic tape transport, Varian Statos 31 plotter, interfaces to the geochemical and single-channel seismic systems, and hardware bootstrap loaders. The processor receives basic data from the PDP-11 logger via the synchronous communications link and writes redundant 1-min logger records on magnetic tape. Additionally the processor logs and records on magnetic tape data from the geochemical system. The second magnetic tape unit is used to record single-channel seismic data as backup for the EMR computer. Pertinent information is transmitted to the EMR for real-time plotting.

In an off-line mode of operation the processor reads back data previously recorded for a line and does further processing to obtain corrections to the potential field data, peak and parameter information from the geochemical records, and generates results to be output to various plotters.

FIGURE C.18 Data logger and data processor magnetic tape units.

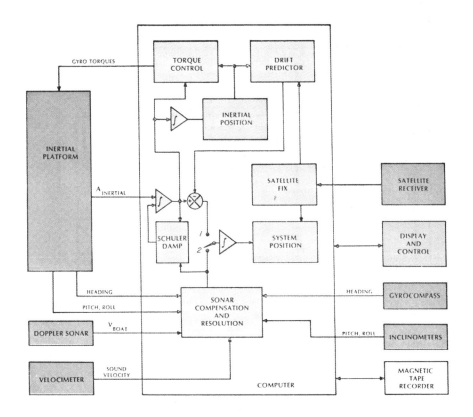

FIGURE C.19 R/V *Hollis Hedberg* navigation system (WINS).

Navigation

Accurate navigation (real time) and positioning (post-cruise) on a worldwide basis are essential to successful offshore exploration surveys. For some detailed work, there is the possibility of having to reoccupy positions for subsequent surveys or well locations, and for all surveys, proper Eötvös corrections for the gravity data reduction require that the speed be known to 0.1 knot and the true course to 0.5°. For some surveys with only regional geological objectives, the navigational requirements are less demanding.

Gulf's early experience with Doppler satellite navigation and a Doppler sonar/gyrocompass system led, in 1970, to the installation of Western's Integrated Navigation System (WINS) aboard the R/V *Gulfrex*. The system consists of a Doppler satellite navigation receiver, a Doppler sonar that measures speed over the bottom along the ship's true course, a gyrocompass, an inertial platform for accurate deepwater navigation, and a dedicated computer. The system on the R/V *Hollis Hedberg* also includes a modern solid-state Loran C receiver with an atomic frequency standard and a special computer. Figure C.19 contains the system information flow of WINS excluding the Loran C subsystem. The detailed specifications of the navigational sensors and computers are given in Table C.1. Figures C.20 and C.21 show the inertial system schematic and the Litton LN-15G inertial platform, respectively.

In almost five years of operation, WINS has proven to be superior to other systems using the Navy satellite system because of its clearly more accurate deepwater navigation and its reliability, which is a function of system redundancy.

Realistic positional accuracy achievable with this system (without Loran C subsystem) on a worldwide basis is as follows:

In shallow water (water depth 600 feet or less):
rms Position Accuracy better than 120 m (400 ft)
"95% CEP" (radius of a 180–275 m (600–900 ft)
circle containing
95% of the data)

In deep water (water depth greater than 180 m; 600 ft):
rms Position Accuracy 305 m (1000 ft) or better
"95% CEP" (radius of a 550–700 m (1800–2300 ft)
circle containing
95% of the data)

These values represent the realistic accuracy achieved under operating conditions (including average downtime of certain sensors) as has been determined from thousands of miles of tests.

The range in accuracy is related to changes in geographical area, especially latitude, and to the geometrical configuration of a survey. The above-mentioned accuracy refers to postcruise processed positions. The real-time position accuracy, which is important as well, is not quite

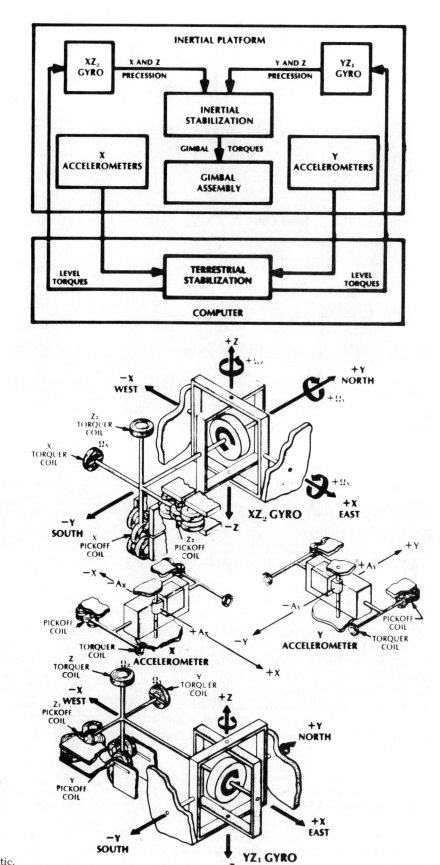

FIGURE C.20 WINS inertial system schematic.

FIGURE C.21 WINS inertial platform.

as good as the postcruise processed accuracy as is to be expected for any dead-reckoning system. However, the real-time position accuracy in shallow water is quite acceptable as is illustrated in the right-hand panel of Figure C.22, which also shows a typical error distribution of postcruise processed positions in shallow water. Figures C.23 and C.24 show typically encountered position errors for a survey line in shallow water (less than 200 m depth) and deep water.

The goal of complete independence from local shore-based navigational aids has not yet been fully achieved, primarily because of sensitivity of Doppler systems to ship's velocity errors. In stationary mode (such as at dockside) the position accuracy of the Doppler satellite system used on the R/V *Hollis Hedberg* is around 5 m, which is exploited in establishing transformation parameters for geodetic datums. However, for surveys requiring best possible position accuracy, high-precision radio location systems such as RAYDIST, CORAC, MOTOROLA, and AUTOTAPE are still used onboard the R/V *Hollis Hedberg*. The range or range difference data from these systems can be recorded digitally on Western Geophysical Company of America's WINS magnetic-tape system through a special interface for subsequent postcruise processing. Each time such systems are employed, simultaneously recorded and processed WINS data are evaluated in order to continuously quality control the system and implement possible improvements. New developments in the field of navigation such as the U.S. Defense Navigation Satellite System are being observed closely in order to achieve increased accuracy.

Side-Scan Sonar and Geological Sampling

The R/V *Hollis Hedberg* can be readily equipped to take piston cores from the ocean bottom and to dredge for bottom samples.

Side-scan sonar equipment of any manufacture can also be readily installed aboard the vessel.

Test Equipment, Spare Parts, and Personnel

Electronic test equipment and spare parts are carried onboard in a quantity sufficient to ensure operation in remote regions with a minimum of downtime. Highly skilled personnel who have received advanced training in all aspects of the operation are aboard at all times. Approximately 80 percent of the technical personnel either have college or university degrees or have completed some work toward a degree.

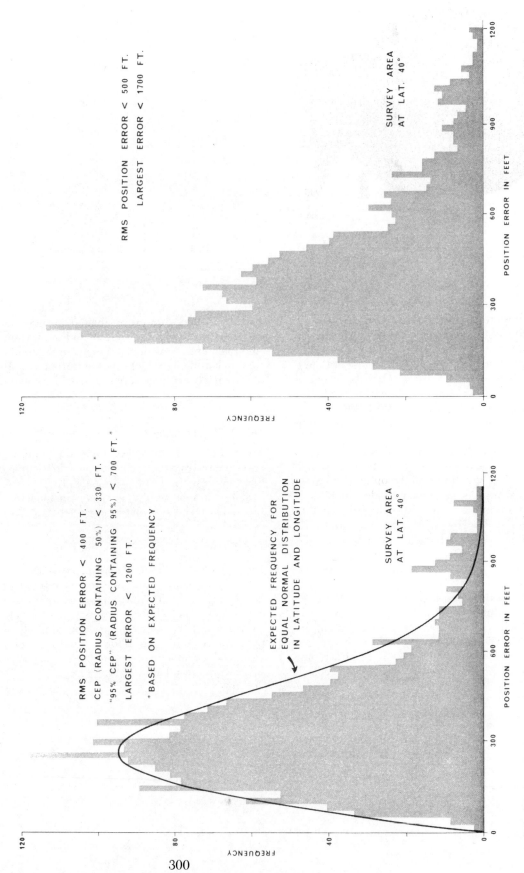

FIGURE C.22 *Left:* Error distribution of postcruise processed WINS positions (water depth 600 ft or less). *Right:* Error distribution of real-time WINS positions (water depth 600 ft or less).

300

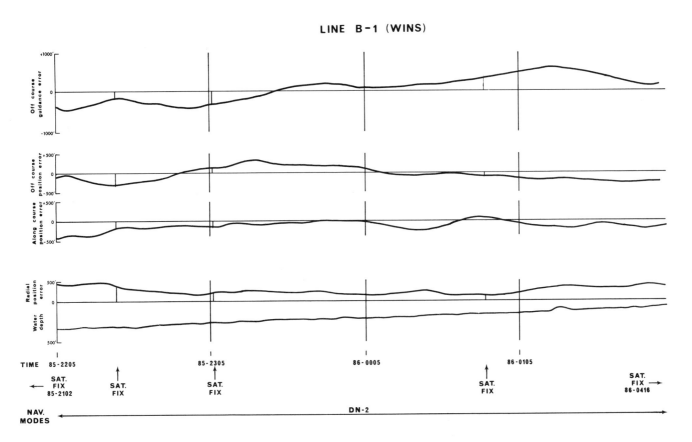

FIGURE C.23 Typical position errors for a shallow-water line.

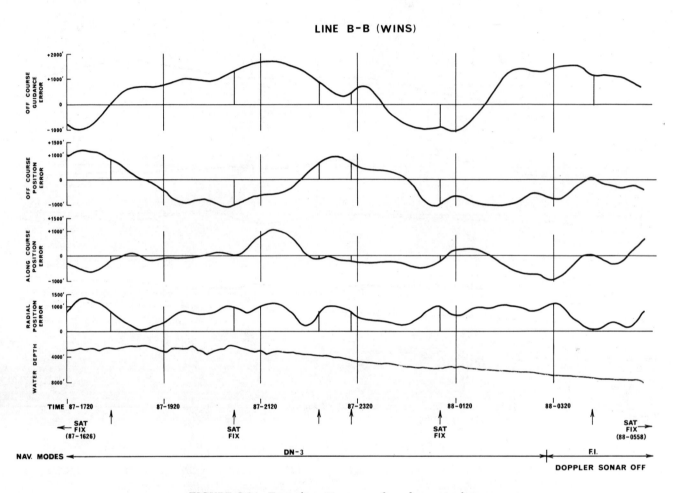

FIGURE C.24 Typical position errors for a deepwater line.